Foundations of Natural Resources Policy and Management

Foundations of Natural Resources Policy and Management

Edited by Tim W. Clark,
Andrew R. Willard, and
Christina M. Cromley

Yale University Press

New Haven and London

Printed in the United States of America.

ISBN 0-300-08356-4 (cloth : alk. paper).——ISBN 0-300-08144-8 (pbk. : alk. paper)
Library of Congress Card Number: 00-035924
A catalogue record for this book is available from the British Library.

The paper in this book meets the guidelines for permanence and durability of the Committee on Production Guidelines for Book Longevity of the Council on Library Resources.

10 9 8 7 6 5 4 3 2 1

Contents

Preface

Foundations of Natural Resources Policy and Management introduces a sound, practical way to understand, analyze, and improve policy and decision making in the management of natural resources toward the goal of sustainability. This book draws on the theory and framework of the policy sciences, a genuinely interdisciplinary problem-solving method, to improve policy and management, and it provides in-depth, detailed, and varied case study examples of how the policy sciences can be applied. The policy sciences are a set of integrated conceptual tools, an analytic framework, and a set of fundamental propositions designed to explore any policy problem and devise practical intervention strategies. This powerful method is currently underutilized in natural resources policy and management. The policy sciences, invented over fifty years ago, anticipate the growing demand in natural resources for a truly interdisciplinary approach, one that places increasing emphasis on understanding the role of humans in shaping policy and management directives and in affecting ecosystem processes. In fact, the policy sciences offer a comprehensive and integrative approach which addresses the need for interdisciplinary problem solving that students, instructors, and practitioners (in and out of government) now seek with growing intensity and concern.

This book is largely an outcome of courses on improving natural resources policy and management taught at the Yale School of Forestry and Environmental Studies in New Haven, Connecticut. It is also based on years of professional use of the policy sciences by the editors, their students, and colleagues, and on a series of workshops and field trips.

The flexibility and resolving power of the policy sciences are demonstrated through ten case studies in conservation and natural resources management in the United States and abroad and in both urban and rural settings. These case studies, prepared by students in the Foundations of Natural Resources Policy and Management courses, also provide specific examples of how policy sciences can be applied to improve policy and decision making in the management of natural resources to reach the goal of sustainability.

Objectives of this book are:

• To introduce the policy sciences theory and framework and provide a practical reference, a how to, for practitioners, students, researchers, and instructors
• To show how this theory and framework have been taught, learned, and used
• To demonstrate, through a range of insightful and skillfully analyzed case studies, the power and flexibility of the policy sciences for understanding and improving complex natural resources problems
• To clarify how a common approach can both highlight differences in social, natural, and political contexts and provide insight into patterns that exist across cases

The audience for this book includes students and instructors in the fields of policy sciences, conservation biology, political and social ecology, sustainable development, and environmental management. The range of cases presented in the book will also appeal to resource and land managers, policy scientists, researchers, activists, as well as members of the lay public interested in improving and understanding natural resources policy and management. Domestic and international agencies, nongovernmental organizations (NGOs), and universities will also find the functional, pragmatic approaches taken in the cases useful and insightful.

The contributors to the book come from varied backgrounds, experiences, countries, and expertise. Authors of the case studies, mostly master's and doctoral students, bring both practical experience (in government, NGOs, and consulting) and an analytical, reflective insight to their cases. Taking a functional approach (as opposed to a conventional one) that transcends traditional, disciplinary analyses, the authors examine complex and sometimes sensitive topics such as human rights and democratic participation in policy-making processes within natural resources arenas. They go beyond identifying what is wrong with current policy, suggesting practical alternatives to the way resources are currently managed. All the authors seek ways to improve policy and management outcomes.

The book is organized into two parts. Part 1, Improving Natural Resources Policy and Management, includes two chapters. Chapter 1 introduces the policy sciences approach and gives a description of how the approach or method can be applied to problems in natural resources policy and management. It also presents the approach by describing a Yale course in which the policy sciences are taught. The course itself can serve as a model or starting point for people who wish to develop courses or learn how to use policy sciences to understand and improve natural resources policy and management. Other means to learn this method such as workshops, field trips, and individual study are also mentioned. Chapter 2 offers a detailed outline, a guide to students, teachers, practitioners, and researchers for applying the policy sciences to natural resources policy and management in a comprehensive and systematic manner.

Part 2, which includes Chapters 3 through 12, consists of ten case studies. Each case study is organized around the problem orientation of the policy sciences and follows the outline presented and described in detail in Chapter 2.

Each of the ten chapters in Part 2 focuses on different policy problems, and each recommends specific, practical ways to improve social and decision processes and problem orientation given the substance and context of the problem at hand. For example, Alejandro Flores discusses natural resources issues particular to urban residents of the Baltimore metropolitan area, some of whom die from exposure to high levels of ozone air pollution. His study examines policy to improve current ozone abatement efforts. It recommends a strategy that targets key decisions affecting both rates of ozone removal and human exposure.

Jessica Lawrence also considers natural resources within an urban environment and addresses social inequalities. Her study focuses on the social and decision processes concerned with Beaver Ponds Park management in New Haven, Connecticut. She recommends improving park management by continuation of the ongoing planning process and strengthening democratic participation.

Christopher M. Elwell seeks to understand and resolve some of the controversy that has arisen from the proposed construction of two pumped-storage hydroelectric power plants in the Sequatchie Valley, Tennessee. He presents four alternatives for citizens' groups that feel excluded from the legally mandated public decision process.

David Kaczka examines sea turtle conservation worldwide and finds that the greatest threat to sea turtles is shrimp trawling. He recommends the use of turtle excluder devices, which are an inexpensive technology already available, to reduce incidental capture of sea turtles in shrimp trawls with little reduction in shrimp catch. He offers suggestions to diffuse the use of this technology.

Katherine Lieberknecht describes the social and decision processes that guide current management of the endangered Barton Springs salamander in Texas. She critiques current management and recommends practical improvements.

Christina M. Cromley analyzes the killing of grizzly bear 209 in Grand Teton National Park, Wyoming, to gain a better understanding of the expectations, demands, and claims of participants in the grizzly bear policy and management debate. The social and political complexities of carrying out endangered species recovery plans are described and a complex of strategies are recommended to meet the expectations and demands of participants.

Eva J. Garen examines ecotourism, a popular strategy for gaining local, national, and international support for natural resource conservation. She explores why many ecotourism programs may not be accomplishing the goal of biodiversity protection, examining how decisions are made during program development and application. She identifies a number of lessons and makes recommendations for improving how ecotourism programs are carried out.

David Lyon examines and appraises the decision functions involved in setting up and managing conservation activities in and around Kyabobo Range National Park in Ghana. His study investigates the roles of different participants in established arenas of power and access to those arenas, issues that are currently controversial. He argues that in several circumstances there is a middle ground between conservation and development that would ameliorate many problems surrounding the park.

Peter R. Wilshusen explores why the majority of international conservation and development projects have not fostered a greater degree of participation by local people. He contrasts local projects that have failed to generate participation with examples of interventions that appear to have succeeded in this respect in order to uncover constraints and lessons regarding local involvement. He concludes with recommendations for increasing the degree and improving the quality of local participation in conservation and development projects.

Gus Le Breton explores the social context within which the Office of the Senior Coordinator for Environmental Affairs (OSCEA)—part of the United Nations High Commissioner for Refugees (UNHCR)—operates and examines the decision processes through which OSCEA tries to improve environmental management. He recommends how OSCEA can best improve UNHCR practices to minimize the adverse environmental effects of concentrating refugees in confined geographic areas.

The final chapter, Improving Natural Resources Policy and Management, is a summary. It discusses the significance of the policy sciences interdisciplinary approach for natural resources policy and management by reviewing the book's contents. All ten case studies show that although natural resources problems often become the focus of attention because of their physical dimensions, the fundamental causes of resource degradation are social, involving human dimensions. The cases in this volume go beyond making claims about the importance of combining the social in its broadest sense and natural sciences in resolving these problems: they actually apply the problem-oriented, contextual, interdisciplinary tools found in the policy sciences to suggest workable improvements in natural resources policy and management. As more natural resources cases are examined using the policy sciences as a stable frame of reference, patterns may begin to emerge about why problems in natural resources policy and management recur and what reasonable, politically workable, and justifiable solutions are already available or could be developed. The policy sciences are not easy to master, but the cases demonstrate that they are practical for developing policy alternatives.

We wish to thank the several hundred students and professionals who participated in our courses, in workshops, and in field trips over the years.

The support and guidance of Jean E. Thomson Black, Nancy Moore Brochin, and Denise Casey are very much appreciated.

Part 1

Improving Natural Resources

Policy and Management

1 Learning about Natural Resources Policy and Management

Tim W. Clark and Andrew R. Willard

Natural resources challenges are growing more complex as the human population expands, the reach of technology grows, and the stocks of many natural resources are drawn down. The best conceptual and applied tools are needed to address these challenges. The policy sciences offer a set of tools that are unparalleled in this regard. There are many ways to learn this set of tools including courses, workshops and short courses, field trips (Clark and Ashton 1999), and on one's own. Perhaps the best and easiest way to learn is through a formal course which systematically introduces and illustrates the tools and their practical utility in improving natural resources policy and management. Our "Foundations" course teaches the policy sciences approach to understanding and addressing natural resources policy and management problems. Used with insight, judgment, and skill, these tools can improve policy making in the management of natural resources toward the goal of sustainability. This chapter describes our course, introduces the policy sciences approach, and gives an initial evaluation of the course. This course offers a way to learn how to improve natural resources policy and management practically. The analytic outline in Chapter 2 and the case studies (Chapters 3–12), each of which follows the organization of the outline, will help newcomers as well as experienced professionals learn the policy sciences and apply them with skill.

Course Overview

We first offered "Foundations of Natural Resources Policy and Management" at the Yale School of Forestry and Environmental Studies in the spring semester of 1996. It is taught by Tim Clark, a conservation biologist, policy scientist, and professor at the school, and Andrew Willard, a policy scientist, an anthropologist, and associate research scholar at the Yale Law School. We were both interested in the problem-oriented, contextual, and multimethod discipline for problem solving known as policy sciences and agreed that it would be a useful addition to the university's curriculum. This method is genuinely interdisciplinary. This chapter is based on three years' experience with this course and over two decades of experience with related courses, workshops, and field trips.

Background

The course grew from the conviction that sustainable human use of the biosphere is critical, but that efforts have been inadequate in clarifying what is meant by sustainability and what actions are needed to secure it. This was hardly a novel observation. Walters (1986), Lee (1993), and Gunderson et al. (1995) have detailed many pathologies in natural resources policy and management. Ludwig et al. (1993:17), for instance, reflecting on diverse policy and management experiences, suggested that many of the current plans for sustainable development "reflect ignorance of the history of resource exploitation and misunderstanding of the possibility of achieving scientific consensus concerning resources and environment." They concluded that "resource problems are not really environmental problems: They are human problems that we have created at many times and in many places, under a variety of political, social, and economic systems" (p. 36). This stimulating insight directed attention to alternative ways of thinking about natural resources policy and management and about sustainability in particular. When natural resources problems are seen in this new light as epiphenomena of human practices, it becomes necessary to apply new methods to identify, understand, and reshape those practices that generate, condition, and respond to natural resources problems. Presenting and applying such methods is the central task of the "Foundations" course.

The notion of sustainability is illuminated by the policy sciences theory central to the course (see Lasswell 1971; Lasswell and Kaplan 1950; Lasswell and McDougal 1992). For example, generally speaking, achieving sustainability in policy and management requires responding to the changing conditions that shape the national and global community and guiding them to the extent possible. It is desirable to have a smooth social transition from current short-term, exploitive policies to more long-term, adaptive and sustainable approaches. Thus, one challenge for scholars and natural resources practitioners is to look broadly in society and its institutions for rigidities in outlook and material conditions that might inhibit a smooth adaptation to a more sustainable future.

Many components of policy sciences can facilitate the student's ability to identify the relevant context of a problem reliably. Regardless of the subject—biodiversity, genetics, endangered species, population growth, oceans, fisheries, forests, parks and reserves, the atmosphere, outer space, or land and water use—every discipline and standpoint has something to contribute in the examination of problems and solutions. Learning how to integrate and synthesize specialized knowledge and contributions is central to the course.

When the policy sciences approach is applied to particular problems, it results in a realistic, reliable, operational "map" of the past and potential future of the policy system under consideration. In turn, this practical map can be used to assess the potential contributions of more specialized views and findings, determine overlooked issues for research, and locate the most likely sites of intervention to facilitate adaptation of the policy system to actual and future conditions. It offers its skilled users a means to contribute to clarifying and securing (achieving) the common interest.

Course Format

The purpose of the course is to help students improve skills in thinking more effectively and acting more responsibly. It is designed for students in any subfield of forestry, environmental studies, or other disciplines. Particular emphasis is placed on practical improvements in policy processes and on clarifying what is meant by sustainability in natural resources policy and management. Students are expected to discuss theory, applications, and cases. Once they gain familiarity with the core methods of problem identification, clarification, and resolution, they apply them to specific issues in natural resources policy and management. The class, fourteen weeks (one semester) in length, meets for three hours once or twice a week. The instructors have open office hours and meet with students individually and frequently throughout the semester. The course is designed to complement other courses in the curricula of the school and the university.

In the first eight weeks, comprehensive and integrated methods of thinking about problems and proposing solutions are explored. Instructors lecture and students summarize and critique readings. Discussion is both extensive and intensive, with students contributing subject matter (often from their case studies) to illustrate the methods in addition to case materials presented by the instructors. Each week's topic is introduced by a one-page description prepared by the instructors, and each week students read about five papers totaling approximately 100 pages. An additional bibliography is provided on the world environment, natural resources issues, and the policy sciences. Over the course of the semester, each student is asked to lead the class in discussing a few readings. The student is asked to answer three questions concerning the readings (Brunner 1997a,b,c): (1) What did you already know about the material covered? (2) What important information (if anything) did you learn from it? (3) What are the most important questions you have about it?

Subsequent weeks are devoted to student presentations applying the methods of the policy sciences. This not only gives individuals a chance to analyze a case of their choice in depth but also gives all members of the class exposure to a wide variety of issues and applications. The students thus learn by doing. Each one-hour presentation is based on a draft of one student's term paper. The paper and presentation are expected to demonstrate explicitly the interdisciplinary problem-solving method studied in the course. (An outline for preparing individual papers is described in Chapter 2.) A concluding class session reviews the policy sciences and the semester's work.

A View of Natural Resources

The course is based on a functional view of natural resources. In the broadest sense, natural resources include the earth and its atmosphere, soils, minerals, energy, plants, animals, including humans, as well as the solar system, the galaxy, and beyond. The term *natural resources* is used to designate the biophysical environment in which social interactions are carried on and which may be directly involved in such interactions (Lasswell and McDougal 1992). The policy sciences can be used to study this social process and the effects of resources use in any society.

The significance or meaning of the biophysical environment is influenced by people's goals at specific times and places and by their expectations about the utility of particular resources for achieving their goals. For example, a forest is a resource that can be used for well-being (e.g., building homes), enlightenment (e.g., knowledge about ecosystem structure), wealth (e.g., increased financial income for users), or other values; a thousand years ago oil and gas, DNA, and the geosynchronous orbit were not resources as they are now.

Humans manipulate their environments, including other humans, and such operations are interwoven with the entire social and cultural context. Manipulating flint for stone tools, atoms for hydrogen bombs, and soils, plants, animals, or ecological processes (such as fire) are all examples of operations. Technology is the operational means people use to manipulate their material culture, which is the ensemble of these operations and the physical objects on which they are performed. When operations are associated with symbols of meaning, which is usually the case, we call them practices. For example, cutting down a tree may be associated with feeding one's family (i.e., well-being), making money (i.e., wealth), or some other mix of values. Operations plus these perspectives are practices. Sound natural resources policy and management call for operations, perspectives, and practices that serve the common interest—that is, those that provide people the mix of values they hold and offer a sustainable future (see Ascher and Healy 1990).

The objects of manipulation in natural resources policy and management are either culture or raw materials. Culture materials include axes, chain saws, bulldozers, domestic livestock and crops, rockets, and computers. Even some soils and whole landscapes may be culture materials if they have been altered significantly by human cultivation over centuries. Accordingly, culture materials change over time. Raw materials are potentially usable resources, such as unexploited grasslands, forests, or oceans, or even Antarctica. Of course, the largest reservoir of raw materials is composed of the not yet discovered realms and phenomena of the universe. Thus the material culture of a society or community, which consists of its technology and of all the operations performed on biophysical objects, is rarely static. Different cultures carry out operations differently. Some operations are intended for short-term value accumulation, which we call "exploitive" practices, while others are long-term, which we call "sustainable" practices.

The Policy Sciences Approach

Students and practitioners need a means to understand and participate in social processes pertaining to natural resources realistically, comprehensively, practically, and constructively. Such a means or method —policy sciences—already exists and has been applied to problems in

natural resources as well as other policy arenas, including social services, science policy, communications, war, revolution, human rights,
international law, and national defense (e.g., Arsanjani 1981; Ascher
and Healy, 1990; Brewer 1988; Brewer and Kakalik 1979; Brunner
and Ascher 1992; Brunner and Byerly 1990; Brunner and Clark 1997;
Clark et al. 1996, in press; Johnston 1987; Lasswell 1948, 1965b; Lasswell et al. 1979–1980; McDougal and Burke 1987; McDougal and Feliciano 1994; McDougal and Reisman 1981; McDougal et al. 1963; Mc-
Dougal et al. 1980; McDougal et al. 1994; Murty 1989a,b; Pielke 1997;
Reisman 1971, 1981; Sahurie 1992). This method is part of a professional movement that, in the largest sense, is a global effort to address
the increasingly complex problems of our time. Its basic aim is to improve the quality of life and policy decisions through genuinely comprehensive and integrated inquiry, and practical action.

From a policy sciences perspective, the ongoing interaction of people in their efforts to achieve what they value is the foundation of all
policy, including that of natural resources. Public policy making is a
never-ending process whereby people attempt to clarify and secure
their common interests. Management is the actual manipulation of
people and resources through programs.

Policy Sciences Method

The power of the policy sciences to generate practical and theoretical insight comes from its unique methodological integration and
specification of the full range of problem-solving tasks. The main conceptual categories of policy sciences are illustrated in Figure 1.1. The
framework has four principal dimensions: problem orientation, social
process mapping, decision process mapping, and observational standpoint. In actual problem-solving situations, empirical data about each
of these dimensions must be gathered, organized, and interpreted.

The policy sciences approach to problem solving is well documented
in the social sciences literature (Lasswell 1971) and has been widely
used with respect to international natural resources issues (e.g., Ascher
and Healy 1990). The approach is directly and immediately applicable
to any and all contexts wherein people interact (Lasswell and McDougal 1992). The approaches' integrated concepts make up a logically
comprehensive set of elements that function as a "stable frame of reference." This stable frame of reference allows analysts and other participants to look not just at the technical particulars of any natural resources
management issue, but, more important, at the functional relationships
that animate them and all of human interaction. Policy sciences is not

Figure 1.1. The principal dimensions and elements of the policy sciences. Drawing by Denise Casey.

simply a "cookbook" approach that adds social science data to biological data in order to analyze natural resources problems or find solutions to them. The power of policy sciences resides, in part, in its highly flexible approach, which is both comprehensive and selective, to mapping human dynamics and their implications in and for policy (see Dewey 1933). The approach offers insights into social process that simply cannot be achieved by using conventional views and terms. Through the use of policy sciences, data—hard and soft—that otherwise would likely be overlooked or misconstrued can be fully discerned, appreciated, and used practically in natural resources management. The four policy sciences dimensions are described briefly below.

Problem Orientation

Problem orientation is a strategy to address problems and invent solutions (Lasswell 1971; Lasswell and McDougal 1992). Five tasks are involved (Figure 1.1). The goals that people seek to achieve must be practically clarified and specified in relation to the problem at hand; hence, careful attention must be given to problem identification and delimitation. Historic trends must be described to discern whether events are moving toward or away from the clarified goals. Factors or conditions that have influenced these trends must be understood. Projections or future trends can be offered in light of past trends and conditions. Finally, potential solutions or alternative actions must be invented, evaluated, and selected if projections are seen as undesirable or less than optimal.

A more thorough examination of these five tasks better illustrates what is involved in problem solving (Wallace and Clark 1999). Problem orientation permits a person to carry out sound, integrated research, management, and policy and avoid common pitfalls. These five tasks direct an analyst, advisor, or decision maker to ask questions and seek out information in a fashion conducive to learning as much as possible about a specific problem.

First is the task of clarifying goals. "The goal-clarifying task is indicated by the blunt question, 'What ought I to prefer?'" (Lasswell 1971:40). Because natural resources problems occur in a context, it is vital to always maintain a view on the problem and its context at the same time. The goal question as posed here can best be answered for a problem after considering the problem's context or social process (Clark and Wallace 1998). Considering the social process (as described below) means analyzing a particular problem or situation including: (1) the actual or desired participants involved, (2) their various perspectives on the issue, (3) the situations in which they interact or might interact, (4) the values (or assets or resources) they use in their efforts to achieve their goals, (5) the actions or strategies they use to achieve their goals, (6) the outcomes they will or might achieve, and (7) the real and potential effects of their actions (Lasswell 1971; Willard and Norchi 1993). Once these factors have been considered, it becomes much easier to determine the utility of desired goals while aiming to reduce unknowns and the possibility of introducing additional problems into the decision making process. Social process mapping should continue over the life of every natural resources issue.

Second is the task of describing (historical) trends. Describing trends means identifying the status of the key elements of the problem with respect to the goals that have been set for them, perhaps provisionally, and also which participants and perspectives in the natural resources management effort have met or fallen short of goals through their actions (Lasswell and McDougal 1992). "The immediate aim is to suggest that much can be accomplished in a problem-solving strategy that gives full weight to asking and answering the questions, 'Where are we? How far have we come in achieving what we are aiming at? Where are the positive and negative instances of success or failure?'" (Lasswell 1971:48).

Third is the task of analyzing conditions that have influenced observed key trends. For each trend identified in a particular natural resources policy or management problem and its human context, there is a set of conditions influencing that trend. To research and understand the trends in natural resources policy and management and also whether and how participants have met or fallen short of goals, it is necessary to analyze the factors that account for observed trends. This task focuses on scientific inquiry, not only about technical matters but also about human, organizational, and policy behavior in the social process.

Fourth is the task of making projections on the basis of past trends and conditions. In part, this task demands that we suspend our beliefs and conventional views of what actions participants should take in the future. Instead it asks that we take a current situation and project it, free of the effects of possible future interventions, to its likely outcome. For example, if current legislation severely reduces an agency budget for a program, the projected outcome is likely to be failure to meet goals. This example illustrates simply that by projecting current circumstances into future outcomes, we gain insight into how those circumstances will affect natural resources problems. Such projections also indicate where interventions or other alternatives are needed to address the problem and produce acceptable future outcomes. The other goal of making projections is to try to foresee the consequences of choosing alternatives.

Last is the task of inventing, evaluating, and selecting alternatives. This task requires creating, reviewing, and choosing objectives and strategies for achieving goals. Key questions include: What approaches do we use to realize the goals we set for sustainablility? What alternatives can we implement to change conditions so that future trends will be favorable for the natural system and the human system involved? On the basis of all we have learned about a given problem in the four previous tasks, what decisions should we make to reach preferred goals?

In natural resources policy and management, these five tasks must be carried out repeatedly as time permits over the lifetime of a policy or program. At any point in carrying out the five problem orientation tasks, problem solvers may be required to return to other tasks.

Social Process

Social process mapping is a way to comprehend the particular social context(s) in which all problems are embedded (Lasswell 1971;

Lasswell and McDougal 1992). To cultivate awareness of the larger context as well as the details of particular situations, policy sciences recommends a set of conceptual categories to describe or map any social process or problematic situation (Figure 1.1). These categories include the participants, with their particular perspectives interacting in particular situations. Participants use whatever values they have (i.e., assets or base values) through various strategies to achieve desired outcomes that have effects (Clark and Wallace 1998). Values—both the things people strive for and the assets they use to acquire them—are created and exchanged (shaped and shared) through social interactions to gain more values. Policy scientists have identified eight functional value categories: power, wealth, enlightenment, skill, well-being, affection, respect, and rectitude. Social and decision processes have outcomes and effects that may be characterized as indulgent or deprivational in terms of various values for participants.

Social process concepts and mapping of a given context can be more easily understood and more effectively formulated if we remember that problems do not exist in a vacuum. Problems are the products of the interaction among people in given biophysical situations. Therefore, it is essential to understand the social process or context of natural resources policy and management problems. Social process is the interaction of people as they influence the actions, plans, or policies of other people, even if they are unaware of one another. Social process is the process wherein people create and sustain the human community and the environment that makes it possible.

In trying to understand social process in general or in any particular natural resources case, most people merely impose conventional classification systems extrapolated from everyday life. The terms used in the conventional approach and the scope permitted by that kind of analysis is usually weak in analytic comprehensiveness and insight. For example, how many times have you heard someone invoke words like "politics" or "personality" to explain away troublesome social dynamics and outcomes in natural resources policy?

A much more practical way to view social process has been devised based on functional anthropology and the policy sciences that enhances understanding of complex policy problems (Lasswell and Kaplan 1950). Briefly, this model encompasses participants with perspectives interacting in particular situations. Using whatever base values they have, people employ strategies to pursue particular valued outcomes, which have effects on all future interactions. These terms and social dynamics are discussed in depth in theoretical and applied works by Lasswell and Kaplan (1950), Lasswell (1971), and Lasswell and McDougal (1992), and are discussed in the context of conservation biology by Clark (1997) and others (see references above). Another way to focus on social dynamics in any context is to ask seven questions: Who participates? With what perspectives? In which situations? Using which power bases? Manipulating them in which strategies? With what outcomes? And with what longer-term effects?

In all human interactions, people tend to act in ways they perceive will leave them better off than if they had acted otherwise. Because of the subjective character of perceptions, people perceive themselves, their environments, and other participants differently and sometimes erroneously (Berger and Luckmann 1987). The differences among people—in identities, expectations, demands, values, strategies, and other variables—may be vast and irreconcilable. Even if it were a desirable goal, no amount of "cold, hard fact," collected by "objective, neutral" scientists and made "equally accessible" to all participants in a policy arena, and no amount of "education" and "exchange of ideas," can completely neutralize basic inherent differences among people. However, this observation does not disaffirm our common interests or obviate the need to strive to clarify and secure them.

One basic feature of the social process mapping heuristic is the interplay of human values. People's values underlie their perspectives of the world. Values are the key medium of exchange in all interpersonal interactions. Values are embodied in the goals people strive for—whether it be for education, skill in performing their jobs, good health, good government, healthy environments, security, among many others.

The policy sciences divide human values—everything that people in all societies in all times at all ages at all levels strive for—into eight categories: power, wealth, enlightenment, well-being, affection, skill, respect, and rectitude (Lasswell and Kaplan 1950; Lasswell and McDougal 1992). These are functional terms and can be used across all communities and cultures to describe how people's behavior shapes the weal and woe of society. For instance, any natural resources policy case can be analyzed functionally in terms of these eight values. All eight values are involved in social transactions. These values are "traded," influencing one another in social transactions. Who is indulged and who is deprived in value terms, and in terms of which values? How are values shaped and shared through natural resources policy and management processes? Continuously mapping the social process of any policy system or management program can reveal trends (and why trends are taking place) that may not be clear otherwise. In turn, future developments might be anticipated better and any foreseeable problems avoided and opportunities capitalized upon. Knowledge about value trends, conditions, and projections is a central element in social process mapping and can be used to aid in the invention, evaluation, and selection of alternatives to solve particular problems in natural resources policy and management.

Values also constitute the individual or group assets or resources on which people draw to achieve their goals. Wealth is used, for example, to support well-being and the pursuit of skill, and it may be used as well to "buy" power and influence. People use the respect of their friends and peers or the power of their status to build support for a cause they believe in. Or they may draw on the mental well-being they feel from knowing that other life forms are sufficiently secure and healthy to build strong communities or political alliances.

Values are also evident in the outcomes of social interaction. Outcomes distribute values in certain ways to the benefit or harm of certain people and groups. Long-term outcomes or effects of social process change institutional practices. For instance, a new law rolling back environmental protection may advance the power or wealth of a special interest group while diminishing public health or it may benefit all humans by enhancing practices to sustain their environment.

In summary, the social process analytic approach offers: (1) a practical way of accounting for differences among people, (2) a device for understanding human dynamics, and (3) practical insights for preventing or correcting weaknesses in the ongoing process of clarifying and securing common interests. This kind of knowledge and the skill in obtaining and using it can improve significantly the effective practice of all people involved in natural resources policy and management.

Decision Process

Decision process mapping is the description and analysis of the decision-making process that is a part of all policy problems (Lasswell 1971; Lasswell and McDougal 1992). Decision processes consist of seven interlinked functions, which are often carried out simultaneously in actual cases (Figure 1.1, Table 1.1). Intelligence must be gathered about the specific problem and its context. The intelligence must be debated and discussed, and problem solutions recommended and promoted. Rules or guidelines must be established to solve the problem. These must be initially specified or enforced. In turn, the rules or guidelines must be applied. Whether successful or not in solving the problem in context, the decision process or particular prescriptions gain greater specificity through application and evolve or eventually

Table 1.1. The Seven Decision Functions Essential to All Natural Resources Policy and Management Efforts

Function[a]	Activities	Examples	Standards	Examples of Questions to Ask
Intelligence	Information relevant to decision making is gathered, processed, and distributed. Planning and prediction take place. Goals are clarified.	Field work, social surveys, models, pluralistic discussion	Dependable, comprehensive yet selective, creative, available	Is intelligence being collected on all relevant components of the problem and its context and from all affected people? To whom is intelligence communicated?
Promotion	Active advocacy debate about what to do takes place. Different alternatives are promoted. Resources, data, and opinion are mobilized to secure preferred outcomes. Expectations begin to crystallize, and demands are clarified.	Forums, pluralistic discussion, recommendations	Rational, integrated, comprehensive, effective	Which groups (official or unofficial) urge which courses of action? What values are promoted or dismissed by each alternative, and what groups are served by each?
Prescription	Policies or guidelines for action are formulated and enacted. Demands are crystallized. Facts and their contexts must be examined, rules clarified, and implications of the rules examined. Rules must be specified, communicated, and approved by the partnership, government officials,	Recovery plans, other written and and verbal agreements	Comprehensive, rational, effective	Will the new prescriptions harmonize with rules by which the agencies already operate, or will they conflict? What rules does the partnership set for itself? What prescriptions are binding (these may be easier to determine if they are written down)?

(continued)

Table 1.1. continued

Function[a]	Activities	Examples	Standards	Examples of Questions to Ask
	and others concerned, i.e., those with authority (full support of officials and people involved) and control (a means to encourage compliance with rules).			
Invocation	General rules are put into practice. They begin to be applied in actual cases.	Programs are organized, teams set up, and work (research, management, public relations, analysis, etc.) begins in field, lab, and office	Timely (prompt), dependable in characterizing facts, rational, not open to abuse by individuals, effective	Is implementation consistent with prescription? Who should be held accountable to follow the rules? Who will enforce the rules?
Application	Administration, including resolving differences or deviations from the rules based on peer review, authority, or other mechanisms. Participants must interpret rules (prescriptions), supplement them if needed, and integrate old and new prescriptions into a working program. There must be enforcement as well as continuous review and approval or disapproval of behavior.	Open, pluralistic forums; internal and external means; the courts may figure prominently, but many resolutions take place formally or informally	Rational (conforming to common interest prescriptions), uniform (independent of special interests), effective (must work in practice), and constructive (mobilizing consensus and cooperation)	Will disputes be resolved by people with authority and control? How do participants interact and affect one another as they resolve disputes?

Table 1.1. continued

Function[a]	Activities	Examples	Standards	Examples of Questions to Ask
Termination	This is the cancellation of past prescriptions and frameworks and the compensation of people who are adversely affected by termination. This function is often overlooked or underappreciated.	Stopping practices that are not working as well as those that have accomplished their goals, moving to a new beginning	Prompt, respectful, and consistent with human dignity, comprehensive, balanced, and ameliorative	Who should stop or change the rules? Who is served and who is harmed by ending a program?
Appraisal	Efforts are evaluated, and continuous assessment is made of success and failure in terms of goal achievement and responsibility and accountability for what happened. This requires gathering information on how well past decision functions worked, assessing the quality of performance, and disseminating findings and recommendations to appropriate people and publics.	Formal and informal, internal and external evaluations	Dependably realistic, ongoing, independent of special interests, fully contextual (taking many factors into account, including matters of rationality, politics, and morality)	Who is served by the program and who is not? Is the program evaluated fully and regularly? Who is responsible and accountable for success or failure? By whom are one's own activities appraised?

[a]The terms *intelligence, promotion, prescription, invocation, application, termination,* and *appraisal* are described in detail by Lasswell (1971).
Source: Adapted from H. D. Lasswell, *A Pre-View of Policy Sciences.* New York: American Elsevier, 1971; and T. W. Clark and R. D. Brunner, "Making Partnerships Work in Endangered Species Conservation," *Endangered Species Update* 1996, 13:1–4

terminate. Finally, the policy or program must be appraised to determine if the problem was correctly understood originally, if it was solved, and if all decision functions were carried out well.

Some people conceive of decisions as occurring at the precise time when commitments to rules are made. This view is misleading because many interrelated decisions both precede and follow that moment. Altogether this complex of decisions makes up a decision process. In other words, decision making is not an event, it is a process.

By knowing how a decision process works or doesn't work, people in natural resources policy and management can maintain good practices or correct a poorly functioning process (Clark and Brunner 1996). A decision process can be a means of reconciling or at least productively managing competition among policies through politics. Politics is part and parcel of human society. It will always be with us because people develop and seek different policies that reflect their particular interests (special interests). Yet in many cases, like sustainability management, people must reconcile interest differences to clarify and secure their common interest. A fundamental outcome of all decision processes should be a working specification of the common interest in particular contexts. The common interest is manifest in rules or norms, both substantive and procedural. Rules or norms clarify what is to be achieved and how it will be achieved. Many different kinds of rules or norms exist for different kinds of policy problems and human communities. They may be formal or informal. Informal and formal rules include diverse community norms, norms of discussion in meetings, population viability analyses, professional standards, laws passed by representatives at local, state, or national community levels (e.g., the Endangered Species Act), and rules about rule making (e.g., the U.S. Constitution). Rules or norms are absolutely necessary for any team, group, or community to coordinate its members' expectations and actions, however imperfectly. An action by a single individual is appropriate to the degree that it meets the appropriate rules or norms in place in the relevant community; actions are inappropriate when they do not conform to community rules or norms. Of course, community rules or norms themselves may be inappropriate for the achievement of important goals. Hence the need to appraise them and make adjustments when necessary or desirable. A large body of theory and experience exists about decision processes—all seven functions—that can be used practically in natural resources policy and management (see Lasswell 1971).

Standards exist for the adequate performance of each of the decision functions, and preferred outcomes have been specified for each function to aid people in decision making (Lasswell 1971; Lasswell and McDougal 1992; Table 1.1). Table 1.1 (adapted from H. D. Lasswell 1971) describes the seven functions. Each function is evident in every complete decision process. The functions can be introduced by asking seven general questions: How is information about a problematic situation gathered, processed, and brought to the attention of decision makers? Based on this information, how are recommendations promoted and made? How are general rules prescribed? How are the rules invoked against challengers in specific cases? How are disputes in specific cases decided or resolved? How are the rules and the decision process appraised? How are the rules and the process terminated or modified?

Table 1.1 also offers examples as well as standards each function should meet, and suggests some basic questions that decision makers and other participants/observers need to ask themselves in any decision process. In many decision processes, it is not always easy to understand how each function is being carried out and to ascertain the people/groups who are responsible for how well each function is performing. By using the social process mapping categories (i.e., participants, perspectives, situations, base values, strategies, outcomes, and effects) to map each decision function, analysts, advisors, and decision makers can make judgments about the hows and whys for particular decision processes. Moreover, it is also possible to become directly involved and improve one or more decision functions so that natural resources policy and management is enhanced and the policy process runs better.

Although it is possible to see which agencies and organizations specialize in a given decision function, all groups/organizations may carry out all functions to some extent. For example, the U.S. Fish and Wildlife Service (FWS) carries out all seven functions as directed by the Endangered Species Act, but many other organizations are involved in species restoration too. The National Biological Service and university and other researchers are directly involved in gathering intelligence and planning. Nongovernmental conservation groups and businesses are often highly involved in promoting one plan over others, although it should be evident that all groups and people, regardless of their claims of objectivity and neutrality, promote decisions they hope will meet their own interests. Rules are set not only by legislative bodies, but also by agencies and groups of all kinds. These have enormous influence in the design and operation of actual conservation efforts down to the field level. The FWS and other agencies and organizations are involved in implementing wildlife conservation programs. Agencies and other diverse people and groups are also involved in dispute resolution, as are the courts. The media are also involved in reporting on conflicts and all other aspects of the decision process. Many others are involved as well in appraisal or review and evaluation of natural resources policy and management efforts. The decision to terminate a policy or program is usually made by government. However, termination involves many other organizations. In any decision process, we may expect to find several official and unofficial participants involved.

Natural resources policy and management will be much more effective and efficient when and if it develops and maintains high-quality decision processes. In part, this depends on whether participants learn explicitly about how the decision process works and how they can improve decisions. The decision functions introduced in Table 1.1 can be used to ask specific questions about any decision-making process. Concerning values, fair trading and mutual exchange must take place among participants in order for a community, organization, or team to improve policy. However, in some and perhaps most decision processes, participants do not share similar priorities, and little individual or group effort is expended on developing a common, well-grounded perspective. As a result, a decision process can be easily distorted by using power or wealth to centralize, concentrate, or legalize certain decision functions for the benefit of special interests and to the detriment of other people's interests.

The decision process of a natural resources policy and management case should be an open, flexible, and fair means to produce operational rules or norms for all participants in achieving their common interest (Table 1.1). Obviously, efforts to resolve policy problems will not work if some people seek their own special interests at the expense of the common interest as determined by the community. After rules are specified in a fair way and agreed to, then rules or norms must be enforced against any and all people who challenge them. Rules or norms can be appraised by the community and changed as needed— that is, assuming they are clear enough to be evaluated.

Decision making must be grounded in actual contexts. It must be comprehensive yet manageable given time and other resources constraints. The decision-making framework within the policy sciences is a tool for building a map of each particular decision process. The resulting map can be used by people interested in improving natural resources policy and management. Building and using such decision process maps will help bring about adequate intelligence and appraisal and will improve all functions as well. Good decision making requires a certain pattern of thought and action. The central challenge to improved natural resources policy and management is the crafting and maintaining of good decision processes.

Observational Standpoint

Observational standpoints are held by all, including those who are engaged in policy processes (Lasswell 1971; Lasswell and McDougal 1992). Standpoint consists of an individual's value orientations and biases resulting from personality, disciplinary training, experiences (parochial/universal), epistemological assumptions, and organizational allegiances, for example. All observers and participants, especially professionals and analysts, should seek to be clear and realistic about their own standpoint and those of others, using both selective and comprehensive "lenses" (Figure 1.1).

Aside from the individual or collective work to be done in any sustainable policy and management effort, managing oneself constructively is essential (Clark and Wallace 1999). Knowing one's own behavior and role in any natural resources policy and management process requires knowing not only about the biological challenge and the organizational environment but also about oneself. All too often, people assume, perhaps unconsciously, that they know what they are doing and why, and that other people will see and appreciate their good intentions and works. Clarifying one's standpoint relative to both the natural resources policy and management challenge and other participants is important to achieving goals, just as knowing the technical, substantive issues is an important variable. Being clear about one's standpoint can aid successful teamwork and successful conservation. Being unclear can lead to problems, conflict, and failure.

Policy participants can be more or less aware of both themselves and other people, depending on how self-reflective and observant they are (Schon 1983). To be as effective as possible, participants must be clear about their standpoint—how they fit into the process—and seek to understand and discount biases. This is possible only by being self-aware and using that knowledge of self when making professional judgments and engaging in interpersonal relationships (Clark and Wallace 1999). Most individuals have at one time or another analyzed their actions and role in relationships, whether with a spouse, partner, parent, sibling, or friend. We do this in order to know where we stand with someone concerning expectations, demands, trust, and many other aspects of relationships. This self-analysis is at the heart of clarifying one's standpoint, and the process, in general, is the same in both professional and nonprofessional relationships. All people have biases as a result of experience, personality, interests, and education, among other things. Learning about one's own standpoint and the perspectives of others is not easy, but it is essential to effective professional practice. Over time, as participants gain experience they improve their understanding of self and others. For policy participants to reach their potential for effectiveness either technically or as leaders, they must be able to look at and understand themselves and others involved or interested in the natural resources policy and management issue of concern.

To clarify your standpoint as a participant in any policy process, you should start by asking yourself five questions about your roles, involvement, work, shaping factors, and orientation (Willard 1988). First, what roles are you and other people engaged in while working in the policy effort—scientist, technician, manager, student, teacher, advocate, advisor, reporter, decision maker, scholar, facilitator, concerned citizen, or others? Second, what problem-solving tasks do you carry out when performing your roles—clarifying goals, determining historical trends, analyzing conditions, projecting trends, and inventing and evaluating alternatives? Third, what factors shape how you carry out your tasks and roles—culture, class, interests, personality, and previous experience? Fourth, what conditioning factors shape your personal and professional "approach" in general and in reference to any particular natural resources policy case? Which approaches or roles are you predisposed toward or against, and how are you predisposed to conduct your own analytic work from each? Fifth, how does your approach shape how you carry out the intellectual tasks associated with your roles? For example, what is the impact of your own "reflective approach" on the goals you clarify for yourself, and how you specify them, on the trends you identify and describe, on the conditions you analyze and how you analyze them, on the projections you make and how you make them, and on the alternatives you invent, evaluate, and select?

You can assume four observational standpoints with regard to policy (Lasswell 1938): participant, spectator, interviewer, and collector. Aside from standpoint, it is important for all to be clear about whether you intend to be descriptive and/or prescriptive in thought and action. Simply describing a natural resources policy and management process is different from seeking prescriptive ways to improve it. Participants in policy processes should always be alert to their standpoint and any biases that may be operating.

The four dimensions of the policy sciences (i.e., problem orientation, social process, decision process, and observational standpoint) are discussed in more detail in Chapter 2 and are applied to diverse natural resources policy problems in Part 2 of this book. Because grasping the policy sciences in the abstract may be difficult, it is best to examine them in light of their application to actual problematic settings.

The Course

The policy sciences introduced above are elaborated in the "Foundations" course, which begins with an introduction to natural resources policy and is followed by an overview of the policy sciences approach. Over six weeks the policy sciences are examined in some detail in this order: social process, decision process, problem orientation, standpoint and method, case illustrations, and sustainability. This sequence of topics begins with people in social process (i.e., the context) and moves to decision making, problem solving, standpoint, and method. The course recognizes that many natural resources professionals are confronted by complex, policy-relevant problems. They need a practical guide, which must rest on an inclusive conception of the social process in which all natural resources problems occur, because all problems are affected by interaction with their contexts.

An outline intended to help guide thinking about particular problems is distributed (described in Chapter 2). This outline, organized around policy sciences concepts, is used to structure the papers written for the course. The papers, some of which are included in Part 2 of this book, are applications or demonstrations of the policy sciences.

One feature of the "Foundations" course is the importance instructors give to the students' presentations of their case studies. Learning how to communicate effectively to a group is a valuable skill that tends to be underemphasized in formal education. A form, entitled "Appraisal of Presentations" (Table 1.2), helps presenters to improve their performance and allows course members to evaluate constructively their colleagues' presentations. The similarity between the categories of the appraisal form and those of the paper format or outline is intentional. The idea is to show how key concepts of the policy sciences can be used for different purposes.

Although the paper format and the appraisal form have been used only in the "Foundations" course, we expect that they can be used by readers grappling with their own efforts to improve natural resources policy and management. Among these challenges are how to clarify and secure the common interest, understand and manage contexts (i.e., social process), meet the highest standards in carrying out the func-

Table 1.2. Appraisal of Presentations

I. Introduction
    1. Was the policy problem stated clearly and simply?   1 2 3[a]
    2. Were the purposes of the presentation stated clearly and simply?   1 2 3
    3. Was the presenter's standpoint clarified?   1 2 3
    4. Were the presenter's methods clear?   1 2 3

II. Problem
    1. Was the problem's context (social process) adequately detailed?   1 2 3
    2. Was the problem's status relative to the decision process clear?   1 2 3
    3. Were goals sought in reference to the problem clarified?   1 2 3

III. Analysis
    1. Were relevant trends (history) adequately described?   1 2 3
    2. Were conditions (factors) that shaped trends adequately described?   1 2 3
    3. Were future trends (projections) adequately described?   1 2 3

IV. Recommendations
    1. Were alternatives to resolve the problem adequately described?   1 2 3
    2. Were alternatives adequately evaluated?   1 2 3
    3. Was the selected alternative (strategy) or complex of strategies appropriate to achieve goals and solve the problem?   1 2 3

V. Conclusion
    1. Were the goals and the problems clearly re-stated?   1 2 3
    2. Was the recommendation stated and justified?   1 2 3

VI. Other criteria
    1. How would you rate the overall quality of the presentation?   1 2 3
    2. How would you rate the use of overheads and other visuals?   1 2 3
    3. What is your recommendation to improve presentation and analytic style? Use reverse side to detail your recommendations.   1 2 3

[a]1 = excellent, 2 = good, 3 = needs work.

tions of decision making, understand problems practically and invent and evaluate solutions, and recognize or create and capitalize on opportunities to improve outcomes in substantive and process terms.

Evaluation

The following evaluation is preliminary, based on student performance in the classroom, anonymous student evaluations, and instructors' appraisals. The real utility of the course can be determined only by how students use the policy sciences throughout their professional careers.

Student Presentations and Papers

The student presentations and papers were generally of high quality. In the majority of instances, students employed policy sciences concepts and language skillfully. The paper topics have been diverse. In 1996, student applications included:

• The International Organization for Standardization and its ios 14000 Process: Meeting environmental management standards
• The problem of bull trout conservation in the Pacific northwest
• Linking biodiversity conservation and development through integrated-development projects: Thailand's Khao Yai National Park case
• Soil contamination as an environmental problem
• International mining practices in Quitovac, Sonora, Mexico
• Restructuring risk in decision making
• Appraisal of New Haven, Connecticut, neighborhood beautification programs
• Management of the sea urchin fishery in Maine
• Improving sea turtle conservation via international use of turtle excluder devices on shrimp trawlers
• Appraisal of the international ecotourism industry
• Appraisal of the Seventh American Forest Congress
• Yale Forest Forum: An agent for change in forest policy, science, and management

In 1997, student applications included:

• Decision process appraisal: Water quality in the Quinnipiac River Watershed, Connecticut
• Management of the endangered Barton Springs salamander, Austin, Texas
• Forest conservation and land use planning for sustainable development in eastern Guatemala
• Improving habitat conservation planning
• An integrated approach to conservation and human development in Kyabobo Range National Park, Ghana
• A gold mining project in northern Costa Rica

- Expanding air pollution abatement opportunities in Baltimore, Maryland
- The proposed pumped-storage hydroelectric power projects, Sequatchie Valley, Tennessee
- Management of Voyageurs National Park, Minnesota
- Public production forests in the Brazilian Amazon
- Appraisal of the National Biodiversity action plan for Indonesia's Kerinci Seblat National Park
- Appraisal of the Ngorongoro Conservation Area General Management Plan
- Dilemma of dignity for the people of Pululahua, Ecuador
- Analysis of the Adirondack Park wolf recovery process
- Improving the policy process for the restoration of Beaver Ponds Park, New Haven, Connecticut
- Greening the United Nations High Commissioner for Refugees (UNHCR): Improving UNHCR's environmental management practices in refugee situations

In 1998, student applications included:

- A preliminary appraisal of elk management on the National Elk Refuge, Wyoming
- Hunting for an acceptable bison hunt near Yellowstone National Park
- Incorporating environmental objectives into land use planning and regulation: Connecticut as case study
- Community-based conservation in Kenya: A policy analysis of the Kimana Community Wildlife Sanctuary
- Conservation of biodiversity in Comoros: Analysis of the United Nations Development Program's Global Environmental Fund Project
- An environmental health policy problem: The continued use of DDT
- Public participation in the planning and management of Lake Umbagog National Wildlife Refuge: A case study
- One canal, two parks, no effective policy process (yet)
- Endangered species recovery on private land: A case study of the Peregrine Fund's Habitat Conservation Plan for the recovery of the Aplomado falcon in south Texas
- Candidate conservation in Idaho: An analysis of two programs
- Analysis of the proposed land swap between the Town of Branford and the Queach/Vigliotti Corporation for the residential/golf course development
- Analysis of the decision-making process operating in the management of contaminated dredged materials in Hudson-Raritan Bay
- The values of trees and forests: Maine's forest management debate
- Asiatic black bear conservation in Korea
- Reintroduction of the Mexican wolf to part of its former range in the southwestern United States
- Tigers, Takin, and district administrators: In situ conservation policy issues in northern Bhutan
- The planning committee for Brownfields redevelopment in the Naugatuck River valley of Connecticut: A decision analysis
- An analysis of the grizzly bear reintroduction process in the Bitterroot ecosystem, Idaho
- Analysis of visitor regulation in Acadia National Park, Maine

• Grizzly bear recovery in the Greater Yellowstone Ecosystem: The problem of human-caused mortality
• Dead Sea development and the possibility of cooperation between Israel, Jordan, and the Palestinian National Authority
• Cows in the cathedral: Sacred or sacrilege? An analysis of grazing in Grand Teton National Park

Each year numerous other case examples were brought into classroom discussion based on student, instructor, and guest speaker experiences. All the cases provided a rich array of examples to explore the utility of the policy sciences.

Student Evaluations

Students were required to complete a course evaluation form. All students indicated that the course content was "good" or "excellent," and course goals were clear and well met. The requirements for presentations and papers were rated "highly relevant" by nearly all students. Overall course work load was judged as "manageable" to "heavy." The course size of about sixteen students was ranked "just right."

The students also made useful recommendations for improving the class. Many felt that more discussion in the beginning and more case studies throughout would help them develop their thinking more quickly. One recommendation was for the instructors to use an active project throughout the semester to illustrate and debate concepts. Working together in groups to prepare case studies was also suggested. Another suggestion was for the instructors to give a capstone lecture at the end of the course. Some thought that the volume of reading, especially the theoretical works early in the semester, was daunting and could be reduced, although there was the admission that rereading it was very helpful, particularly after work on the case studies was under way and the policy sciences framework was understood more fully.

A number of students made very positive comments about the course. Several thought it was an excellent class, especially once they learned exactly what it was about. They thought it was useful to apply the concepts through a case analysis. Some said they would highly recommend the class to prospective students. Others felt that it should be compulsory for all students because it provided a way of conceptualizing natural resources problems in a very clear and comprehensive way. Another said it was an excellent course for everyone, no matter what one's area of specialization. One student described it as the best course he had taken at Yale in his two-year program and that it taught a usable, highly practical skill. Another wrote that it was a "very useful course," great for problem analysis, formulation of solutions, and grasping one's own standpoint in relation to problems and solutions.

Instructors' Evaluation

Policy sciences theory and methods represent a new way of thinking, analyzing, and understanding natural resources problems for students, and the course therefore makes demands on them unlike most of their previous university or professional experiences. The concepts and terminology are largely new, and the readings are often considered dense. At first, many of the students, some of whom come from scientific or activist backgrounds, are not quite ready to accept that natural resources problems are actually policy problems. Many are also not ready to assimilate the concepts and language and, for a host of reasons, are not prepared to meet the demands of policy analysis. As a result, teaching the policy sciences is always a pleasant challenge.

When introduced, the categories in the policy sciences framework are not always fully appreciated by students. As Lasswell (1965a:17) noted, the "function [of the concepts and language] is not to introduce a new cult but to give a sounder general analysis [to social and decision process] than has been possible heretofore." The terms used in the policy sciences are neither empty nor meaningless; they are essential guides to analysis and communication (Eulau 1969). Making this clear to students is a perennial challenge.

Typically, about midway through the semester, before the students begin to understand and connect the concepts and language to their cases in workable, analytic ways, some grow skeptical about the utility of the policy sciences. Resistance develops (see Moore 1968). Unable to embrace or apply the policy sciences fully, many voice uncertainty about the course's benefit to their professional education. In each semester we have taught this class, this has proven to be a temporary phase in students' intellectual development. This uncertainty comes about in part because, as they progressively learn the policy sciences, their original analytical approaches are brought into sharper focus and into question and, eventually, in most cases, are abandoned—at least temporarily—in favor of the policy sciences. For some students, embracing the policy sciences requires them to undergo a fundamental change in their intellectual outlook.

From the instructors' viewpoint, the major challenge is to introduce policy sciences theory, methods, and language in ways that mesh as smoothly as possible with students' previous conceptions about problem solving, policy process, and professional roles. We meet this challenge with explanations, examples, and lectures, but have been differentially successful. We encourage constructive professional and intellectual reflection about different policy models and professional and personal perspectives, and we discuss concerns that students raise openly with the class. This seems to alleviate the anxiety some students experience and further illustrates the practical benefit of the policy sciences approach. As each semester has progressed, students become better able to see the utility of the policy sciences for themselves, and their reservations largely evaporate.

Conclusions

The "Foundations" course is a practical vehicle for examining the problems of sustainable natural resources policy and management and for teaching the policy sciences approach to meet this challenge. The readings, lectures, discussions, case examples, student analyses, presentations, and written papers provide a varied and useful mix to explore the policy sciences and their practical application. The students demonstrate a solid ability to apply the policy sciences approach to diverse, real natural resources policy and management cases. Both instructors and students consider the course successful in meeting learning and skill goals. The policy sciences should be of broad interest to all people interested in improving natural resources policy and management.

LITERATURE CITED

Arsanjani, M. H. 1981. *International Regulation of Internal Resources: A Study of Law and Policy.* Charlottesville: University Press of Virginia.

Ascher, W., and R. Healy. 1990. *Natural Resource Policymaking in Developing Countries: Environment, Economic Growth, and Income Distribution.* Durham, NC: Duke University Press.

Berger, P., and T. Luckmann. 1987. *The Social Construction of Reality: A Treatise in the Sociology of Knowledge.* New York: Penguin Books.

Brewer, G. D. 1988. "Policy Sciences, the Environment and Public Health," *Health Promotion* 2:227–237.

Brewer, G. D., and J. S. Kakalik. 1979. *Handicapped Children: Strategies for Improving Services.* New York: McGraw-Hill.

Brunner, R. D. 1997a. "Raising Standards: A Prototyping Strategy for Undergraduate Education," *Policy Sciences* 30:167–189.

———. 1997b. "Teaching the Policy Sciences: Reflections on a Graduate Seminar," *Policy Sciences* 30:217–231.

———. 1997c. "Introduction to the Policy Sciences," *Policy Sciences* 30:191–215.

Brunner, R. D., and W. Ascher. 1992. "Science and Social Responsibility," *Policy Sciences* 25:295–331.

Brunner, R. D., and R. Byerly, Jr. 1990. "The Space Station Program: Defining the Problem," *Space Policy* May:131–145.

Brunner, R. D., and T. W. Clark. 1997. "A Practice-Based Approach to Ecosystem Management," *Conservation Biology* 11:48–58.

Clark, T. W. 1997. "Conservation Biologists in the Policy Process: Learning How to Be Practical and Effective," in G. K. Meffe and C. R. Carroll, eds., *Principles of Conservation Biology*, pp. 575–597. 2d ed. Sunderland, MA: Sinauer Associates.

Clark, T. W., and M. S. Ashton. 1999. "Field Trips in Natural Resources Professional Education: The Panama Case and Recommendations," *Journal of Sustainable Forestry* 8(3/4):181–198.

Clark, T. W., and R. D. Brunner. 1996. "Making Partnerships Work in Endangered Species Conservation: An Introduction to Decision Process," *Endangered Species Update* 13:1–4.

Clark, T. W., N. Mazur, S. Cork, S. Dovers, and R. Harding. In press. "The Koala Conservation Policy Process: An Appraisal and Recommendations," *Conservation Biology*.

Clark, T. W., P. Paquet, and A. P. Curlee, eds. 1996. "Conservation of Large Carnivores in the Rocky Mountains of North America," *Conservation Biology* 10:936–1058.

Clark, T. W., and R. L. Wallace. 1998. "Understanding the Human Factor in Endangered Species Recovery: An Introduction to Human Social Process," *Endangered Species Update* 15(1):2–9.

———. 1999. "The Professional in Endangered Species Conservation: An Introduction to Standpoint Clarification," *Endangered Species Update* 16(1):9–13.

Dewey, J. 1933. "What Is Thinking? and Why Reflective Thinking Must Be an Educational Aim," in J. A. Boydston, ed., *John Dewey: The Later Works, 1925–1953*, pp. 111–139. Carbondale: Southern Illinois University Press.

Eulau, H. 1969. "The Maddening Methods of Harold D. Lasswell," *Journal of Politics* 30:3024.

Gunderson, L. H., C. S. Holling, S. S. Light, eds. 1995. *Barriers and Bridges to the Renewal of Ecosystems and Institutions*. New York: Columbia University Press.

Johnston, D. 1987. *The International Law of Fisheries: A Framework for Policy-Oriented Inquiries*. New Haven, CT: New Haven Press.

Lasswell, H. D. 1938. "Intensive and Extensive Methods of Observing the Personality-Cultural Manifold," *Yenching Journal of Social Studies* 1:72–86.

———. 1948. "The Structure and Function of Communication in Society," in L. Bryson, ed., *The Communication of Ideas*, pp. 37–51. New York: Institute for Religious and Social Studies.

———. 1965a. *World Politics and Personal Insecurity*. New York: Free Press.

———. 1965b. "The World Revolution of Our Time: A Framework for Basic Policy Research," in H. D. Lasswell and D. Lerner, eds., *World Revolutionary Elites: Studies in Coercive Ideological Movements*, pp. 29–96. Cambridge, MA: M.I.T. Press.

———. 1971. *A Pre-View of Policy Sciences*. New York: American Elsevier.

Lasswell, H. D., and A. Kaplan. 1950. *Power and Society: A Framework for Political Inquiry*. New Haven: Yale University Press.

Lasswell, H. D., D. Lerner, and H. Speier, eds. 1979–1980. *Propaganda and Communication in World History*, vols. 1–3. Honolulu: University Press of Hawaii for the East-West Center.

Lasswell, H. D., and M. S. McDougal. 1992. *Jurisprudence for a Free Society: Studies in Law, Science, and Policy*. New Haven: New Haven Press.

Lee, K. N. 1993. *Compass and Gyroscope: Integrating Science and Politics for the Environment*. Washington, DC: Island Press.

Ludwig, D., R. Hilbon, and C. J. Walters. 1993. "Uncertainty, Resource Exploitation and Conservation: Lessons from History," *Science* 260:17, 36.

McDougal, M. S., and W. T. Burke. 1987. *The Public Order of the Oceans: A Contemporary International Law of the Sea*. New Haven: New Haven Press.

McDougal, M. S., and F. P. Feliciano. 1994. *The International Law of War: Transnational Coercion and World Public Order*. New Haven: New Haven Press.

McDougal, M. S., H. D. Lasswell, and L. Chen. 1980. *Human Rights and World Public Order: The Basic Policies of an International Law of Human Dignity*. New Haven: Yale University Press.

McDougal, M. S., H. D. Lasswell, and J. C. Miller. 1994. *The Interpretation of International Agreements and World Public Order: Principles of Content and Procedure*. New Haven: New Haven Press.

McDougal, M. S., H. D. Lasswell, and I. Vlasic. 1963. *Law and Public Order in Space*. New Haven: Yale University Press.

McDougal, M. S., and W. M. Reisman. 1981. *International Law Essays: A Supplement to International Law in Contemporary Perspective*. Mineola, NY: The Foundation Press.

Moore, J. N. 1968. "Prolegomenon to the Jurisprudence of Myres McDougal and Harold Lasswell," *Virginia Law Review* 54:662–688.

Murty, B. S. 1989a. *The International Law of Propaganda: The Ideological Instrument and World Public Order*. New Haven: New Haven Press.

———. 1989b. *The International Law of Diplomacy: The Diplomatic Instrument and World Public Order*. New Haven: New Haven Press.

Pielke, R. A., Jr. 1997. "Asking the Right Questions: Atmospheric Sciences Research and Societal Needs," *Bulletin of the American Meteorological Society* 78:255–264.

Reisman, W. M. 1971. *Nullity and Revision: The Review and Enforcement of International Judgments and Awards*. New Haven: Yale University Press.

———. 1981. "International Lawmaking: A Process of Communication," *Proceedings of the American Society of International Law* 1981:101–120.

Sahurie, E. J. 1992. *The International Law of Antarctica*. New Haven: New Haven Press.

Schon, D. 1983. *The Reflective Practitioner: How Professionals Think in Action*. New York: Basic Books.

Wallace, R. L., and T. W. Clark. In press. "Solving Problems in Endangered Species Conservation: An Introduction to Problem Orientation," *Endangered Species Update*.

Walters, C. J. 1986. *Adaptive Management of Renewable Resources*. Macmillan, New York.

Willard, A. R. 1988. "Incidents: An Essay in Method," in W. M. Reisman and A. R. Willard, eds. *International Incidents: The Law that Counts in World Politics*, pp. 25–39. Princeton, NJ: Princeton University Press.

Willard, A. R., and C. Norchi. 1993. "The Decision Seminar as an Instrument of Power and Enlightenment," *Political Psychology* 14:575–606.

2 Analyzing Natural Resources Policy and Management

Tim W. Clark and Andrew R. Willard

To improve problematic natural resources policy and management, an analyst, citizen, or professional first needs to understand the processes involved, and on that basis come up with a likely solution, a realistic intervention strategy that will improve matters. Carrying out a practical analysis of policy and management processes, including mastering the substance or technical aspects of an issue, is not always straightforward or easy. A problem-oriented, contextual, multi-method approach is needed, and the policy sciences interdisciplinary approach is recommended. This method focuses on the problem at hand comprehensively, whatever it is, and its social and decision context. This approach serves as a "stable frame of reference" from case to case and over time, and it grounds one's understanding of a particular problem in a conceptually and practically solid way. This chapter describes and illustrates an outline for guiding analysis into any policy and management process. It is developed from policy sciences theory and methods, and from considerable experience applying them to actual natural resources policy and management cases.

Overview of the Outline

It is often said that if you can write out your argument or analysis in a clear, intelligible fashion, then you understand it and can communi-

cate your understanding to others. Other people may be persuaded by
your analysis. The following outline offers a sound way to go about structuring your analysis and recommendations. The outline or guide uses
the concepts and categories of the policy sciences introduced in Chapter 1. The policy sciences may be used more or less explicitly, or the
outline may be converted to different, more conventional language
given the topic and the intended audience. The ten case studies in Part
2 of this book show considerable diversity in the use of the recommended outline and terminology. Some case studies (e.g., Chapters 5 and
10) follow the outline closely and use policy sciences terms, whereas
other studies (e.g., Chapters 8 and 11) are guided by the outline and include all the necessary components but do so in creative ways, with little use of technical policy sciences terms. The flexibility of the outline
is evident as applied in the ten cases. Consult the cases to explore and
study this flexibility and diversity.

Short Outline

The outline in short form using only the major headings is as follows. The outline is problem-oriented and contextual, as introduced in
Chapter 1. It can be used to describe, analyze, and recommend practical solutions to policy problems.

> Title
> Abstract
> Introduction
> I. Problem (Description of the problem(s))
> II. Analysis of the Problem (trends, conditions, projections)—
> The what, the why, and what's likely to happen?
> III. Recommendations (Alternative(s) promoted)—What to do?
> Justify your recommendation
> Conclusion
> Acknowledgments
> Endnotes (or use footnotes; keep short)
> Literature Cited

Rationale for the Outline

It is necessary to look at the rationale for this outline before we examine the outline itself in detail, section by section, and before we illustrate how it was used by the ten case study authors. The outline is
structured following the interdisciplinary problem-solving approach
of the policy sciences. As such, it is problem-oriented and contextual in

terms of the social and decision process involved in the natural re-sources policy and management case under investigation. This ap-proach permits a comprehensive, yet selective, systematic examination of the policy problem and possible solutions (Willard 1988). The foun-dation for this outline comes from policy sciences theory, especially Lasswell and Kaplan (1950), Lasswell (1971), and Lasswell and Mc-Dougal (1992). Remember that the four principal dimensions of the policy sciences are: problem orientation, social process mapping, de-cision process mapping, and observational standpoint.

This approach makes special demands on its users. It asks you, the analyst, to enter into a frame of reference that may be quite different from your own modus operandi, especially if you have traditionally re-lied on convention (i.e., commonplace assumptions and explanations for human interactions). As such, the approach invites a certain intel-lectual flexibility, a capacity for viewing the world through a system-atic, comprehensive framework, which is unlike conventional viewing. People who are good at abstracting from experience quickly pick up the "logic" of the outline and become skilled in its practical use.

The outline also asks you, the analyst, to clarify your standpoint rel-ative to the issue under examination and to the standpoints of the par-ticipants in the policy process. It asks you to acknowledge precon-ceived outlooks and limited conventional notions about what people do and why. The outline, and policy sciences in general, recommends that you clarify your standpoint as an ideal worth striving for if you want to account for your biases in analysis and improve policy processes and outcomes. Conventional approaches are simply not up to the task of systematic analysis. In using convention, people may fall back on their individual experiences and everyday notions, encompassing such clichés as "personalities, power, and politics," their biases, including their out-look or ideology, to explain why a process happens as it does. Or they may focus only on the technical aspects of a problem as best they can and conclude that if only more positivistic science were available or if decision makers would simply listen to scientists, everything would be fine. Such conclusions are misleading and superficial. Moreover, con-ventional approaches often conclude that matters would be made bet-ter if the people involved "would just do the right thing." Such conclu-sions indicate that analysts have not clarified their standpoint relative to that of other people involved in the process or come to appreciate the likelihood that they themselves are trapped in conventional outlooks. The conventional approach typically falls far short of providing the kind of systematic process analysis and practical insight required to un-derstand and improve policy processes in the real world.

Using the policy sciences interdisciplinary approach introduced in Chapter 1 and outlined and elaborated in this chapter for case analysis puts you, the analyst, in a good position to engage the task of improving natural resources policy and management with the necessary intellectual and imaginative flexibility required to move society closer to sustainable conservation and development.

Full Outline

The problem-oriented, contextual outline is flexible and can be used in creative and tailored ways, depending on the author's preferences, the topic under investigation, and the intended audience. In the next section below, the outline is described in more detail and illustrated with the ten case studies. Part 2 shows in detail how the outline was used by different authors in their studies of diverse natural resources policy and management problems.

Title
Your Name
Affiliation
Address (current and permanent, including phone numbers)
Abstract (1 paragraph of less than 200 words)
Introduction
    A. 1st paragraph: Be problem-oriented = goals, problems, alternative(s); say "the policy problem is . . ."
    B. 2nd paragraph: Specify the purposes of your paper (3 purposes—describe, analyze, and recommend)
    C. 3rd & 4th paragraphs: Clarify your standpoint in reference to the problem(s) of concern, including specification of the methodology to be used in your study/paper
       1. Clarify your role, including motivation for interest in the problem(s) of concern and roles you performed (e.g., scientist, advocate, advisor, decision maker, etc.)
       2. Contemplative orientation (i.e., intensive and extensive observational standpoints)
          a. Participant-observer, interviewer, spectator, collector
          b. Case studies
          c. Prototypes
I. Problem (Description of the problem(s))
    A. Provisional statement of the problem(s) in natural resources policy and management that is (are) the subject of your study and provisional clarification of goals

B. Specify contextually and in some detail the problem(s) and goals in natural resources policy and management that is (are) the subject of your study

1. Identify which decision function(s) is (are) to be emphasized in your treatment of the problem(s) of concern. Explain why. Decision functions include: intelligence, promotion, prescription, invocation, application, termination, appraisal

2. Describe how each relevant decision function is performed in reference to the problem(s) of concern by using feature analysis. Feature analysis permits you to give a more detailed specification of the context of performance. Feature analysis requires that you address the following questions in reference to each relevant decision function:

 a. Who participates?
 b. With what perspectives?
 c. In which arenas?
 d. Using what base values?
 e. In what strategic ways?
 f. To generate what outcomes?

3. In light of the description you develop above (by addressing the questions listed in item B.2), make a "final" determination and more detailed specification of the problem(s) and goals in natural resources policy and management that are your primary focus of concern

II. Analysis of the Problem (trends, conditions, projections)— The what, the why, and what's likely to happen?

A. Description of trends in decision making that have had an impact on the problem of concern, including identification of particular impacts and their relation to the achievement of goals

B. Identification and examination of the factors that have shaped the trends and impacts described in item II.A

C. Projection of future trends in decision and accompanying impacts, with an emphasis on exploring the relation between projected impacts and the achievement of goals

III. Recommendations (alternative(s) promoted)—What to do? Justify your recommendation.

A. Invention of alternative(s) for resolving the problem given the projections described in II.C above

B. Evaluation of alternative strategies proposed for their potential contribution towards reaching goals

C. Selection and justification of particular strategy or complex of strategies to resolve the problem as defined

Conclusions
   A. Very brief re-statement of goal(s), problem
   B. Your recommendation(s) and justification to solve the prob-
lem as defined
Acknowledgments
Endnotes (or use footnotes; keep short)
Literature Cited

Description and Illustration of the Outline

This discussion explores each major heading in the full outline, giving
more complete guidance—a how-to—in applying it. The ten cases in
Part 2 demonstrate how various authors used the outline and also the
flexibility possible in analyzing cases. These cases can help you see
how to use the outline to address a range of complex natural resources
issues from around the world (also see Willard 1988).

Supplying the Title, Author, and Abstract

Your paper should have a clear, concise, descriptive title. It should
be approximately ten words. Two good examples are Lieberknecht's
(Chapter 7) "How Everything Becomes Bigger in Texas: The Barton
Springs Salamander Controversy" and Wilshusen's (Chapter 11) "Local
Participation in Conservation and Development Projects: Ends, Means,
and Power Dynamics."
   The title should be followed by your full name, affiliation, and address including contact numbers (phone, fax, email).
   The abstract is very important, as potential readers often make a
judgment about whether or not to read a paper based on its title and
abstract. The abstract should be one paragraph of less than two hundred words. Tell what the policy problem is, give details of your analysis, and list your specific recommendations. The abstract should be
highly readable. It should not only broadly describe what is in the paper
but also summarize its substantive contents. The ten cases generally
follow this recommendation and can be consulted for examples.

Introducing the Policy Problem at Hand

The introduction should be attempted early on in writing but should
be finalized when your paper is largely finished. Be concise, interesting, and clear. The introduction is brief and should be composed of
roughly four paragraphs, as recommended by the outline.

The first paragraph is a key to the paper, as it sets up the policy problem and potential solutions for the reader. It introduces goals, problems, and alternatives. This paragraph also introduces the policy problem at hand. Be sure to say explicitly "The policy problem is. . . ." For example, Bowes-Lyon (Chapter 10), in his study of integrated conservation and human development in Africa, notes that the policy problem is that even though great benefits are possible from the establishment of a national park in Ghana, West Africa, the park can also produce value losses to villagers resulting from the loss of traditional lands and limited future access to resources historically available. Also, if the perceived benefits from the park, such as rural development, tourism, and employment, are not realized in a reasonable time frame, then local support will most likely decrease or disappear. The policy problem is how to establish the park and ensure that local development takes place simultaneously in a clear and timely way that benefits all concerned.

The second paragraph states the paper's three purposes, which are to describe, analyze, and recommend. The three major parts of each paper correspond with these three purposes. For example, Kaczka's (Chapter 6) introduction to his investigation of sea turtle conservation policy states that his paper's purposes are to "describe, contextually and in detail, the threat to sea turtles from shrimp trawlers, and current policies to control those impacts outside the United States. . . . Second, the policy sciences are used to analyze current policy defects. . . . Finally, I recommend alternative measures to increase TEDs [turtle excluder devices] use and improve sea turtle conservation." All ten cases follow this basic three-purpose formula.

The third paragraph of the introduction includes the clarification of your standpoint in reference to the policy problem(s) of concern, including specification of your contemplative orientation as participant-observer, interviewer, spectator, or collector (Lasswell 1938; Chapter 1 in this volume). Your motivation and interest in the problem(s) of concern and the roles you performed in carrying out your investigation (e.g., scientist, advocate, advisor, decision maker) are described as well. For example, Cromley, in her study of grizzly bear management in the Greater Yellowstone Ecosystem (Chapter 8), states that she "acted as a collector, spectator, interviewer, and participant observer." She carried out both intensive and extensive observational activities, and her conclusions and recommendations are based on a synthesis and integration of the information obtained from these differing standpoints. Elwell (Chapter 5) spends nearly two pages detailing his standpoint. He notes that he is an academic and a professional forester, and that he has a personal interest in the topic of concern as he lives in the region. He admits that he feels that the Tennessee Valley Authority (TVA) proposed project would be detrimental to the values he places on the area. Therefore he aligned himself with the citizens' opposition to the project and sought to aid their efforts.

The fourth paragraph summarizes your methods briefly so that readers can understand how you went about your work. Your methods should be described in sufficient detail so that anyone else could duplicate them. For example, Cromley (Chapter 8) began her study by "conducting an attitudinal survey in the summers of 1996 and 1997." She "began studying the Bear 209 case by surveying books and articles on grizzly bear management written by academics and researchers. These provided [her] with analysis and a historical context." She goes on to more fully detail her methods for the reader.

When you complete the introduction, ask yourself the following questions: Is the policy problem stated clearly and simply? Are the purposes of the paper stated clearly and simply? Is my standpoint clarified? And, are my methods clear?

Describing the Policy Problem

This major section introduces the policy problem in more detail including how it figures in the ongoing decision and social process under way. It begins by asking you to state provisionally what you think is the policy problem in your subject of study. This requires that you identify and state the goal of the policy process you are investigating. For example, Elwell's (Chapter 5) examination of a water power project in Tennessee first notes the policy problem is "[citizen] exclusion from the decision process [led by the TVA]." He goes on to note that the goal is "authentic inclusion in the decision process." His later and more complete analysis leads him to stick with this problem statement and goal. However, quite often analysts change or refine their problem statement and goal after conducting their analysis and learning more about the problem and its context.

Next the outline asks you to identify which decision function or functions are the focus of your study. Decision functions include: intelligence, promotion, prescription, invocation, application, termination, and appraisal (see Chapter 1, Table 1.1). Explain why. For example, Flores's (Chapter 3) study of ozone pollution in Baltimore, Maryland, researches all seven decision functions. He groups some of the functions and addresses them together, rather than each separately. Flores describes intelligence and promotion together, then prescription, invocation and application, and finally termination and appraisal. He describes how each function worked for better or worse to appreciate and resolve the policy problem. By contrast, Lawrence (Chapter 4) focuses almost exclusively on the intelligence function because she was an active participant and observer in this activity and knew it best. However, she does address the other functions briefly. You can see from these two examples that there is flexibility in using the outline according to your purposes.

Finally, you are asked to describe how each relevant decision function is performed in reference to the problem(s) of concern by using feature analysis. Feature analysis permits you to give a more detailed specification of the context (social process). It requires that you address the following questions in reference to each relevant decision function: Who participates? With what perspectives? In which arenas? Using what base values? In what strategic ways? To generate what outcomes? For example, Lawrence's (Chapter 4) case on urban park management thoroughly describes and analyzes the intelligence function by answering these questions. By addressing these questions, Le Breton's (Chapter 12) piece on United Nations programs for refugees identifies and examines the context or social process involved in the entire decision process, though he does not focus on particular decision functions. In Lyon's study (Chapter 10), all seven decision functions are explored using these questions.

In light of the results of carrying out the tasks outlined above, you are asked to make a "final" determination and more detailed specification of the problem(s) and goals in natural resources policy and management that are your primary focus of concern. For example, Garen's study (Chapter 9) of ecotourism and biodiversity conservation stays with her preliminary problem definition after her description of the "problem" and its context. She concludes that "ecotourism programs often fail to protect the natural area upon which they depend. Decisions regarding programs frequently are monopolized by a minority of participants interested in personal gains in wealth and power." Lieberknecht's paper (Chapter 7) on endangered species recovery first proposed that the policy problem was that there were few concrete steps taken by authorities toward reducing short- and long-term threats to the survival of the endangered salamander she studied. But after mapping the social process and the decision functions, she decided that the initial problem statement needed to be refined. She concluded that the problem is one of "goal substitution" by key participants, the influence of developer and private property interests on state government, and an absence of citizen power in the decision process. She went on to elaborate the refined problem definition in terms of decision process and noted that the policy problem in her case is similar to those seen in other endangered species cases.

When this section of your paper is complete, ask the following questions: Is the problem's context (social process) adequately detailed? Is the problem's status relative to the decision process clear? Are goals sought in reference to the problem as defined clarified?

Analyzing the Policy Problem

Analyzing the policy problem in more detail requires focusing on problem orientation, that is, what happened (trends), why (conditions), and what's likely to happen in the future (projections). The three subheadings in this major section address these three elements, respectively. First is the trend question (what happened), and answering it requires the analyst to describe the historical trends in decision making that have had an impact on the problem of interest, including identification of particular impacts and their relation to the achievement of goals. Second is the conditions question (why), which focuses the analyst's attention on why trends occurred. The analyst must identify and examine the factors that have shaped the trends and impacts described immediately above. Third is the projection question (what's likely to happen), which focuses on the possible future given trends and conditions. Will events move closer to or further away from goals? In other words, will the problem become smaller or larger if no one intervenes?

In the previous section of the outline, other elements of problem orientation were addressed; specifically, goals and a problem statement were offered. In the following section, the last element in problem orientation—that is, alternatives or potential solutions—is attended to systematically. The kind of contextual problem analysis required by policy sciences method in the form of the outline increases the likelihood that the problem and the factors that brought it about will be well enough understood so that practical, realistic solutions can be found, evaluated, selected, and implemented successfully. For example, Flores's study (Chapter 3) on ozone pollution in Baltimore tracks key changes in decision making and in the well-being value outcomes (e.g., human health) for the city's population. He organizes trends into three periods demarcated by milestones in both value outcomes and prescriptions (potential solutions) to the problem. Key conditioning factors include adoption of emission control and public education programs. These factors have conditioned trends by not alleviating the problem fully. Ultimately, the causes of pollution are a host of social, political, economic, and ecological factors that go beyond the scope of any particular current decision process. Finally, his projection section looks at what is likely to happen if business continues as usual. Flores suggests that health problems will continue and possibly worsen for Baltimore's citizens if nothing is done.

Le Breton's case study (Chapter 12) on the environmental impact of certain types of refugee circumstances identifies key trends including the formation of a United Nations program in 1993 followed by the establishment of the Office of the Senior Coordinator for Environmental Affairs (as part of the United Nations High Commissioner for Refugees) to address refugee problems. Conditioning factors include the growing, diverse number of refugee-related problems and programs, including the 1991 Honduran government's refusal to accept Guatemalan refugees, the 1992 Rio Earth Summit, the 1994 Rwandan refugee crisis (which produced several million refugees and threatened the mountain gorilla), and many more conditioning factors. Projections include continuing problems, such as host government reluctance to accept refugees and a lack of donor support to fund required work, and the growing, but highly bureaucratic capacity of the United Nations to address the problem of maintaining the ecological integrity of environments overrun by refugees, especially in crisis situations.

When this part of the outline is completed, ask the following questions: Are relevant trends (history) described adequately? Are conditions (factors) that shaped trends described adequately? Are future trends (projections) described adequately?

Recommending Solutions to the Policy Problem

Once goals, problems, trends, conditions, and projections have been investigated contextually as called for by the outline, then it is time to recommend potential solutions to the problem(s). The outline calls for the specification of practical recommendations to solve the problem(s). What alternatives are to be promoted? What should be done and why? Your recommendations should be fully justified. Once you invent your alternatives for resolving the problem given the projections described previously, you need to evaluate the alternative strategies proposed for their potential contribution toward reaching goals. Finally, you must select a strategy or complex of strategies to recommend.

For example, in Flores's case study (Chapter 3), the goal is to reduce significantly or eliminate harmful human health effects from ozone pollution. There appear to be limits to reducing the amount of ozone produced, but other measures are possible, and Flores proposes a number of strategies and justifies them well. Lawrence's study (Chapter 4) on urban park management recommends improving matters by including more community leaders in the policy process, carrying out a "decision seminar" designed to engage key people in examining the problem and finding solutions, and hiring an outreach coordinator to work either with the city or independent of it to facilitate communication among participants. She also suggests organizing a "Friends of Beaver Ponds" park group. Finally, she recommends having university students help gather the kinds of intelligence leaders and friends groups need. Garen's analysis (Chapter 9) of ecotourism recommends carrying out systematic appraisal of the ecotourism industry, case by case, to determine if in fact it is aiding conservation or not. It is impossible to have informed policy making without good appraisal, she argues. And Bowes-Lyon (Chapter 10) offers fifteen recommendations to improve the decision process that currently makes up park management. He methodically evaluates each recommended alternative in practical terms with respect to each decision function. He evaluates alternatives in terms of his goal (having an integrated human development and conservation project in the common interest). Finally, he concludes that the set of intervention strategies he outlined could practically aid matters and move the process and its outcomes closer to the goal. In Wilshusen's study of local participation in conservation and development projects (Chapter 11), the author recommends specific strategies for enhancing such participation, especially in the formative phases of project development.

After completing this section of the paper, ask the following questions: Are alternatives to resolve the problem(s) described adequately? Are alternatives evaluated adequately? Is the selected alternative (strategy) or complex of strategies appropriate to achieve goals and solve the problem(s)?

Concluding the Paper

In conclusion, the outline asks you to make a very brief re-statement of the goal(s) and policy problem. Following that, you are asked to list briefly your recommendation(s) and justification to solve the problem as defined. The conclusions should be short—a single paragraph or at most two paragraphs. Quickly look at the ten cases to see how each succinctly and clearly concludes.

On completing this last section, ask the following questions: Are the goals and the problem re-stated clearly? Is (are) the recommendation(s) stated and justified?

Acknowledgments and Documentation

It is important to acknowledge the support and information you received in preparing your paper. Be generous in recognizing people who aided your work. If further documentation is needed on the many points in your study, these can be added as endnotes (or use footnotes; keep them short). In the Literature Cited section be sure to include all relevant source references. Make sure all your citations are complete and accurate. Be sure to check this list against the text to verify that the two match.

Conclusions

An interdisciplinary, problem-oriented outline is presented, described, and illustrated with ten case studies that can guide an analyst, decision maker, or concerned citizen in investigating any natural resources policy and management issue. The outline calls for a clear statement of the goals and policy problem, a thorough and clear analysis of the problem, and practical recommendations to resolve the identified problems. The outline can be used as a checklist of sorts to make sure that no part of the problem orientation is left out or covered insufficiently. The questions offered at the end of each section of the outline are helpful in this regard. Skill in the use of this approach as embodied in the outline comes with practice.

LITERATURE CITED

Clark, T. W. 1997. "Conservation Biologists in the Policy Process: Learning How to Be Practical and Effective," in G. K. Meffe and C. R. Carrol, eds. *Principles of Conservation Biology*, pp. 575–598. 2d edition. Sunderland, MA: Sinauer Associates.

Lasswell, H. D. 1938. "Intensive and Extensive Methods of Observing the Personality-Cultural Manifold," *Yenching Journal of Social Studies* 1:72–86.

———. 1971. *A Pre-View of Policy Sciences*. New York: Elsevier.

Lasswell, H. D., and A. Kaplan. 1950. *Power and Society: A Framework for Political Inquiry*. New Haven: Yale University Press.

Lasswell, H. D., and M. S. McDougal. 1992. *Jurisprudence for a Free Society: Studies in Law, Science and Policy*. New Haven: New Haven Press.

Willard, A. R. 1988. "Incidents: An Essay in Method," in W. M. Reisman and A. R. Willard, eds. *International Incidents: The Law that Counts in World Politics*, pp. 25–39. Princeton, NJ: Princeton University Press.

Part 2

Case Studies: Analyzing Problems and Finding Solutions

3 Protecting Human Health from Ozone Pollution in Baltimore, Maryland

Revising the Current Policy

Alejandro Flores

According to the United States Environmental Protection Agency (USEPA), residents of Baltimore, Maryland, continue to be exposed to unacceptable amounts of ground-level ozone, one of the most damaging air pollutants affecting human communities in the United States. While there is evidence of improvements in the overall quality of Baltimore's air since the 1970s (Energy and Environmental Analysis 1995, hereafter "EEA"), today's ozone pollution continues to take a toll on the health and well-being of its residents (USEPA 1995a; Maryland Department of the Environment 1996a and 1997, hereafter "MDE"). In fact, a recent study by researchers from the Harvard School of Public Health ranked the Baltimore region as second—just after Los Angeles—in the percentage of respiratory problems linked to ozone exposure during the 1994 season (American Lung Association 1996, hereafter "ALA"). Evidently, failure to protect residents from the health risks posed by ambient ozone represents a serious public policy problem in the area.

This study stimulates an examination of a broad range of policy alternatives that could be pursued in an effort to further reduce the health risks posed by ambient ozone in the Baltimore metropolitan area (BMA). The first section describes violations of national standards and the role of key actors and practices driving ozone dynamics in the

Research for this project was conducted between 1996 and 1997.

area. An analysis of historical trends in decisions and conditioning factors leading to the current state of affairs along with a projection of these trends into the future follows in the second section. The final section recommends a policy strategy based on the evaluation of three policy alternatives and suggests specific actions for implementation.

The primary source of motivation for this work is the basic premise that improvements in the air pollution abatement policy process could significantly accelerate the transition of the BMA from a severe nonattainment area to a place where no resident is exposed to federally defined unacceptable levels of ambient ozone. Working under this premise, I rely on the policy sciences analytic framework of Harold Lasswell and colleagues (Lasswell and Kaplan 1950; Lasswell 1971; Lasswell and McDougal 1992) to identify key areas for improvement and suggest ways of implementing changes in a problem-oriented and contextually sensitive manner. Given my primary concern with impacts of decisions on local human victims of ozone exposure, my analysis is limited to the human health dimension of the ambient ozone problem in the BMA. In studying this dimension, however, I take an ecosystem science approach to account for some of the critical, yet often neglected, physical, chemical, and biological connections driving ozone dynamics and human exposure in the area.

While my primary biases stem from my current affiliation as a doctoral student of the Yale School of Forestry and Environmental Studies (a source of identity) and in particular of the school's Urban Resources Initiative (a source of ideology), the role sought for this work is that of an independent policy analyst. The contemplative orientation can be described as a balance between an interviewer and observer-collector. Direct contact has been established with several participants in the policy process through interviews and past personal interaction. Additionally, a variety of available records including publications from participant institutions, newspaper articles, scientific literature, and publicly accessible databases for the area have been collected and analyzed.

The Ozone Problem in Baltimore

Ozone and Human Health

Residents of the BMA continue to suffer significant morbidity from exposure to federally defined unacceptable levels of ground-level ozone, the preeminent member of a family of secondary air pollutants termed photochemical oxidants. The current unacceptable level of ambient ozone concentration as defined by the federal government is 0.080 ppm (eight-hour average). Human response to exposure to high levels of ozone is expressed primarily in terms of reduction in lung functioning, impairment of immune defenses to respiratory illnesses, and inflammation of lung tissue throughout the population at large but especially among children, residents with a history of respiratory diseases, and outdoor enthusiasts (World Health Organization 1992, hereafter "WHO"; USEPA 1986, 1996a, 1996b). This statement does not imply, of course, that exposure to ambient ozone is solely responsible for prevailing respiratory ailments in the area. For instance, the contribution of ozone exposure to total respiratory hospital admissions in the BMA was estimated to be approximately only 8% during the 1994 ozone season (ALA 1996). Other causative agents of respiratory ailments may include, for example, exposure to particulate matter and sulfur dioxide.

Neither does the above statement regarding human response to ozone exposure imply that residents' respiratory ailments are the only sources of concern for observed levels of ambient ozone in the BMA. Exposure to high levels of ambient ozone is also known, for instance, to induce morbidity in other organisms such as trees and highly cherished crops (Guderian 1985; Smith 1990; USEPA 1996a) as well as physical damage to materials such as rubber, textile fibers, and dyes (Haynie and Spence 1984; Lake and Mente 1992; USEPA 1996a). Moreover, concern over high levels of ozone is often coupled with concern for high levels of its precursor pollutants (i.e., volatile organic compounds and oxides of nitrogen), which have important human and environmental health implications of their own.

From a common interest perspective, the significant and wide-ranging morbidity resulting from people's exposure to high levels of ozone represents an undesirable outcome of historical and ongoing decision processes shaping and sharing the health values of BMA residents. A more desirable outcome of the process would be one in which no resident is involuntarily deprived of a minimum degree of protection against unacceptable levels of morbidity. From a common interest perspective, therefore, the pollution abatement policy goal becomes one of striving toward having no human exposure to unacceptable levels of ozone in the BMA.

Ozone in the Baltimore Metropolitan Area

The Baltimore metropolitan area, Maryland, United States of America (76° E; 39° N), is home to approximately 2.5 million people. Of these, close to 700,000 live within the limits of Baltimore City (Maryland Office of Planning 1996, hereafter "MOP"), while the rest live in suburban and formerly rural communities radiating outward from Baltimore's downtown area. On a more regional scale, people living in the BMA are also residents of the "northeastern megalopolis," a large and highly industrialized urban corridor extending from Washington, D.C., to Boston, Massachusetts, along the northeastern coast of the United States. The regional context is relevant to ozone dynamics in the BMA in two important ways. First, large emissions of both volatile organic compounds and oxides of nitrogen in the BMA are common to metropolitan areas in this region, which are characterized by a high rate of energy and material throughput along with a strong dependence on the combustion of fossil fuels for transportation. Second, given the distinctively close proximity between urban centers in the region, local ozone dynamics in the BMA can be highly influenced by dynamics in other northeast coast metropolitan areas and vice versa. This relationship can be quite significant, especially because the BMA is rather close (20 to 50 miles) and generally downwind from the Washington metropolitan area (3.5 million people).

Human exposure to high levels of ozone in the BMA is evidently not regarded as a high-priority problem by health officials in the area. There is not even a program in place to address the problem, for instance, in the local Department of Health itself (Don Torres, personal communication 1996). Compared, however, to ozone problems in other areas of the United States, the BMA's ozone problem is nevertheless one of the most severe. Indeed, the BMA is currently classified as a "severe" nonattainment area based on violations of National Ambient Air Quality Standards (NAAQS) set by the EPA (USEPA 1996c). Moreover, when compared to other environmental issues in the state of Maryland, ozone pollution is perceived among the highest-priority problems faced by its residents (MDE 1996b).

Baltimore's Ozone Pollution Abatement Policy Process

Key Participants

The participants of a policy process can be thought of as encompassing all those personal and institutional actors involved in the shaping and sharing of values of concern. Analyses of participants have conventionally focused on "official" or conspicuous actors at the expense of many others who, although unofficial, inconspicuous, or unconscious, also have a significant impact on value outcomes. This lack of attention has been particularly true with regard to the conducting of promotional activities by semiofficial or private actors (see Lasswell 1971:38). In light of such chronic limitation, an analyst is best advised to adopt a broad conceptualization of participants. Following such broad conceptualization, a list of institutional participants in the process of ozone pollution abatement in BMA is presented in Table 3.1. This list, of course, is not meant to be exhaustive but rather representative of the diversity of relevant participants involved.

The policy sciences framework offers a comprehensive and systematic scheme to further describe participants through the characterization of their perspectives in terms of demands and expectations and identifications on the one hand and, on the other, doctrine, formulas (norms that guide behavior), and miranda (highly expressive elements of doctrine) (Lasswell 1971). While a description of the perspective of each and every participant would be an overwhelming task, much practicality can be gained by grouping perspectives into meaningful clusters. Table 3.2 describes the contrasting perspectives of three major clusters of participants having a considerable impact in the policy process. A description of each cluster's base values and strategy (Lasswell 1971) is also included. This overall description is helpful in answering three basic questions about participants: who they are (identity, doctrine, formula, miranda), what they want (value demands and expectations), and how they get it (base values and strategy).

The first cluster is composed of environmental protection agencies and other partner government agencies that are expected to protect the air quality of the region. Included in this group are the Maryland Department of the Environment (MDE), its analogous agencies in neighboring states, the EPA (Region III—Delaware, Maryland, Pennsylvania, Virginia, and West Virginia), and, to some extent, transportation agencies such as the Maryland Department of Transportation. These participants generally use power and wealth (base values) in an economic and sometimes coercive way to gain respect and power (demands) in their expectation to meet ambient ozone standards. Salient aspects of their identity are typified by a doctrine of command and control. Not surprisingly, their behavior is generally guided by norms related to the measurement of pollution concentrations and the control of emission sources. The publication of an air quality index and the imposition of fines or suspensions on local polluters are highly expressive elements of their doctrine. Care should be taken, however, in not inferring that all other participants expected to work toward the protection of the air quality of the region necessarily share a similar perspective. Examples of such participants include the Sierra Club and the Yale School of Forestry and Environmental Studies. Although these

Table 3.1. List of Participants According to General Kind of Institution

Governmental Institutions	Private Institutions

Environmental Protection Agencies
  Maryland Department of
    Environment
  MD Neighboring States' EPAs
    (Region III)
  EPA Region III
  National EPA
	Nonpoint Source Polluters
	  Maryland Petroleum Council
	    (fuel providers)
	  Service Station Associates
	    (fuel providers)
	  Foundation for Clean Air Progress
	    (automakers)

Health Agencies
  Baltimore City Department of
    Health
  Maryland Department of Health
	Marine Trade Association (nonroad
	    vehicles)
	  AAA—Potomac (motor vehicle
	    drivers)

Transportation Agencies
  Maryland Department of
    Transportation
	Point Source Polluters
	  Chemical Industry Council
	  Baltimore Chamber of Commerce
	  Power generators

Planning Agencies
  Transportation Steering
    Committee
  Baltimore City Department of
    Planning
  Maryland Office of Planning
	Area Hospitals
	  Johns Hopkins University Medical
	    Center
	  University of Maryland at
	    Baltimore Medical Center

Natural Resource Agencies
  Baltimore City Dept. of Rec. and
    Parks
  Maryland Department of Natural
    Resources
  US Forest Service NE Exp. Station
	Media Broadcasters
	  Baltimore Sun newspaper
	  CBS-WRC TV station (Channel 4
	    in DC)
	  WJZ TV station (Channel 3 in
	    Baltimore)

Housing Agencies
  Baltimore City Department of
    Housing and Community
    Development
  Maryland Department of Housing
    and Community Development
	Nongovernmental Institutions
	Health Associations
	  American Lung Association,
	    Maryland chapter

Other Agencies
  Ozone Transport Commission
  Metropolitan Washington Air
    Quality Committee (Wash. Council
    of Gov't.)
  Maryland Department of
    Economic and Employment
    Development
  Maryland Energy Office
	Environmental Groups
	  Sierra Club, Appalachian region
	  Urban Resources Initiative
	
	Other
	  Air Quality Control Advisory
	    Council
	
	Academic Units
	  UM (Baltimore): School of Public
	    Health
	  UM (College Park): meteorology
	    department

Table 3.1.   continued

Governmental Institutions	Private Institutions
	JHU: School of Public Health
	JHU: Department of Environmental Engineering
	Harvard: School of Public Health
	Yale: School of Forestry and Environmental Studies

Note: EPA = Environmental Protection Agency; JHU = Johns Hopkins University; UM = University of Maryland.

organizations have similar demands, their strategies are quite different from those used by environmental protection agencies.

The second cluster includes many regulated polluters or those participants known to expose BMA residents to higher levels of ozone. Automakers, fuel providers, power generators, chemical manufacturers, and large employers of commuters and their representative nongovernmental institutions such as the Maryland Petroleum Council (fuel providers) and the Chemical Industry Council are included here. Unlike participants of the first cluster, their identity is better typified by a capitalistic doctrine. Their conduct is guided rather by norms related to producing surplus from the local market, and their doctrine is clearly symbolized by product performance standards (e.g., vehicle miles per gallon) and publicity campaigns (e.g., television commercials). Through the deployment of wealth and skill (base values) in a decisively economic and ideological strategy, these participants strive to protect and increase their market share/profitability (expectations) in fulfilling their demands for increased wealth. A notable group of polluters whose practices do not necessarily conform to this profile are the vehicle drivers in the area.

The third cluster of participants in the policy process is associated with human health protection organizations. Notable participants in this group include local hospitals (e.g., Johns Hopkins), professional health associations (e.g., ALA, Maryland chapter), and a handful of public health schools. This group relies primarily on rectitude and respect (base values) in their ideological strategy to protect human health (demands) and to reduce the incidence of respiratory ailments in the BMA (expectations). Salient aspects of their identity are associated with social well-being doctrine. Their conduct is guided by norms related to the integrity and scientific soundness with which residents are educated or

Table 3.2. Examples of Contrasting Clusters of Participants in Baltimore's Ozone Abatement Policy Process

Cluster	Env. Protection Agencies, Some Partner Agencies	Automakers, Fuel Providers, Some Polluters	Hospitals, Other Health Organizations
Doctrine	Command and control	Capitalism	Social well-being
Formula	Norms related to controlling emissions from the local polluters	Norms related to producing surplus from selling energy, cars, or other manufactured products	Norms related to treating or informing residents with respect to the health effects of ozone
Miranda	Air quality index Closure of manufacturing plants	Product (fuel, vehicle, or manufactured good) performance standards Local auto commercials and sale events	Emergency room visits and other expressions of morbidity Endorsement studies from "authoritative" schools of public health (e.g., Harvard)
Demands	Respect; well-being	Wealth; respect	Well-being; respect
Expectations	Have no unacceptable levels of ambient ozone	Protect share and profitability of local markets	Slow the growth in incidence of respiratory ailments
Base Values	Power; wealth	Wealth; skills	Rectitude; respect
Strategy	Coercive; economic	Economic; ideological	Ideological; economic

treated in relation to the health effects of ozone. Highly expressive elements of their doctrine include publications of hospital room visits and study endorsements from authoritative medical professionals. Other participants having a direct impact on the protection of human health, yet not conforming to this pattern, include some participants working to reduce human exposure such as local weather forecasters (e.g., University of Maryland at College Park's meteorology department) and local media broadcasters (e.g., *Baltimore Sun* newspaper).

The Decision-Making Process

The policy sciences framework recognizes seven basic functions of decision processes: intelligence, promotion, prescription, invocation, application, termination, and appraisal (Lasswell and McDougal 1992). A description of such functions allows for a process-oriented analysis of participant dynamics. This section describes the most salient aspects of each decision function as it is performed in the BMA ozone abatement policy process. A subsequent appraisal of the decision process organized by function is presented in the problem analysis section of this paper.

Intelligence and Promotion. The intelligence function deals with the way information is gathered and processed, and the way it comes to the attention of all who participate in the decision process (Lasswell 1971). In the BMA ozone abatement policy process, this function is dominated by technical measurements, estimates, and projections of emissions and concentrations of pollutants in the troposphere (lower atmosphere). Most of the data pertaining to the concentration of ambient ozone and the emission and concentration of precursor pollutants (oxides of nitrogen and volatile organic compounds) is gathered by the MDE through source emissions reports and a network of air quality stations in the area. While some participants occasionally engage in the processing of raw air quality and meteorological data, information on pollutant emissions and concentrations comes to the attention of most participants through official agency annual reports or some third-party interpretation of these reports. In addition to data on emissions, concentrations, and air quality dynamics, some data regarding the environmental and human health effects of ozone is gathered, processed, and disseminated among participants. Examples of human health related information include public education on the health effects of ozone exposure, public warnings on upcoming ozone episodes, and a few estimates of hospitalizations and resident morbidity following exposure.

Closely related to the intelligence function is the promotion function, which deals with the way recommendations are formulated and promoted. The objective is to transform promoters' demands into group prescriptions (Lasswell 1971). Promotion of policy alternatives for ozone abatement in the BMA seems neither intensive nor frequent. There seems to be little activity and continuity in terms of participants' involvement in this decision function. A closer examination of the policy process reveals, however, that considerable promotional agitation with respect to ozone abatement alternatives for U.S. metropolitan areas takes place in Washington, D.C., in the form of reports or direct lobbying. Promotional activities also take the form of propaganda or campaigns for or against stricter environmental protection on different media including magazines (e.g., Exxon 1997).

Prescription. Prescription is a key decision function in the policy sciences framework. A clear and elegant way to understand prescription is through its outcomes: "Prescription outcomes are characterized by the stabilizations of expectations concerning the norms to be severely sanctioned if challenged in various contingencies" (Lasswell 1971:45). Conventionally, prescriptions are codified or crystallized in written form, yet "stabilizations of expectations" need not to be manifested in that fashion. Prescriptions also take the form of "customary law," or informal norms with no statutes or text associated with them. The thrust of air pollution statutory prescriptions in the BMA echoes national ones. Ozone pollution statutes prescribed by the MDE for the BMA are adopted directly from federal statutes, notably the 1970 Clean Air Act and its 1990 amendments. Stabilizations of participant expectations in the tradition of "customary law," on the other hand, while not explicit or highly transparent, are nevertheless evident in general expectations with regard to, for example, the freedom of residents to put their private land to use as they wish or of vehicle drivers to drive as many miles as they want.

Invocation and Application. The invocation and application decision functions can be thought of as complementary phases of the actual process of implementing a prescription. Invocation is the initial act, and application the final one, "of characterizing a concrete situation in terms of its conformity or nonconformity to prescription" (Lasswell 1971). In the BMA these concrete characterizations take the final form (application) of permits or fines applied by the MDE to polluters and approval or sanctions applied by EPA Region III to the MDE. The MDE, with the help of other state agencies (notably the Maryland Vehicle Administration), may issue emission permits (application) upon request by local polluters (invocation) or may fine local polluters (application) after inspecting them (invocation). Similarly, EPA Region III may approve State Implementation Plans (SIPs) (application) submitted by the MDE (invocation) or otherwise sanction the MDE (application) following an audit (invocation) of compliance with federal prescriptions (i.e., Clean Air Act). The invocation and application functions are not limited to statutory prescriptions. An example of conformity with a customary prescription is when, for instance, residents decide to mow their lawns on hot ozone summer days despite advice to do otherwise in order to reduce ozone formation.

Termination and Appraisal. A description of these decision functions addresses the general questions of how prescriptions are terminated and appraised. The most notable terminations of prescriptions in ozone abatement policy in the BMA take the form of phase-outs or finalizations to pollution control programs. These terminations typically follow important changes to federal air quality statutory law (e.g., 1977 and 1990 amendments to the Clean Air Act). However, terminations may also occur after changes in customary law, as when, for example, participants cease to be involved actively in the policy process as a result of discontinuity in interest or funding.

Appraisal outcomes "characterize the aggregate flow of decision according to the policy objectives of the body politic, and identify those who are causally or formally responsible for successes or failures" (Lasswell 1971:29). Evidently, appraisal outcomes in the BMA ozone abatement decision process are not very systematic, comprehensive, or clear in their identification of the impact of decisions on successes or failures. Partial appraisals of the decision process, however, take the form of technical/scientific reports presenting a particular interpretation of air quality and, to some extent, health information suggesting particular trends in some value outcomes. Examples of these appraisals include: (1) a report conducted by Energy and Environmental Analysis on behalf of the Foundation for Clean Air Progress (Energy and Environmental Analysis 1995) highlighting past success in reducing mobile source emissions (cars and trucks) in the context of improvements in local air quality; (2) a study conducted by the Harvard School of Public Health on behalf of the ALA highlighting the role of ambient ozone in human respiratory ailments in the BMA and other U.S. ozone nonattainment metropolitan areas (ALA, 1996); (3) reports prepared by environmental protection agencies in the region (e.g., USEPA 1995a) describing improvements in air quality and identifying remaining challenges.

Analysis of the Problem

Evolving Threads of the Policy Process

It has been stated that the ozone abatement policy goal in the BMA is to strive toward having no involuntary exposure of humans to unacceptable levels of ozone. In the policy sciences framework this goal represents an intention to arrive at an acceptable level of a particular kind of value outcome (i.e., well-being) among all residents of the BMA. A meaningful analysis of trends, therefore, should track not only key changes in decisions but also key changes in value outcomes. Toward that end, milestones in both prescriptions and value indicators are useful delineators of the evolution of the ozone abatement policy process in the BMA.

The following description of trends in the policy process is organized into three periods demarcated by milestones in both value outcomes (included in the description of the perception of the problem) and prescriptions (included in the description of policy response). A synopsis of the full trend from the discovery of the pollutant in the 1950s to the present is presented in Table 3.3. It should be noted that the ideal indicator of trends in value outcomes is some expression of ozone-induced morbidity among residents. The traditional measurement of progress in the BMA, as in most other cities, however, has been ambient concentration of ozone; although limited, this is the closest indicator available.

1950–1970: The Discovery of Anthropogenic Ozone,
and Local Response to the Problem

The problem of ozone pollution in the BMA was first perceived as a threat to crops in the area (Went 1950; MaGill et al. 1956). At that time there was little concern with human morbidity and the issue was approached primarily as a phenomenon of great scientific interest because anthropogenic ozone was only recently discovered in Southern California. As in virtually all other areas of the United States, the first policy response to the problem in the BMA was local and nonaggressive. Because ozone pollution was not of significant concern and there were city ordinances that dealt with air quality, there was little incentive for the adoption of new prescriptions and involved policy programs: "At the present time it is considered that the Baltimore Nuisance Ordinance of the city code of 1927 is of sufficient regulation for the control of air pollution" (MaGill et al. 1956:2–12). The early policy

Table 3.3. Major Phases of Ozone Abatement Policy Process in Baltimore Metropolitan Area (1950–Present)

Period	Perception of Problem	Policy Response
I. Discovery and early signs of concern: 1950–1970	Some threat to crops and incipient threats to health Scientific interest	1. Local and passive Baltimore city ordinances (1927, 1946) 2. Federal research grants and failed state policies Motor vehicle acts (1960, 1965) Air quality acts (1963, 1967)
II. Growing problem and federal involvement: 1970–1990	Serious threats to human and environmental health (average of 22 exceeding days/yr, 1983–1989) One of many air pollution problems	3. 1970 CAA - Birth of federal and state environmental protection agencies - Establishment of NAAQS 4. 1977 CAA amendments - Revised ozone standard - Extensions for noncompliance
III. 1990 CAA amendments to present: 1990–1997	Persistent threats to human and environmental health (average of 7 exceeding days/yr, 1990–1996) Focus on air pollution abatement in BMA and US	5. 1990 CAA amendments - Regulation of lesser sources - Inventory and monitoring of VOCs - More extensions for noncompliance 6. 1997 revised ozone standard

Note: BMA = Baltimore metropolitan area; CAA = Clean Air Act; NAAQS = National Ambient Air Quality Standards; VOCs = volatile organic compounds.

reaction to air pollution in the area involved, however, some invest-ment in research. In 1952 the Division of Air Pollution Control was set up in the Bureau of Industrial Hygiene of the Baltimore City Health Department to conduct studies in atmospheric pollution (MaGill et al. 1956:2–12). Similarly, during the 1950s and 1960s the role of the fed-eral government in air pollution abatement was limited initially to in-vestment in research (Percival et al. 1992:103–108). Indeed, even by 1960 when Congress adopted the similarly research-oriented Motor Vehicle Act, it explicitly emphasized that air pollution control was the responsibility of states (Percival et al. 1992).

The quality of the air, however, continued to deteriorate in many areas in the country and soon the U.S. government started to shift its policy response toward federal control of state air pollution. Congress first provided for direct regulation in the adoption of the Motor Vehicle Air Pollution Control Act (MVAPC) of 1965, which established the first national auto emissions standards. The adoption of the 1963 Clean Air Act (CAA) and the Air Quality Act of 1967 (AQA) was also justified by the need for federal involvement in efforts to protect interstate air pollu-tion (Percival et al. 1992: 761-762). Yet, it was not until the amendment of the 1963 Clean Air Act in 1970, marking the beginning of the next phase of the policy process, that the federal government became in-volved decisively in direct regulation of the air quality of the BMA.

1970–1990: A Growing Problem, and Direct Involvement of the Federal Government

In the BMA, as in many other areas of the nation, the problem of air pollution gradually evolved from a "nuisance" in the early 1950s to a se-rious human and environmental health threat by the late 1960s. The perception of a growing and a widespread problem led eventually to con-siderable changes in prescriptions with the introduction of the United States Clean Air Act in 1970. By replacing the 1963 CAA and the 1967 AQA, the 1970 Clean Air Act established a new framework for federal reg-ulation of air pollution throughout the country. Among other provisions, the act set deadlines for the EPA (established by executive order also in 1970) to promulgate air quality standards to be implemented by every state (Percival et al. 1992). National Ambient Air Quality Standards (NAAQS) were promulgated in 1971 for six air pollutants (criteria pollu-tants) based on ambient concentrations for a given unit of time (Federal Register 1971). The standard for photochemical oxidants (ozone in-cluded) was set at 0.80 ppm, not to be exceeded for more than 1 hour per year. The photochemical oxidant standard was later changed to an

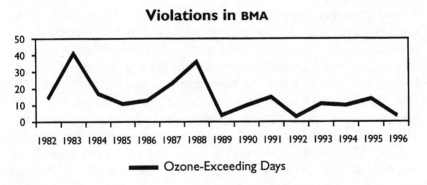

Violations in BMA

── Ozone-Exceeding Days

Figure 3.1. Trend in indicator of value outcome (ozone-exceeding days) from 1982–1996 (EEA 1995, MDE 1997a). 1990–1997: The 1990 CAA Amendments to the present.

ozone standard and relaxed from 0.08 to 0.12 ppm (1 hour) in 1977 when the act was amended (Federal Register 1977). These amendments also extended the compliance date for all nonattainment areas to December 31, 1982, and authorized further extensions for the pollutants most difficult to control up to December 31, 1987.

Reliable official indicators of value outcomes in relation to ozone pollution in the BMA did not become available until the early 1980s when systematic monitoring of its ambient concentrations started in the area (Figure 3.1). The average annual number of ozone-exceeding days in the BMA in the period from 1983 to 1989 was twenty-two. The year-to-year number of ozone-exceeding days was noted to vary considerably, with some years recording only a handful of exceeding days and others recording over twenty. Of particular significance in raising awareness of the ozone problem was the extremely hot year of 1988, in which thirty-six ozone-exceeding days were recorded in the area. It is important to note, however, that ozone was only one of several criteria air pollutants seriously affecting air quality in the BMA through most of this policy phase. Indeed, the perception of the overall air quality problem in the country did not change much through this period because although emissions of some air pollutants, notably lead, started to decline rapidly, the emissions of others, like oxides of nitrogen and sulfur dioxide, increased or only slowly declined.

1990–1997: The 1990 Clean Air Act Amendments to the Present

In light of the continued widespread failure to meet ambient standards for several pollutants in the 1970s and 1980s, demands for considerable changes in national policy were being increasingly articulated and promoted during the second part of that period. Ozone pollution, in particular, proved to be a rather elusive pollutant to control in many metropolitan areas. Proposals for change, however, faced gridlock during the two Reagan administrations, and it was not until 1990 that significant changes in prescriptions were introduced in the form of the most recent amendments to the 1970 CAA (U.S. Code 1991). These changes mark the beginning of the current phase in the policy process. Three of the major eleven titles of the 1990 CAA amendments are directly relevant to ground-level ozone: titles I, II, and IV. Title I concerns new controls on emissions of oxides of nitrogen and hydrocarbons. It also provides for the reclassification of ozone nonattainment areas into a new range that spans from "marginal" to "extreme" nonattainment (the BMA is classified as "severe") and assigns new attainment dates (2005 for the BMA). Title II details requirements for all aspects of new regulation over mobile sources including new emission and fuel standards for vehicles. Finally, title IV (Acid Deposition Program) details a plan to reduce (starting in 1995) the annual emissions of oxides of nitrogen from fossil fuel–fired electric utilities by approximately 2 million tons per year from the 1980 level.

A scheme of emission permits, fines, and suspensions as mandated by the 1990 CAA amendments has been slowly put in place in the BMA through the adoption of several emission control programs. Most of these air quality control programs target ozone as it has become recently the focus of concern in the BMA. Examples of these control programs are the Vehicle Emissions Inspection Program (VEIP) and the more recent Vapor Recovery Nozzle Program (VRNP). Other policy programs such as FORCAST, aimed at predicting ozone episodes, and Ozone Action Days (OAD), aimed at reducing emissions during forecasted ozone episodes, have also been adopted as part of the overall policy response although these are not mandated by the 1990 CAA amendments.

Meanwhile, violations in the current phase of the policy process are proving to be less frequent and less severe than in the previous phase. The average number of ozone-exceeding days in the period from 1990 to 1996 is only seven compared to twenty-two for the previous seven years. As discussed below, care should be taken in correlating recent changes in value outcomes with recent changes in prescriptions. Moreover, care should be taken in interpreting these changes as an unequivocal signal that the ozone problem is no more, for one must keep in mind that ground-level ozone standards have been (and continue to

be) violated in the BMA ever since the federal government first established these standards in 1973.

Examination of Factors that Have Shaped Trends in Outcomes

In the policy sciences framework, an examination of factors that
have shaped trends would fall under the rubric of an analysis of conditioning factors. This intellectual task is the scientific pattern of thought,
the purpose of which is to understand the cause-effect relationship of
factors shaping trends en route to policy goals (Lasswell 1971:50).

The adoption of emissions control and public education programs
along with other decisions of key participants in the policy process
have no doubt had a significant impact on the historical trends. This
statement does not imply, however, that changes in decision making
are always sufficient or even necessary to explain variations in observed value outcomes for any given period of time. As it will be elaborated below, the process leading to the formation, accumulation, exposure, and, ultimately, the health impact of ground-level ozone is a
rather involved process in which several social and ecological factors
(respectively, human and nonhuman factors beyond the scope of any
particular decision process) play a significant role in shaping value
outcomes (Table 3.4).

Ozone Formation

The formation of ozone results from several precursor pollutants with
multiple human and natural sources involved in a dynamic sequence of
chemical reactions that is highly dependent on meteorological conditions (World Meteorological Organization 1991, hereafter "WMO"). Thus,
in addition to emission control programs certain social and ecological
factors condition the formation phase of ozone dynamics.

Key social factors are the potential for emissions of both oxides of
nitrogen (NOXs) and volatile organic compounds (VOCs) in the area,
which, of course, have not remained constant throughout the history
of the policy process. Many factors beyond the scope of local emission
control programs have conditioned emission levels of NOXs and VOCs in
ozone nonattainment metropolitan areas. Notable among these are
changes in the number of circulating vehicles and the average distances traveled per vehicle, locally and in relevant upwind areas. Indeed, many participants have maintained that considerable gains in
fuel and combustion efficiency (again, not necessarily as a result of

Table 3.4. Key Factors Conditioning Value Outcomes of Ozone Problem in Baltimore Metropolitan Area

Type	Formation	Removal	Exposure
Decision Process Situation: social process	Emission control programs Potential for NOX emissions in BMA and upwind areas (vehicles, utilities, manufacturers)	Land use programs Demand for land use in BMA Soil/forest cover (land-atmosphere exchanges & meteorological effects)	Public education programs Demographics Pop. density Age structure Lifestyle Outdoors attitude Place and time of outdoor activity
	Potential for VOC emissions in BMA and upwind areas (vehicles, land fills, manufacturers)	Demand for land use upwind of BMA	
Situation: ecological process	Meteorology Temperature Wind speed Potential for biological emissions of VOCs (soil & vegetation dynamics)	Meteorology Temperature Wind speed Potential for biological removal of ozone and NOX (soil & vegetation dynamics)	Meteorology Temperature Precipitation

Note: BMA = Baltimore metropolitan area; NOX = oxides of nitrogen; VOC = volatile organic compounds.

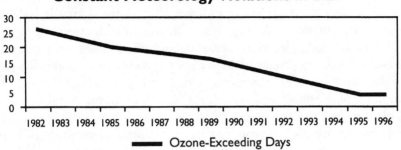

Figure 3.2. Estimation of trends in indicator of value outcomes (ozone-exceeding days, former 0.12 ppm standard) if key meteorological conditioning factors had remained constant from 1983–1996 (EEA 1995, 1995; and 1996 are inferred).

local policy programs or even national programs) have been offset by an increasing number of vehicle miles traveled (VMT) in metropolitan areas across the United States. On the other hand, ecological factors also play a key role in the formation of ozone. This is particularly true with respect to fluctuating meteorological conditions like solar irradiation and wind speed, upon which chemical reactions leading to the formation of ozone and, ultimately, ambient concentrations so highly depend (Figure 3.2). A comparison of this trend with that displayed in Figure 3.1 reveals the considerable effect of key ecological factors in shaping trends in value outcomes.

Ozone Removal

The accumulation of ozone in any given area remains checked by a number of "sinks" that prevent it from concentrating indefinitely. These "sinks" act as forces that remove some of the ozone being formed. As with the case in the formation phase, certain social and ecological factors condition the accumulation phase. A pivotal ecological sink entails the transport of ozone into adjacent areas by wind currents. The lack of wind currents that is characteristic of "stagnation" situations is rather highly conducive to the accumulation of ozone. Yet another significant ecological "sink" entails the sequestration of ozone by non-human biological systems such as forests and soils (Smith 1990). The fact that this latter kind of ecological sink is often neglected as a conditioning factor in most analyses does not mean that its impact is negligible (e.g., Nowak 1994). While humans may have no direct control over the rate of transfers of atmospheric gases to specific landscape surfaces, they do exert some indirect control by virtue of induced changes in land cover through time. Change in land cover represents both a social and a decision type of conditioning factor—the latter by virtue of land use programs being part of ozone abatement decision making, and the former by virtue of demands on land use shaping the evolution of the landscape configuration in a given metropolitan area. Evidently, the land cover configuration in the urban, suburban, rural, and upwind areas of the BMA NOX has not remained constant through time, and observed changes can hardly be ascribed to the ozone abatement policy process.

Human Exposure to Ozone

Finally, in a similar fashion, trends in the human health impact phase of ozone dynamics cannot be explained by public education programs alone. Two obvious social factors conditioning the magnitude of ozone-induced health deprivations in the BMA are the potential for human exposure and the susceptibility of the population to health damage following exposure. The potential for human exposure has evolved during the policy process if for no other reason than changes in population size (since 1950, the BMA population size has gradually increased, and that for Baltimore City itself has considerably decreased). Residents' attitude toward spending time outdoors (higher ozone concentrations) is yet another factor that is unlikely to have remained constant since 1950. Similarly, in light of differential exposure-response curves of younger versus older people, susceptibility to damage after ozone exposure has changed if for no other reason than changes in the age distribution of the BMA population (the percentage of people under 19 years of age in Baltimore City, for instance, decreased from approximately 35% in 1970 to 25% in 1990) (MOP 1996). Ecological factors such as temperature and precipitation conditioning the time and place of outdoor activity may also be significant factors in the exposure phase of ozone dynamics.

Examination of Factors That Have Shaped Trends in Decisions

The analysis of conditioning factors explaining trends in value outcomes can be complemented by a decision process appraisal providing insight into the factors that have shaped trends in decisions themselves. Of course, trends in decisions have impacts on the shaping and sharing of all values in the BMA. Accordingly, decisions in the decision process can be regarded as the ultimate constraints on the attainment of policy goals. This section presents the most salient results of an appraisal of the ozone abatement decision process in the BMA based on normative criteria established for each decision function (Lasswell 1971).

Intelligence and Promotion. The intelligence function is performed with adequate dependability and openness but inadequate comprehensiveness. The main intelligence-gathering practice, the ambient ozone monitoring network in the BMA, is a dependable one because data gathering is done systematically and in agreement with professional standards. The processing of air quality may take some time, and meteorological data may not be readily accessible but it is ultimately available to all participants at a reasonable cost. Intelligence gathering and processing, however, is not sensitive enough to consider actual victims of ozone. It is considerably narrow in scope, the bulk of it being limited to the measurement and simulation of emissions and concentrations of pollutants. The function is particularly blind with regard to the human exposure phase of ozone dynamics and with regard to spatial and temporal heterogeneity in all three phases (more detailed intelligence could be gathered, for instance, with regard to the spatial distribution of populations that are most vulnerable and less protected). In light of these deficiencies in the intelligence function, one can expect only a proportionate lack of comprehensiveness and a magnified lack of context in the promotion function, especially when the bulk of promotion takes place in another location (e.g., Washington, D.C.; see the section on description of the decision process).

Prescription. Stability of expectation, rationality, and comprehensiveness are the main criteria for the prescription function (Lasswell 1971). Ozone abatement prescriptions for the BMA are relatively stable and transparent but not comprehensive or fully rational. Air pollution abatement prescriptions for the whole state of Maryland are directly adopted from those delineated by the federal government in the CAA, which is rather explicit in both the air quality goals to be pursued by states and the measures necessary to achieve such goals (Percival et al. 1992). From a local human health perspective, however, this outcome of the prescription function is neither comprehensive nor fully rational. Indeed, the goodness of fit between the BMA's air pollution abatement prescriptions and its air pollution health problems is compromised by the adoption of narrow and highly uniform prescriptions set for the entire country. The adoption of uniform national standards can be criticized, among other things, as inefficient, for it disregards marked differences in policy opportunities and constraints resulting from variability in key conditioning factors including local demographics, meteorology, and land use.

Yet undoubtedly, the most consequential factor influencing the national prescription as well as its local application is the fact that the entire prescription process has been shaped by a major goal substitution. Dr. Ronald Brunner's comments on the issue are useful in explaining this common deficiency of decision processes (Brunner, personal communication 1997): "In effect, health problems were diagnosed as the result of emissions; the goal became control of emissions according to the federal standards; and this framing of the problem restricted the consideration of policy alternatives relevant to the health problems in question—for example, removal of air pollutants by various means, and reduction of length of [human] exposure or exposure during transient peaks of air pollution—this seems to me a textbook case of goal substitution: First there is a single-factor explanation of trends and projections relevant to a common interest goal (e.g., better health). Then control of that factor (e.g., emissions) becomes a goal in itself. And finally, alternatives based on multiple-factor explanations of the original problem are neglected." Direct adoption of national air quality prescriptions has resulted, therefore, in a decision process that is not only insensitive to the local context but also severely confining in its potential course of action.

Invocation and Application. In the BMA, prescriptions have been implemented in a relatively nonprovocative and uniform manner, yet the process has been slow and unreliable. Initial and final characterizations of conformity and nonconformity with respect to prescriptions (invocation and application, respectively), such as those resulting in the issuing of permits and fines to polluters by the MDE, seems neither controversial nor selective. The implementation process, however, has been unrealistic and slow—characteristic of implementation initiatives throughout the history of the CAA. The BMA has been slow in implementing programs mandated by the CAA and has failed regularly to comply with various deadlines including those set for the attainment of ambient ozone standards.

Similarly, the EPA (Region III) has been slow and passive in policing the state's performance and rather lenient and reluctant in the application of sanctions despite unequivocal and continued violations (in the case of ozone, the imposition of mandatory sanctions has been postponed until the year 2005 for the BMA). Moreover, the decision process is considerably undermined by the fact that many participants (including actual victims of ozone exposure) have no practical means of invoking noncompliance and rarely participate in final characterizations of conformity with prescriptions. It is important to note, however, that this has been the case for many other metropolitan areas in the United States, leading many analysts to describe the CAA as a potpourri of postponements, revisions, extensions, and suspensions.

Termination and Appraisal. The termination function is somewhat less problematic for the BMA ozone abatement policy process. Key criteria for the termination function include timeliness, balance, and ability to ameliorate (Lasswell 1971). Although terminations have historically been slow they have gradually become more balanced and ameliorative. This trend is reflected, for instance, in terminations related to the national ozone ambient air quality standard, the most recent of which was promulgated in July 1997. While it could be argued that changes to the ozone standard continue to be undertaken too late or too soon, this most recent termination has been unprecedented in terms of participant awareness and explicit analyses of implications (e.g., re-mapping of nonattainment areas under different termination scenarios).

The most consequential deficiencies in the BMA ozone abatement decision process, however, undoubtedly relate to the appraisal function. Dependability (reliability), comprehensiveness, independence, and continuity are some the main criteria for the appraisal function (Lasswell 1971). Appraisal efforts in the BMA are continuous but hardly comprehensive, independent, or dependable. Continuous appraisal efforts include, for example, yearly reports of BMA violations of NAAQS (including ozone) and yearly analyses of aggregate trends in emissions and concentrations for the larger Region III area. These appraisal efforts, of course, are not comprehensive in terms of the phases of ozone dynamics (e.g., relatively little is done with regard to the removal and human exposure phases), the multiplicity of participants (e.g., unofficial or indirect yet influential participants such as health and planning agencies are not given due consideration), or the wide range of policy activities (e.g., most appraisal efforts focus on intelligence and prescription at the expense of other decision functions). Moreover, these appraisals are rarely independent because most are undertaken individually by the key participants themselves. Finally, these and other less continuous appraisal efforts are certainly far from being dependable, for how can an appraisal function that has failed to identify lack of context, goal substitution, and insensitivity to human health victims for more than twenty years be thought otherwise?

Projection of Trends in Decisions and Outcomes

This section of problem analysis addresses the question of what is likely to happen if business continues to be done as usual. The purpose of this exercise is to render implicit maps of expectation more explicit and dependable. The immediate task is to project trends in decision, and with a complementary projection of social and ecological conditioning factors, one can ultimately project trends in value outcomes. My projection in value outcomes is considerably less optimistic than those of other analysts. Whereas, for instance, a recent study done for the Foundation for Clean Air Progress (EEA 1995) suggested the year 2000 as the deadline for the BMA to solve its ozone problem (in terms of ozone-exceeding days), I project this will not happen at least until 2008. Key trends in conditioning factors supporting a less optimistic projection include the following four: (1) the recent adoption of a stricter national ozone ambient air quality standard changing the unacceptable level to 0.08 ppm (8-hour average) from 0.12 ppm (1-hour average); (2) a more gradual decrease in the rates of emissions of precursor pollutants; (3) an actual net decrease in the rates of ozone removal resulting from overall loss of permeable land cover that could sequester ozone; and (4) lack of evidence supporting a significant revision in the current ozone abatement policy approach. Each of these four trends is discussed in the context of evaluating policy alternatives in the following section.

Recommendations

Characterization and Evaluation of Policy Alternatives

In light of the previous description and analysis of the ozone exposure health problem in the BMA, three courses of action for improvement on the basis of the targeted phase of ozone-exposure dynamics can be visualized. Each one is characterized and evaluated in turn.

Improve Efforts to Control Emissions

The first alternative is to improve current efforts to control emissions, which are the focus of the prevailing policy approach as characterized in the problem description section. However, as suggested in the problem analysis section, it is unlikely that policy goals will be achieved soon under this course of action. The first salient reason for a less optimistic projection of trends pertains to a recent decision to terminate a prescription. As alluded to before in the appraisal of the termination decision function, the national ozone ambient air quality standard of 0.125 ppm (1 hour) has been revised to 0.080 ppm (8 hours). This was done in light of accumulating evidence of human health risks at lower ozone concentrations (USEPA 1996a; Bascom, personal communication 1997). An updated and more realistic trend projection should be based on the new standard. It should be noted that while actual human health damage is independent of revisions in standards, the latter do have an impact on the trend toward the attainment of a policy goal that is defined in terms of exposure to federally unacceptable levels of ambient ozone.

The second reason for the less optimistic projection is the fact that reducing emissions of precursor pollutants in the BMA is becoming an increasingly expensive and difficult enterprise. Controlling emissions is becoming more expensive because of decreasing returns in emission prevention per unit of resource investment. Indeed, many analysts (e.g., Krupnick and Portney 1991) have concluded that the truly cost-effective gains in air pollutant emission controls took place in the early phase of the current policy period (i.e., 1970s and 1980s). Moreover, new proposals for further increases in energy and material use efficiency (to reduce emissions) will face increasing resistance from polluters across the board. Finally, improvements in, for example, vehicle fuel combustion efficiency do not necessarily translate into lower overall emissions of pollutants. Gains in efficiency may be offset by increasing vehicle miles driven in the context of increasing affluence and population size. Indeed, the gradual anticipated decreases in future rates of emissions of NOxs and VOCs in the BMA are suggested by similar emission projections at the national level (USEPA 1995b). Increasing difficulties in curbing total precursor pollutant emissions will translate into slower than expected improvements in ozone ambient concentrations for the BMA.

Begin Efforts to Offset Losses in Rates of Ozone Removal

One of the reasons for anticipating slower decreases in future ozone ambient concentrations in the BMA is the likely decrease in the net rate of ozone removal in the area. While the trend projection in meteorological factors conditioning ozone removal is highly uncertain, a projection of trends in land-cover configuration indicates a net loss in the potential for ozone sequestration. It is anticipated that permeable surfaces will continue to be lost in areas that are generally upwind of the BMA (portions of the landscape between Washington, D.C., and the BMA). The conversion of forest and agricultural lands to less permeable surfaces (e.g., those in residential and commercial land uses) in southern and central Maryland is rather significant (MOP 1996) and unlikely to suddenly stop in the next decade. A policy effort can be initiated, however, to influence such rates of land conversion in strategic areas upwind of the BMA.

Losses in rates of ozone removal can also be ameliorated by a complementary effort to accelerate the conversion of impermeable landscape surfaces to more permeable ones within the BMA. Within the limits of Baltimore City, a staggering number of abandoned-building lots, which could be converted to soil or vegetated land, along with the availability of a community forestry infrastructure (local residents to enable and help maintain such conversion) makes this option unusually viable. Indeed, it is anticipated that more permeable surfaces will soon be created in the BMA. Grove and Cock (1997) estimate that at least a thousand new vacant lots will be created this year as the Baltimore City Department of Housing and Community Development prepares to demolish properties that are unsalvageable for housing (the number of vacant houses in Baltimore City is now estimated at ten thousand).

Begin a New Effort to Minimize Human Exposure and Impacts on Victims

A third alternative is to increase current efforts to minimize the potential for human exposure to ambient ozone and impacts on victims. This alternative is very attractive in light of the fact that human health damage from ozone exposure is a highly spatially and temporally specific phenomenon. As suggested in Figures 3.3 and 3.4, the severity of impact on value outcomes varies markedly according to the month of the year and time of day. Moreover, as suggested in Figure 3.5, the severity of potential health deprivations varies markedly according to location in the BMA. Residents living in areas registering a high number of ozone-exceeding hours (e.g., the Ft. Mead area) are less "protected" than those residing in areas registering a considerably lower number of ozone-exceeding hours (e.g., the Essex area). Finally, not all residents living in a given location are equally "vulnerable" to health damage after exposure. Children, residents with a history of respiratory diseases, and outdoor enthusiasts have been identified as being at greater risk (WHO 1992; USEPA 1986, 1996a, 1996b). Considerable deprivation of healthful conditions can thus be prevented by marginal decreases in the personal exposure time of particularly "vulnerable" residents in the less "protected areas" during "ozone episodes."

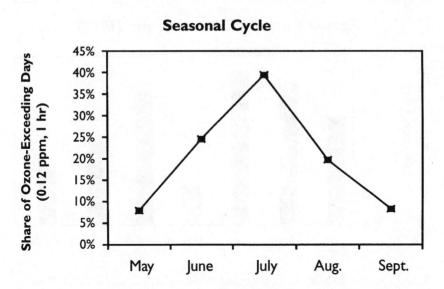

Figure 3.3. Seasonal cycle of ozone in the Baltimore metropolitan area based on ambient concentration data from 1982–1994 (EEA, 1995). Number of ozone violations vary markedly according to month of the year with close to 85% of exceeding days taking place in only three months (June, July, and August).

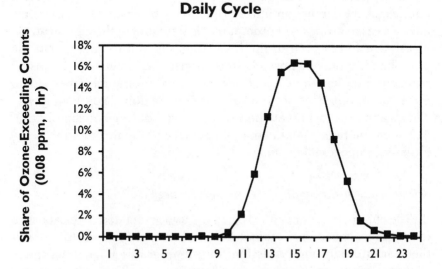

Figure 3.4. Daily cycle of ozone in the Baltimore metropolitan area based on 1994 ambient concentration data. Ozone concentrations vary markedly according to the time of the day with over 80% of exceeding hours taking place between 1:00 and 6:00 in the afternoon.

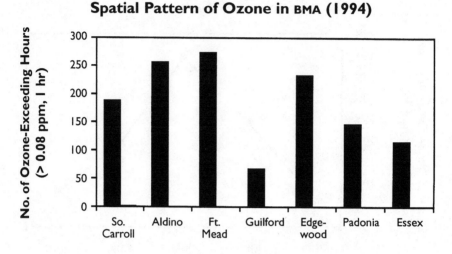

Spatial Pattern of Ozone in BMA (1994)

Figure 3.5. Spatial pattern of concern of ozone concentrations in the Baltimore metropolitan area based on 1994 data. Number of violations of the standard vary markedly according to location in the BMA with residents in less protected areas being potentially exposed to more than twice the number of exceeding hours than residents in the more protected areas.

A complementary effort within this course of action would focus on minimizing the post-exposure impact on victims. Impacts on health after exposure can be minimized right away by ensuring that victims have access to proper and prompt medical treatment through strategic investment in research to improve diagnosis and treatment. On the other hand, it may be possible to minimize the impact on victims by compensating health deprivations with wealth or some other value. The latter option does not reduce morbidity (a policy goal) but nevertheless attempts to mitigate imposed damages during noncompliance. Although the policy goal cannot be achieved fully through the latter initiative, important benefits could be attained.

Selection and Justification of a Particular Strategy

The elimination of health risks in the BMA associated with exposure to federally defined unacceptable levels of ozone is unlikely to occur in the near future through emission control programs alone (alternative 1, or the current policy approach). This is especially true in light of a more strict ozone standard and decreasing returns on emission control investments. Evidently, there is a need to take a revised approach to problem resolution.

Policy goals could be redefined to better address the BMA's ozone problem, and the range of solution alternatives to explore could be widened. From a human health perspective, pollution abatement goals should aim at reducing residents' morbidity from ozone exposure and not at reducing the number of days in which ambient concentrations are exceeded. On the other hand, starting efforts to accelerate ozone removal (alternative 2) as well as new efforts to reduce personal exposure and impact (alternative 3) should prove strategic in complementing current emission control strategies. In such a context, a policy strategy combining aspects of all three alternatives is recommended. Such a policy strategy would involve two new initiatives aimed at more effective interfacing with two key intersecting policy processes: land-use allocation decision processes having a considerable impact on ozone sinks; and health prevention and care decision processes having a considerable impact on preventing or treating ozone exposure.

Several key actions are suggested for the implementation of an initiative to influence land-cover change decisions that would lead to increases in the BMA's rates of ozone removal. *Project upcoming and future land-use changes in the BMA and upwind areas.* The Baltimore City Department of Planning and the Maryland Office of Planning could be particularly helpful in this regard. *Conduct studies to simulate the impact of potential preventable losses (suburban and rural) and simulated gains (urban) in vegetation cover in the context of projected land-use changes throughout the BMA.* These studies could be spearheaded by the USDA Forest Service—Northeastern Experimental Station. Interested research scholars in pertinent academic units could also participate in this effort. *Advocate for specific land-use allocations to optimize the impact on health risks in appropriate land-use decision processes.* Effectiveness of promotional activities in upwind areas could be greatly enhanced through coordination with the state's Smart Grow Program, aimed at curbing urban sprawl. Conversely, effectiveness of promotional activities in the inner city area could be greatly enhanced through coordination with housing demolition efforts by the City Department of Housing and Community Development and complementary community forestry efforts by the Yale University School of Forestry and Environmental Studies.

The following key actions are suggested for the implementation of a complementary initiative to influence decision processes having an impact on both residents' time of exposure and residents' morbidity after exposure:

• Map the spatial distribution of groups at risk and the incidence of respiratory ailments (including those induced by ozone) for the BMA. There is an incredible amount of intellectual infrastructure in the local schools of public health that could be very useful in this regard.

• Educate residents in targeted areas about measures to prevent/minimize ozone exposure. Perhaps the ALA would have an interest in leading this effort. Participation of local schools could be helpful in this regard.

• Increase effectiveness in alerting residents of ozone episodes. More strategic and intensive warning efforts could be undertaken with the active participation of all local print, audio, and video media broadcasters.

• Increase the effectiveness of medical treatment through access and research. This effort would offer an opportunity to both the Baltimore City and Maryland departments of health to become active participants in the process. Both health departments could coordinate hospitals in the BMA to ensure that potential victims would have easy access to the best available treatment following exposure. Some resources could also be invested in diagnosis/treatment research that could offer results in the near future.

• Initiate a fund to compensate victims of ozone exposure. This fund could be created through matched contributions from various participants. Money could be distributed through the respiratory departments of local hospitals at the time of treatment.

Conclusions

From a common interest perspective, the significant and wide-ranging morbidity resulting from BMA residents' exposure to high levels of ozone represents an undesirable outcome of historical and ongoing decision processes. A more desirable outcome would be one in which no resident is deprived of a minimum degree of protection to his or her health as a result of exposure to unacceptable levels of ambient ozone. From this perspective, thus, the policy goal becomes one of striving toward the minimization of human morbidity resulting from ozone exposure in the Baltimore metropolitan area.

Three policy alternatives are identified and evaluated in this study. These are differentiated on the basis of the targeted phase of ozone-exposure dynamics: (1) improve efforts to control emissions; (2) begin efforts to offset losses in rates of ozone removal; and (3) begin a new effort to minimize human exposure and impacts on victims. It is concluded that the achievement of the policy goal is unlikely to occur in the near future by following alternative 1 alone (the thrust of the current policy approach). A policy strategy that additionally integrates key intersecting decision processes affecting rates of ozone removal on the one hand (alternative 2) and residents' health procurement (alternative 3) on the other is recommended instead.

Successful implementation of the proposed strategy would involve considerable commitment that certainly goes beyond relabeling of current ozone abatement programs. Critical resources needed to successfully implement the proposed strategy are the enlightenment and skill necessary to reformulate prevailing policy goals and widen the range of alternative solutions considered, and the respect, skill, and wealth necessary to support effective participation in key "external" decision processes in the BMA. Shifting from a focus on ozone monitors to a focus on potential victims' faces might be challenging, but can we afford to do otherwise if we are striving to secure basic values to all BMA residents?

ACKNOWLEDGMENTS

This paper represents the culmination of the work started as a project for the course "Foundations of Natural Resources Policy and Management" taught at Yale University's School of Forestry and Environmental Studies. Drs. Tim W. Clark and Andrew R. Willard, course mentors, were extremely helpful in introducing me to the policy sciences analytic framework of Lasswell and colleagues. I thank both of them for their dedication and spirit. Along the same lines I would also like to thank Dr. Ronald Brunner from the Center for Public Policy Research at the University of Colorado, Boulder, for his helpful and enjoyable conversations regarding decision process appraisals.

LITERATURE CITED

American Lung Association. 1996. *Breathless: Air Pollution and Hospital Admissions and Emergency Room Visits in 13 Cities*. Washington, DC: American Lung Association.

Energy and Environmental Analysis. 1995. "Emission and Air Quality Trends in the Baltimore Metropolitan Area," final report prepared for the Foundation for Clean Air Progress, Washington, DC.

Exxon. March 1997. "EPA's Proposed Air Regulations Promise Uncertain Benefits and Higher Costs," *Exxon Perspectives*.

Federal Register. 1971. "National Primary and Secondary Ambient Air Quality Standards," *Federal Register* (April 30) 36:8186–8201.

————. 1977. "Review of the Photochemical Oxidant and Hydrocarbon Air Quality Standards," *Federal Register* (April 20) 42:20493–20494.

Grove, M. J., and J. Cock. 1997. "Baltimore Reports on Plans for Neighborhood Revitalization Through Open-Space Restoration," *Urban Issues* (Spring 1997). A newsletter of the Urban Resources Initiative, Yale School of Forestry and Environmental Studies, New Haven.

Guderian, R., ed. 1985. *Air Pollution by Photochemical Oxidants: Formation, Transport, Control, and Effects on Plants*. Berlin: Springer-Verlag.

Haynie, F. H., and J. W. Spence. 1984. "Air Pollution Damage to Exterior Household Paints," *Journal of the Air Pollution Control Association* 34:941–944.

Krupnick, A. J., and P. R. Portney. 1991. "Controlling Urban Air Pollution: A Benefit-Cost Assessment," *Science* 252:522–528.

Lake, G. J., and P. G. Mente. 1992. "Ozone Cracking and Protection of Elastometers at High and Low Temperatures," *Journal of National Rubber Research* 7:1–13.

Lasswell, H. D. 1971. *A Pre-view of Policy Sciences*. New York: American Elsevier.

Lasswell, H. D., and A. Kaplan. 1950. *Power and Society: A Framework for Political Inquiry*. New York: Yale University Press.

Lasswell, H. D., and M. S. McDougal. 1992. *Jurisprudence for a Free Society: Studies in Law, Science and Policy*. 2 vols. New Haven: New Haven Press.

MaGill, P. L., F. R. Holdren, and C. Ackley, eds. 1956. *Air Pollution Handbook*. New York: McGraw-Hill.

Maryland Department of the Environment. 1996a. "Maryland Air Quality Data Report (1994)," annual report of the Air and Radiation Management Administration, Maryland Department of the Environment, Baltimore.

———. 1996b. "Phase I Report of the Maryland Environment 2000 Project," Internet-published document submitted to the EPA by the Maryland Department of the Environment, August 16, 1996, Baltimore.

———. 1997. "Maryland Air Quality Data Report (1995)," annual report of the Air and Radiation Management Administration, Maryland Department of the Environment, Baltimore.

Maryland Office of Planning. 1996. "Planning Data Services," Internet-published document produced by the Maryland Office of Planning, January 1996, Baltimore.

Nowak, D. J. 1994. "Air Pollution Removal by Chicago's Urban Forest," in E. McPhearson et al., eds., *Chicago's Urban Forest Ecosystem: Results of the Chicago Urban Forest Climate Project*, pp. 63–81. General technical report NE-186. USDA Forest Service, Northeast Forest Experiment Station, Randor.

Percival, R. V., A. S. Miller, C. H. Shroeder, and J. P. Leape. 1992. *Environmental Regulation: Law, Science, and Policy*. Boston: Little, Brown.

Smith, W. H. 1990. *Air Pollution and Forests: Interaction Between Air Contaminants and Forest Ecosystems*. New York: Springer-Verlag.

United States Code. 1991. Clean Air Act. U.S.C. 42:7408–7409.

United States Environmental Protection Agency. 1986. "Air Quality Criteria for Ozone and Related Photochemical Oxidants," EPA Office of Health and Environmental Assessment, report #EPA/600884020, Research Triangle Park, North Carolina.

———. 1995a. "Our Mid-Atlantic Environment: 25 Years of Progress," EPA Region III report #EPA-903-R-95–017, December 1995, Philadelphia.

———. 1995b. "National Air Pollutant Emission Trends (1900–1994)," EPA, Office of Air and Quality Standards, Research Triangle Park, North Carolina.

———. 1996a. "Air Quality Criteria for Ozone and Related Photochemical Oxidants," EPA Office of Research and Development report #EPA/600/AP-93/004abc, Research Triangle Park, North Carolina.

———. 1996b. "Health and Environmental Effects of Ground-Level Ozone—A Fact Sheet," document posted on EPA's home page. Office of Air & Radiation, Office of Air Quality Planning & Standards. USEPA, November 29, 1996, Washington, DC.

———. 1996c. "National Air Quality and Emissions Trends Report (1995)," EPA, Office of Air and Quality Standards, Research Triangle Park, North Carolina.

Went, F. W. 1950. "The General Problem of Air Pollution and Plants," in *Proceedings of the First National Air Pollution Symposium*, pp. 148–149. Los Angeles: Stanford Research Institute.

World Health Organization. 1992. "Acute Effects on Health of Smog Episodes," WHO regional publications, European series #43, Washington, DC.

World Meteorological Organization. 1991. "Tropospheric Processes: Observation and Interpretation," in *Scientific Assessment of Ozone Depletion.* Global Ozone Research and Monitoring Project, pp. 5.1–5.2, report #25. World Meteorological Organization, Geneva, Switzerland.

4 Improving the Policy Process for the Restoration of Beaver Ponds Park, New Haven, Connecticut

Jessica Lawrence

Beaver Ponds Park is one of the largest green spaces in the city of New Haven, Connecticut (pop. 530,240). The park is underused because of its reputation as inaccessible, polluted, dangerous, and uninteresting. The area is relatively flat, with two large ponds, mowed fields, and wetlands bordered by stands of small-diameter black locust and red maple trees. It is bordered on the north by the grounds of Southern Connecticut State University, and residential neighborhoods on the east and west sides. Hillhouse High School borders the southern edge, where the running track, football stadium, baseball diamonds, and parking lot are located.

Neighborhood groups from the bordering communities of Newhallville (pop. 7,798), Dixwell (pop. 6,298), and Beaver Hills (pop. 4,953), as well as New Haven's Department of Parks, Recreation and Trees (DPRT) have expressed interest in improving native species diversity, restoring water quality, and beautifying the park. It is the goal of many groups to make the park safe and attractive for a more diverse range of users. However, the communication dynamics of participants involved in the policy process or park restoration are sporadic at best, exclusionary at worst. Time shortages and value differences among participants also make collaborative planning difficult. Using the policy sci-

Research for this project was conducted in 1996 and 1997.

ences framework to guide interviews with local organizations' leaders, a graduate student research team clarified how to create a more value-inclusive prescription for park management (Clark 1997).

I present an analysis of the social and decision processes concerned with park management over the past decade. I then analyze how the factors that hindered the information-gathering process of the graduate student research team also prevent park restoration. Finally, I make recommendations for continuing the planning process and strengthening democratic management of Beaver Ponds Park.

With limited staff and funding, the New Haven DPRT has frequently sought outside assistance when tackling the planning and management problems of their 193 urban parks, most of which are less than two acres. To analyze policy problems concerning Beaver Ponds Park, graduate students from the Yale School of Forestry and Environmental Studies (YFES) Ecosystem Management class, under the direction of Professor William R. Burch, Jr., were invited by the DPRT to carry out an information-gathering process from January through April 1997. The department requested that the final report be in the form of a draft management plan that could be reviewed, refined, and implemented.

I was a member of the YFES team of seven students, whose goal was to create a document that summarized all available information about Beaver Ponds Park. We hoped to produce a report that would be accessible and useful to the DPRT as well as community groups in their planning processes. We used several ecosystem management frameworks to focus our questions about biophysical and socioeconomic conditions (e.g., Machlis et al. 1995). My task was to interview representatives of all local institutions involved with the park in the three local neighborhoods so that institutional assets and linkages could be mapped. It became obvious that our team could not create a valid draft management plan in four months, so we focused our efforts on gathering ideas about potential local involvement in the creation of a final management plan.

Our team believes Beaver Ponds Park is worth spending time and effort to restore. Although the park is not in crisis condition, we assumed that neighborhood people—whose well-being could benefit from an improved park—would want a voice in the creation of a new set of policies concerning the restoration and management of the park.

The research team reviewed past reports, maps, and historical government and media documents relating to the park (Hengen 1994; Everett et al. 1996). One team member and I interviewed over twenty individual park users and twenty participants in the restoration process, most of whom were representatives of local organizations. We used semistructured interview techniques, incorporating the five questions of policy orientation (Lasswell 1971:39). Each person or group interviewed was asked to clarify the goals of their organization, describe the trends that have had an impact on the park, analyze past and present environmental and institutional conditions, and describe their relationships to other participants. We asked interviewees to make predictions for the park and suggest and evaluate alternatives to the status quo.

Urban Open Space: A History of Park Transformation

A century ago, Beaver Ponds Park was a malarial swamp, complete with resident beavers, surrounded by farmland as well as residential neighborhoods. As New Haven expanded, farms became suburbs and the swamp was dredged and pooled into two large ponds. Malaria was eradicated and policy makers planned a beautiful park for fishing and swimming. The plans never quite materialized, however. The waterways were used as dumping grounds by local slaughterhouses, households, and the nearby Winchester Repeating Arms factory. Despite the waterways' history of pollution, an increased park budget around mid-century allowed for improved maintenance and the park became a favorite area for families, children, fishermen, and local events. As economic decline over the past three decades has affected all of New Haven, park maintenance has not been a priority. Beaver Ponds again has become an illegal dumping ground for locals, outsiders, and businesses. Construction materials from around Connecticut have been dumped in or near the waterways with little official response.

For over twenty years the New Haven Department of Parks, Recreation and Trees has operated with insufficient budget and staff size. Because there has been no mandated format for participatory planning, public involvement in planning and decision making has been sought only when time and personnel have been available to do so. With insufficient staff, all administrative functions, including the planning, monitoring, and appraisal of projects, have had to be curtailed.

Over time, reduction of on-site maintenance has led to trash and weed accumulation. Natural areas have become choked with poison ivy, thorny vines, and dense wetland plant species that have prevented view of or access to the ponds. Lack of interpretive services has prevented outreach and education of park users. The aesthetic beauty of the area has decreased, and public awareness of the name "Beaver Ponds Park" has diminished. Neglected areas have been fenced off to prevent further dumping, which has limited recreational use. As usage has declined, the park has become a haven for illegal drug use and trade and therefore has been feared and avoided by many residents. Because of the site's poor reputation, there has been little public effort or governmental support to invest in restoration.

Despite this history of decline, the park retains wild plants and animals, open spaces, waterways, and free recreational facilities within walking distance for almost nineteen thousand people.

Historical Changes in Local Neighborhoods

The process of decline I have described has happened in many of New Haven's parks. However, both East Rock and Edgewood parks today enjoy high levels of public use and relatively good reputations. What is it about Beaver Ponds Park that has prevented its restoration in recent years? The answer lies partly in the context of the local neighborhoods. Thirty years ago, Newhallville was a working-class neighborhood. Most of its residents were able to raise families on the steady salaries provided by employment in nearby manufacturing industries. With the loss of these jobs when industries relocated, downsized, or closed, the social structure of Newhallville shifted dramatically. Most homes became renter-occupied, and few owners had enough capital to maintain their buildings. When crack cocaine became available to low-income consumers in the mid-1980s, drug activity on the streets caused property values to continue to fall despite unflagging local attempts to keep the streets safe and drug-free. Abandoned houses often became burned-out shells with trashed yards and walls covered with graffiti. Vacant lots became dumping grounds for nonlocal contractors who realized they could get away with trashing low-income areas in order to avoid disposal fines elsewhere. These conditions continue today, though some blocks have experienced a revitalization of community spirit and have beautified outdoor areas by creating community gardens, landscaped areas, and playgrounds. Religious and youth organizations have grown strong in an area with fewer governmental support services than in nearby Dixwell.

Because Dixwell has historically been a poorer neighborhood, many organizations providing social services have long been a part of community infrastructure. The recession affected this neighborhood as well, with accompanying declines in property values and increased levels of drug-related crimes.

Beaver Hill has remained a middle-income neighborhood despite the economic downturns. The neighborhood continues to have higher property values, higher levels of home ownership, and well-maintained streets and yards. Public space in this neighborhood is frequently maintained by block watch groups with organized beautification projects and community gardens. Thus, social divisions among the three neighborhoods have grown from decades of economic disparity.

Community involvement in New Haven planning has been hampered historically by differences in race and class and in levels of wealth, power, and education. Residents we interviewed commonly expressed the sentiment that community loyalty and participation had been stronger within ethnic and class categories in the past than it is now. With transitory populations in neighborhoods where most residents rent apartments for short amounts of time, investments of citizen time and effort into an improved decision process appear to have become rarer. Some residents claim that lack of community organization is a problem, while others express the jaded opinion that no amount of organizing by people in Newhallville and Dixwell makes a difference when the government ignores their demands anyway. For Beaver Ponds Park, lack of government responsiveness has created resentment in residents who feel deprived of power and respect. Wealthier neighbors tend to criticize those living in lower-income neighborhoods because they apparently do not share "common" ideals of how buildings, yards, streets, and public spaces should be treated. However, standards of beauty, cleanliness, and pride in and responsibility toward public spaces may vary much more with local ability to affect government than with economic status or cultural heritage.

The Social Context of 1997

For over a generation, Beaver Ponds Park has failed to live up to its potential to improve community well-being and empowerment in decisions concerning public lands and resources. How can we understand the details of the current situation? To clarify the social processes[1] and decision phases[2] concerning the park, I have grouped participating organizations into three categories: governmental, community-based, and community-assisting organizations (Table 4.1). The values[3] of individuals within these categories are not homogeneous because the categories include participants whose interests may conflict with others within the same category. The groupings are useful for categorizing where each organization stands in the context of others.

Decision Making in Park Management

All decision functions have been the historical responsibility and mandate of the DPRT. However, other participants have been critically

Table 4.1. Institutional Participants in the Beaver Ponds Park Restoration Process

Organizations	Contact Person[a]
Governmental	
Dept. of Parks, Recreation and Trees	Pam Kressman, Pat Rubano
New Haven Riverkeeper	Peter Davis
Alderman for Beaver Hill	Bruce McLenning
Animal shelter	Tim Griffin
Dept. of Public Works	Carmen Mendez
Dept. of Environmental Conservation	Bruce Blair[b]
Livable Cities Initiative	Loreen Larson Oboyski[b]
Community-based	
Friends of Beavers Pond Park	Rob Forbes
Whalley-Edgewood-Beaver Hill management team	Richard Dozier
Dixwell Enterprise Community management team	Ed Grant
Neighborhood specialists	Jerry Tureck
Newhallville management team	Alfreda Edwards[b]
Newhallville Restoration Corporation	Roger Hughes[b]
Pond St. block watch	Hazel Williams[b]
Block watch coordinator: AMERICORPS	Tanya Tolsen[b]
Newhallville Enterprise Community	Richard Spears[b]
Dixwell Neighborhood House	Ted Hogan[b]
Community-assisting	
Urban Resources Initiative/ New Haven (URI)	Colleen Murphy-Dunning
Center for Coastal and Watershed Studies (CCWS)	Emly McDiarmid
Community Foundation for Greater New Haven	Ana Arroyo
Fighting Back	Betty Rawls
New Haven Land Trust	Lauren Brown
Neighborhood Housing Services, Inc.	Tracy Smith, Henry Dynia
Family Alliance	Barbara Tinney
Southern Connecticut State U. groundskeeping	Van Seldon
Local Initiative Support Corporation	Angel Fernandez[b]
Verick African Methodist Episcopal	Lester McCorn[b]
Buelah Heights	Bishop Brooks[b]
Good Shepard Outreach Ministry	
Bethel African Methodist Episcopal	
Masjid Muhammed	
Young Israel	
Church on the Rock	
Dixwell Community/Q House	Jackie Downey[b]

(continued)

Table 4.1. continued

Organizations	Contact Person[a]
Project GREEN/Alternative Sanctions Program	Ralph DiDominic
Citywide Youth Coalition	Dan Newall[b]
Students United for the Return of Excellence (SURE)	
Youth Continuum	
Young Voices Initiative	

[a]Contact people were interviewed January through April 1997.
[b]These contact people were not interviewed despite efforts to contact them by phone.

important in some of the functions. This analysis focuses on the activities of community-assisting groups in intelligence and promotion, community-based groups in promotion, and the DPRT in all decision functions, especially application of value-exclusive prescriptions. I include a complete analysis only of the intelligence function because I was an active participant and observer in this process during data collection. The DPRT maintains authority over the other decision functions; I have analyzed those minimally. A comprehensive policy sciences analysis would perform a feature analysis for every decision function.

Gathering Information: Who Participates with What Perspectives?

The intelligence function is defined as the gathering, evaluation, processing, storage, retrieval, dissemination, and utilization of relevant information (McDougal and Reisman 1981:269–272). The way in which information concerning Beaver Ponds Park is managed is problematic. The DPRT is the primary participant in the intelligence function because traditionally it has been the only permanent institution responsible for information related to the park. Records of park improvements exist within the archives of board meetings and in annual reports and budgets. Because park-specific information is not centralized or compiled, the DPRT remains a relatively poor resource for formal documentation of historic facts. The perspective of the department is that formal information gathering is a low priority that is better left to local student groups or researchers who have the necessary time and funds.

Other government agencies such as City Planning, City Engineers, Department of Public Works, and the State Department of Environmental Conservation make brief reference to the park in their maps and research, but such information is not separate and easily accessible.

Community-assisting organizations (CAOs) such as the Urban Resources Initiative and the Center for Coastal and Watershed Studies (CCWS) at the Yale School of Forestry and Environmental Studies (YFES) work with both government and communities to research specific problems, offer technical assistance, and provide funds or other resources to community initiatives. Other CAOs include governmental social service agencies, private grant-making foundations, housing renovation organizations, and environmental agencies. Researchers from Yale University, Southern Connecticut State University, and other local schools have done preliminary water quality testing, but the data and results are not accessible to the public.

The most active and comprehensive intelligence-gathering processes concerning Beaver Ponds Park as an entity in itself were conducted by teams of students from the YFES Community Forestry and Ecosystem Management classes. Student research teams from YFES operate with the perspective that they can learn ecosystem management theory by analyzing a local case study. Beaver Ponds Park has been one of several areas of interest for these classes in recent years.

How Do the Participants Interact?

Interactions among organizations are limited because the DPRT has no formal arena that encourages public participation or power sharing in park planning and management. Decisions are made sporadically by a few individuals in response to crises rather than in anticipation of them.

Special interest groups (such as local athletic groups) pressure the DPRT to spend most available funds on fields and locker room facilities (Gillesberg et al. 1997). There remains no funding for open space beautification, safety patrols, trash and weed removal, or infrastructure for other interest groups. There are no benches in the park, no obvious walking or jogging trails, and no playground equipment for children. The effect of such policy is that local residents are deprived of a potential place to enjoy and strengthen individual, family, and community well-being, affection, and respect.

How Are Participants Relatively Advantaged?

Basic human values are shaped and shared in the policy process according to the historical interplay of various factors (Brunner 1987:10). The DPRT could be powerful, skilled, and enlightened when it comes to the intelligence process, but it lacks the funds needed to gather and organize both formal and informal knowledge. Without sufficient funds, it is unlikely there will be an adequate staff to gather intelligence or consult with interested participants.

YFES teams hoped to assist the policy process by doing three months of unpaid work in the form of information gathering from as many participants as possible. Student teams drew upon base values of enlightenment concerning ecosystem restoration theory, skill in applying research methods, and wealth through school-subsidized research. As we will see, however, these benefits are not without drawbacks, and can even lead to conflict.

Other CAOs have the strengths of diverse skills, respect of many participants, knowledge of local dynamics, and wealth—because project funds are often grant-supported.

The community-based organizations (CBOs) often include individuals with decades of acquired knowledge about the interpersonal, institutional, and biophysical dynamics of Beaver Ponds Park. Nevertheless, CBOs are relatively inactive in formal aspects of the intelligence function because of lack of time, funds, and the skills necessary to actively gather, process, and disseminate information.

What Strategies Are Used?

Within the DPRT board meeting minutes, formalized information concerning Beaver Ponds Park consists of sporadic paragraphs dealing with pond dredging or athletic field improvements from 1890 to circa 1960. Little mention of the park was found in the minutes from 1960 to 1997. No separate financial records exist for the park, because management costs are subsumed in the financial accounting for all city parks. The most recent formal planning exercises within the DPRT were done in the late 1970s, when a map was created that determined boundaries of future athletic field improvements (P. Kressman, personal communication 1997). Informal information available from DPRT employees is much richer, including undocumented, detailed knowledge of budgets specific to the park, natural history and ecology of the area, history of land and water pollution, and community interest in the planning and use of athletic facilities.

Information is not centralized in one place within the city. The New Haven Historical Society archives historical documents related to the park. Local newspapers keep their own records by date, but not by subject. Scientific research on environmental quality has focused on water quality issues and has been conducted according to the schedules of grants and temporarily funded projects.

Two substantial YFES student reports were produced in the early 1990s (Hengen 1994; Everett et al. 1996). These works focused on gathering, evaluating, processing, and storing any social or biophysical information relevant to the park within a one-semester time constraint. Student teams investigated park and local neighborhood history, plant and animal species diversity, as well as public opinion and behaviors relating to park recreation. The reports contain detailed alternatives for park restoration.

For a functional understanding of the intelligence process, we must ask from what source information was received and to whom it was communicated (Lasswell 1956:5). Student reports were disseminated to many governmental and community-assisting organizations, with fewer to community-based groups. All reports have been archived at the Yale Urban Resources Initiative, New Haven Historical Society, and DPRT.

The third and most recent YFES report, based on research carried out during spring 1997, focused on issues of water quality and non–point source pollution, natural habitat restoration, socioeconomic trends in local neighborhoods, local school involvement in outdoor education, sustainable finance mechanisms for the park, and recommendations for the creation of a collaborative management plan (Gillesberg et al. 1997).

During our information-gathering process with community-based organizations and individuals, we expected that by clarifying our goals and methods, we could reduce the socioeconomic and power tensions so common in Yale–New Haven interactions. Despite our attempts to maintain our strategy strictly within the intelligence rather than promotion function, in two of four large public meetings with community-based organizations, local leaders who had been selected for participation in the promotion function felt deprived of power and respect. In turn they criticized our initiative and withdrew their organizations' support.

We brought to these meetings a summary of past YFES reports in the form of a matrix. One column contained a list of proposed actions for park restoration as defined in the reports. The other columns asked how and where these goals might be achieved, who should be involved, and what had been overlooked. Although the reports stated that there was diverse public involvement in the selection of alternative solutions, we found that some leaders denied involvement and maintained that the recommendations were YFES-created alternatives. We were asking for public appraisal of these alternatives, yet community leaders assumed we were attempting to build on past student reports to move into an already determined promotion phase with restoration activities resulting. Our mistake was to skip the important question: Are these stated goals in line with yours? What do you want to see happen in Beaver Ponds Park over the long run? In jumping to the evaluation of alternatives, we were skipping over the important preliminary step of problem orientation with the participants.

We tried to clarify that we would refrain from simply promoting past goals as stated in the reports, which perhaps had been premature. We agreed to limit our analysis to intelligence gathering that included interviews as well as public appraisal of past planning efforts. Nevertheless, one community leader, who was wary of Yale student involvement in community issues, demanded that participants refrain from sharing any information with us. Many concurred. Respect for Yale students was bluntly withdrawn in order to assert power.

In an effort to maintain good standing in the eyes of their constituents, government and other CAO representatives sided with community leaders. At these large group meetings, no information sought by our team was gathered, although various participants reassured us after the public debate that they would still support our efforts if we shifted our methodology toward "beginning at square one, again." This included contacting key leaders before public meetings and letting them know our agenda. If it became apparent that we were likely to infringe upon someone else's domain, a community representative trusted by the leaders would be prepared to actively facilitate an information-gathering process.

After these meetings we questioned whose power was being asserted: Was it the community as a whole or the one leader who led the initiative that criticized our research ethics? Had past YFES research failed to satisfy some of the people with whom they had worked, or was it that our research team was now treading on the egos of individuals who had not been involved in the policy process in 1996? The answers were not obvious. Whatever the cause, we realized our involvement would be severely hampered by this legacy. Future efforts by student research teams would likely be similarly incompatible with local dynamics if they were not welcomed by politically dominant individuals. One possible reason for public dissatisfaction is the fact that student researchers are constrained by semester or class time frames. The deadline for a final product may not coincide with the time frame of community leaders, and the final reports may not be widely dispersed without outside funding. Furthermore, the nature of student involvement in community issues is sporadic and temporary. A history of these types of interactions has disappointed some participants to the extent that they are no longer interested in having research conducted in this way.

Local political processes in New Haven are hampered by poor communication pathways based on socioeconomic class as well as ethnicity. "I was not informed of this decision" is a legitimate local protest.
These divisions among participants have long been a detriment to democratic planning. Individuals and groups who have relatively little power
and wealth are justified in feeling territorial over their own knowledge
when it has always been documented (and altered) by those with greater
power, wealth, enlightenment, and skill. Indeed, one of the most effective strategies a community leader can employ is to deny respect to the
powerful by stating that the planning process has not been participatory, and is therefore invalid.

It is useful to refer to the maximization postulate here (Lasswell
1971; Clark 1997:12). This postulate states that actors behave in ways
they perceive will leave them better off than if they had acted differently. In this case, if decision makers value fair representation and democratic standards, then residents who claim to have been excluded can
stall the policy process and prevent application from occurring. In my
judgment, this has taken place in reference to the Beaver Ponds Park
restoration effort, with mixed results.

What Are the Effects of These Strategies?

The DPRT did not consider or use previous YFES reports in planning
or management. However, the reports may have been critically important in galvanizing local interest in the park among such CBOs as Friends
of Beaver Ponds Park and the Dixwell Neighborhood Management
Team. These are community-based organizations made up of concerned
residents who volunteer their time to coordinate local programs focusing on public space beautification, safety, economic opportunities, and neighborhood services. The Friends of Beaver Ponds Park
has organized specifically around issues of park restoration and beautification.

All YFES student reports may fall short of being value-inclusive because the studies have been carried out by people the neighborhoods
consider outsiders, rather than trusted, long-standing participants.
Knowledge gathered is inevitably biased by the collector's identity and
social status (Brunner 1987:2–7).

According to McDougal and Reisman (1981), dependability, comprehensiveness, availability, and economy of information are critical.
In our interactions with community-based groups, we learned that
previous student groups may have interviewed many key players, but
they did not reference the identity of informants along with their per-

spectives. Finally, the information gathered was not presented in an easily readable fashion, nor was the availability of the final document known to many. As a result, some community leaders were of the opinion that the reports were neither credible nor comprehensive in their attempt to clarify common interests. Their appraisals of the previous reports were expressed with considerable frustration. Neighborhood residents who felt deprived of power and respect in the intelligence function expressed a strong desire to be the initiators and controllers of social research, rather than depend on outsider involvement. Until community leaders feel respected and involved in how outsiders approach participants, cooperation will remain unlikely.

How Are Restoration Plans Promoted?

It is the goal of the community-based Friends of Beaver Ponds Park to spend the next few years of volunteer effort organizing support and broad local participation for the creation of a formal management plan for the park. The plan is to clarify common interests of the three local neighborhoods and recommend a management prescription for the DPRT to apply.

One reason a common interest was not clarified previously stems from lack of communication, affection, and respect across wealth, power, and ethnicity lines. Populations in Newhallville and Dixwell are composed primarily of Christian, Moslem, and secular African Americans. Beaver Hill includes African Americans as well as European Americans whose affiliations range from orthodox and reform Jewish, Catholic, and Protestant to secular culture groups. Different members within each of these groups fit into low-, medium-, and upper-income brackets. Levels of poverty are highest in parts of Newhallville and Dixwell where housing stock has been greatly devalued.

Friends of Beaver Ponds Park (FBPP) is an emerging organization, formed in December 1996. As with most community-based organizations in the area, the group seeks respect from other participants and well-being for its constituents. As an institution it has relatively little wealth and power in the decision-making process. Participants in the monthly FBPP meetings hail from all three neighborhoods, but representation is skewed toward Beaver Hill, where volunteerism is more common and organized. Almost every organization listed in Table 4.1 has attended at least one FBPP meeting. Participants have included local women with children, gardening enthusiasts from block watch associations, teachers, environmental education specialists, aldermen, and members of the Neighborhood Management Teams and Enterprise Community groups. Representatives from URI, CCWS, and the government agencies of Public Works, DPRT, and Livable Cities Initiative have also attended. FBPP hopes to bring together interested members of all these communities to promote local involvement in the creation of a management prescription. The hope is that such a plan will keep governing agencies accountable for their actions and value-inclusive in their approach. Currently, FBPP prefers to remain a community-based organization that plays the role of activities promoter and volunteer coordinator rather than that of an employing organization with budget, staff, and results-oriented deadlines. Community leaders in the Neighborhood Management Teams and other CBOs are involved in publicizing reform demands with the intent to influence legislative prescription by grassroots efforts.

Prescription: Defining Steps in the Restoration Process

Prescription has been performed mainly by the DPRT, but other city agency agendas have led to the loss of park land for other uses. For example, Hillhouse High School was built on park land in 1956; Jackie Robinson Middle School was constructed two decades later. Park land was further reduced by the construction of the animal shelter, police firing range and junked car lot, and Southern Connecticut State University buildings.

As Lasswell (1956:4) states, "The burden of applying community prescription rests not only upon the official bureaucracies, but upon the leaders and bureaucracies of private organizations (and individuals)." Local residents have had little success at having their visions for the park respected and adopted by decision makers. Despite such discouraging results, there remains the possibility that CAOs can play a key role in strengthening local recommendations and seeing these to fruition.

Invocation: Defining Inappropriate Behaviors

The invocation function is the "characterization of conduct according to prescriptions, including demands for application" (Lasswell 1956:7). The DPRT is the agency through which all prescriptions are invoked. When lands formerly part of the park were appropriated by other city government agencies, the takings were deemed legal according to city judicial processes. Though some developments, such as Jackie Robinson School, were strongly protested by local residents, their invocations of acceptable land management were ignored.

Some residents expressed frustration at an apparent lack of concern shown by the DPRT and the police for the illegal dumping that occurred within the park. When neighbors reported violations, the response was typically delayed.

Application of Restoration Prescriptions

Maintenance of Beaver Ponds Park has long been carried out according to the priorities of special interest groups such as athletic organizations. Since the establishment of FBPP, a different set of priorities for application has begun to develop within the DPRT. One example is the maintenance work along the boundaries of the park. The DPRT has replaced rusted, decrepit fences with guardrails, opened an entranceway in the fence that previously prevented access from Newhallville, and repaved and landscaped the large parking lot during renovations of Pop Smith's Little League baseball field. The department also applied for a $50,000 grant to build a stable platform extending over one pond to facilitate bird-watching, fishing, and canoe access.

Swimming is unlikely to occur in the ponds, as it did decades ago. The ponds have become heavily silted and shallow from erosion in the watershed, which extends for about two miles north and south of the park, and about one mile to the east and west (Gillesberg et al. 1997). The shores and bottoms of the ponds are littered with drug needles, cans, broken bottles, car parts, and decades-old waste. In March 1997, the DPRT hired Riverkeeper Peter Davis, whose job is to remove trash from within and around the ponds. With long-term efforts, the ponds could be restored to swimmable conditions. One constraint to this is that the riverkeeper must also remove trash along the West River, Wintergreen Brook, Mill River, and Quinnipiac River within city boundaries. The riverkeeper also serves as the primary environmental educator, offering guided walks and canoe tours of all waterways. Davis is actively involved with FBPP. His communication skills, knowledge of restoration needs, and enthusiasm for community involvement could provide the link that has been missing for so long between park users and the DPRT administration.

How are participants involved in the application activities of others? In April 1997, a neighborhood cleanup was done of the northwest section of the park, closest to the Beaver Hill neighborhood and Southern Connecticut State University (SCSU). It was organized by an employee of SCSU who is also a member of FBPP. Hundreds of colorful flyers announcing the event were distributed throughout the three neighborhoods and local schools by representatives of community-based organizations. On the day of the event, maintenance staff of the

DPRT and the Department of Public Works pruned trees and hauled away over one hundred bags of trash and leaves collected by volunteers. Most community members who attended were from Beaver Hill block watches and families. Several community leaders from Dixwell attended, but few families. Elementary school students chaperoned by Sierra Club Urban Outings, our YFES research team, SCSU maintenance staff, and SCSU administrators were also present and hardworking. The bulk of the cleanup work, however, was done by over sixty men sentenced to community service for misdemeanors. One such program, Project Green, provides maintenance labor for Connecticut parks using people sentenced to alternative sanctions. Local fishermen, Newhallville residents, and teenagers were noticeably absent from the event.

The demographics of volunteers indicate that Newhallville residents are unlikely to participate in park restoration unless it is actively geared to improving the parts of the park that border their community. As no one from this neighborhood is active in promotion of such activities, it appears unlikely that the eastern side of Beaver Ponds Park will share in the restorative process begun on the north and west sides that border Beaver Hill and SCSU. Friends of Beaver Ponds Park has applied for a green space beautification grant from URI and the Community Foundation for Greater New Haven. With a bias toward the wealthier northwestern edge, park beautification may be achieved primarily by and for the benefit of Beaver Hill residents. The result will be a park with one restored section. Once the more powerful restoration advocates and community organizers are satisfied with their immediate area of interest, overall restoration of the park may stagnate, to the detriment of the ponds, biodiversity, and potential Dixwell and Newhallville users.

Appraisal: Evaluating Past Policy

The DPRT does not conduct appraisals of its policies that are open to power sharing. The "reform wave" (Lasswell 1956:9) has begun in relation to the policy process surrounding Beaver Ponds Park. The information-gathering process of YFES students also included an appraisal of current practices. The student reports attempted to give a formal voice to the evaluations they heard in interviews and discussions. Members of both CBOs and CAOs have criticized the DPRT's special interest–based planning and lack of attention to Beaver Ponds Park.

Appraisals by community-based and community-assisting organizations are informal, with diverse opinions. One outcome of this is that DPRT staff claim not to have enough time to clarify the common inter-

ests of residents and potential park users or to actively open the decision process to wider participation. The DPRT has long been allowed to
remain in what Lasswell (1956:9) calls a "corruption phase." This does
not mean that DPRT staff steal from or cheat the public. Rather, they are
less than strict in their appraisal of their own conduct than is optimal
for collaborative management.

DPRT staff cite the lack of funds for the department as the key reason for park decline. This situation is unlikely to change over the long
term. Even if a reform movement succeeds in creating a candidate
management plan that mandates collaborative planning, the DPRT
may again lapse into the corruption phase by not complying with or
being responsive to the new prescription. Why? The staff of the DPRT
is involved in "triage" and "emergency care" of the 193 parks in New
Haven. In the DPRT's view, there is simply no short-term benefit in seeking extensive local perspectives on issues that are traditionally dealt
with within its exclusive domain. In the long run, park restoration and
management could be done primarily by community volunteers advised and supported by DPRT staff. The transition to such a system will
require much community organizing and commitment.

McDougal and Reisman (1981:285–286) maintain that appraisal is
particularly important in a period in which conditioning factors are
changing rapidly. Appraisal is ideally continuous, impartial, and contextual. Participation in the appraisal function is critical for democ-
racy in the policy process, but it has been unorganized in the case of
Beaver Ponds Park. As with the appraisal of past YFES reports (which
were not publicly discussed until a new group of students arrived on
the scene and community members were able to clarify past frustrations), appraisal of other organizations' actions is rarely communicated in complete feedback loops.

Termination: A Legacy of Unsatisfactory Endings

Some aspects of exclusionary park management have been terminated. With the advent of a proposed fishing platform for one pond and
a plan to lease the northern half of the park to SCSU for landscaping
and maintenance work, it appears that the DPRT will respond to requests for nonathletic uses and facilities. What has not changed, however, is the practice of sporadic communication. In my view, the DPRT
has listened to the loudest voices and is making an attempt to satisfy
them temporarily. No participatory forum or outreach for involvement
in the decision process has been established. Community-based groups
may protest if SCSU leases a large area of the park where neighbors

were planning their own restoration activities. Once again, dissenting voices may be too late to have an effect on the decision process.

Criticisms of previous student research have focused on the poor dispersal of the final reports. This convinced our team of the importance of distributing the information we gathered in a timely manner. Copies of the report have been given to the DPRT and Riverkeeper Peter Davis, who can lend them to interested parties. We also plan to send a personal letter of thanks and a pamphlet summarizing our research, along with the address of where our final report can be accessed, to every participant interviewed. Unfortunately, securing funds for pamphlet printing has taken much longer than anticipated, so many participants are likely to be dissatisfied by the termination of our research.

Projections for Beaver Ponds Park

Without the community's commitment to establish an inclusive planning process that seeks to clarify common interests, problems of power and respect will continue to hamper Beaver Ponds Park restoration. Without an organized community-based coalition that has successfully included disenfranchised local leaders, the park will continue to be managed by an exclusionary decision-making process, which is unacceptable to many local residents.

A brighter projection is that with the mayor's current support of funding for the re-greening of public spaces, there is a window of opportunity for community-based concerns to be incorporated into the policy process for Beaver Ponds Park as well as for all of New Haven's green spaces.

Even without increased funding for park restoration, if Friends of Beaver Ponds Park becomes a respected clearinghouse for information about park activities, the area could become a well-used and more beautiful space once again with enough volunteer coordination in all three neighborhoods. Long-term maintenance and pollution problems, though, will remain the responsibility of public service agencies.

Suggestions for an Improved Policy Process

Current communication among participants is inadequate for collaborative management. If the DPRT continues to neglect pollution, aesthetic, and safety issues while spending the majority of its funds on athletic facilities, most of the park's nineteen thousand neighbors will continue to litter or ignore it.

The DPRT could create a management plan "in house," using ideas and priorities from their agency alone or a collection of government agencies, or through a minimal public input process in which participants appraise a finished document but have no real authority to influence the prescription. This is the least difficult type of reform available to the DPRT, because it is a relatively simple matter to mask deficiencies by declaring that the planning process was participatory. The DPRT estimated that funding requirements for "in house" creation of a park management plan would be about $15,000 (P. Kressman, personal communication 1997). However, the DPRT had no plans to implement this alternative. Though such a plan could turn out to be similar to a plan designed by and agreed upon by all participants, it would likely alienate community leaders and prevent a local sense of pride and respect for the park. Common interests would remain undefined and contentious. The decision process in general and the restoration process for the park in particular would stagnate.

An improved policy process for park restoration is possible, with potentially greater long-term benefits. I recommend the following alternatives:

1. A few community leaders have been calling for an alternative to community-assisting organization or governmental research into park policy; they advocate that local leaders and neighborhood volunteers carry out their own research and planning, despite a longer time frame for completion. Since FBPP has had success in bringing together representatives of almost all participant groups, I suggest that this group organize a series of "decision seminars" (Willard and Norchi 1993:575–582). These seminars, or participatory workshops, should develop and evaluate alternatives to work toward the creation of a prescription that defines common interests as critical goals. To attempt redress of past community disempowerment, open access to decision making must be a central focus of the process. Representatives of interested community organizations from each of the three neighborhoods as well as staff from all levels of the DPRT should be included. The creation of the prescriptive document should provide the appropriate authority signal to participants that it is a document they can use, evaluate, and revise. Policy content (what will be tried) and control intentions (what behaviors will be controlled) should be evaluated rigorously before final selection of policy mandates.

Decision seminars and related activities should be attempted during the next fiscal year or whenever more money and staff are available within the DPRT. Though definition of the common interest will not be a smooth process for such diverse and historically divided groups, the very act of attempting resolution of poor communication processes could go a long way toward promoting local pride and ownership in Beaver Ponds Park. There is likely to be resistance on the part of the DPRT to such community-promoted alternatives as opening access to the Newhallville neighborhood using gates or openings, or removal of the eastern edge fence. The workshops and discussions will surely be contentious, but what will become of a prescription that avoids addressing community perspectives? It seems likely that many of the community-assisting organizations can be brought on board to assist in a participatory planning and management process.

2. Funding could be sought for long-term support of an outreach coordinator (to work either within the DPRT or independent of it) to facilitate communication among participants. With most of the current technical knowledge about Beaver Ponds Park documented in the YFES student reports, participants in the planning effort should be able to move forward and begin park restoration with neighborhood equity issues attended to. With a longer time frame to work with than the YFES team had, the outreach coordinator could organize participatory planning through workshops, committees to appraise past reports, and regular phone communication with all participant groups. The newly formed Steering Committee of the Friends of Beaver Ponds Park represents each of the three neighborhoods. Its members could potentially be counterparts to help facilitate the planning process. Though paying them would offset their personal costs of volunteering, funding directly from the DPRT may bias the counterparts or cause them to lose credibility and respect among the neighborhood residents they attempt to represent. Therefore funding from a nongovernmental source may be the most socially acceptable strategy.

3. If an organization such as Friends of Beaver Ponds Park does not exist, or if it fails to remain functional and representative in the future, I advocate that the outreach coordinator set up a structure of communication with local residents and park users with the goal of democratizing the decision functions. One of the most efficient ways to do so is to have a strong coalition of community-based and community-assisting organizations, with a steering committee that can be contacted whenever a decision concerning the park is taking place. The steering committee would then be responsible for widespread outreach to all interested participants, which would greatly reduce DPRT staff time spent on organizing public outreach and participation. With such a structure in place, future declines in funding within the DPRT should no longer be an excuse to forgo democratic processes in the planning and management of Beaver Ponds Park.

4. Finally, I would like to make recommendations for future YFES or other "outsider" information gatherers. Do not assume that members of community-based organizations are familiar with past research. Rather than initiating contact via interviews or appraisal of past efforts, spend time with informants individually with the explicit goal of developing a shared understanding and acceptance of the objectives, motivations, and methods of the "outsider" before giving presentations at community meetings. Be aware of the fact that past efforts at research termination have not been satisfactory and have left poor foundations for collaboration.

Conclusions

Community groups as well as governmental and nongovernmental organizations are interested in restoring and improving Beaver Ponds Park. However, value conflicts and low communication levels among different organizations hamper collaborative organizing, planning, and management.

In this paper I have analyzed ways in which the policy process is problematic and can be improved. A series of decision seminars is recommended which include the full diversity of participants. Such events should seek to clarify common interests, map institutional assets, and bring participants to agreement on a prescription for park restoration and management. The seminars, prefaced by diverse teams that appraise past reports and recommendations, could be facilitated by an outreach coordinator from Friends of Beaver Ponds Park or New Haven's Department of Parks, Recreation, and Trees. The planning workshops should be useful for all participants to define their common interests and move toward the creation of a publicly acceptable restoration and management plan for Beaver Ponds Park. The legacy of park-related research by Yale graduate student teams may offer useful information for management. However, poor termination of research work may cause future student involvement to be problematic.

ACKNOWLEDGMENTS

I owe a tremendous thank-you to all the residents of New Haven who were willing to share their insights into this issue. Conversations with members of Friends of Beaver Ponds Park, especially steering committee members, were quite illuminating. Within the Department of Parks, Recreation and Trees, I wish to thank Pam Kressman and Riverkeeper Peter Davis. Special acknowledgment is due my colleagues Ammy Gillesberg, Namrita Kapur, Keely Maxwell, Karen Steer, Chado Tshering, and Sasha Weinstein for their attention to the social dynamics of community research. Finally, I thank Drs. Tim Clark, Andrew Willard, and William Burch, Jr., for their helpful suggestions.

NOTES

1. Social context mapping requires analysis of participants, perspectives, situations, base values, strategies, outcomes, and effects.
2. Decision phases include intelligence, promotion, prescription, invocation, application, termination, and appraisal.
3. Basic human values include power, wealth, enlightenment, skill, respect, rectitude, well-being, and affection.

LITERATURE CITED

Brunner, R. D. 1987. "Conceptual Tools for Policy Analysis." Center for Public Policy Research, University of Colorado, Boulder.
Clark, T. W. 1997. "Conservation Biologists in the Policy Process: Learning How to

Be Practical and Effective," in G. K. Meffe and C. R. Carrol, eds., *Principles of Conservation Biology*, pp. 575–597. Sunderland, MA: Sinauer Associates.

Everett, J., C. Kloster, and R. Williams. 1996. "Beaver Pond Park: From Boundary Park to Bountiful Park." Urban Resources Initiative, New Haven, Connecticut.

Gillesberg, A., N. Kapur, J. Lawrence, K. Maxwell, K. Steer, C. Tshering, and S. Weinstein. 1997. "The Restoration of Beaver Ponds: A Park, A Plan, A Process." Urban Resources Initiative, New Haven, Connecticut.

Hengen, M. 1994. "Inventory and Analysis: Beaver Ponds Park." Urban Resources Initiative, New Haven, Connecticut.

Lasswell, H. D. 1956. "The Decision Process: Seven Categories of Functional Analysis." Bureau of Government Research, University of Maryland, College Park, Maryland.

———. 1971. *Problem Orientation: The Intellectual Tasks. A Pre-View of the Policy Sciences*. New York: American Elsevier.

Machlis, G. E. , J. E. Force, and W. R. Burch, Jr. 1995. *Measuring the Social Impact of Natural Resource Policies*. Albuquerque: University of New Mexico Press.

McDougal, M. S., and W. M. Reisman. 1981. "Constitutive Process," in *International Law Essays: A Supplement to International Law in Contemporary Perspective*, pp. 269–286. Mineola, NY: Foundation Press.

Willard, A., and C. Norchi. 1993. "The Decision Seminar as an Instrument of Power and Enlightenment," *Political Psychology* 14:575–606.

5 Analysis of the Proposed Pumped-Storage Hydroelectric Power Projects, Sequatchie Valley, Tennessee

Christopher M. Elwell

The Sequatchie Valley in eastern Tennessee is predominantly a pastoral area with a few small towns. The valley is incised in the Cumberland Plateau and cradles a small river which meanders its way to the Tennessee River. Because of the steep drop between the Cumberland Escarpment and the valley floor, the area has become the focus of two pumped-storage hydroelectric power plants.

Armstrong Engineering Resources (AER) of Butler, Pennsylvania, is the private industry that proposed these projects in 1994. Public speculation arose that the company had entered into an agreement with the Tennessee Valley Authority (TVA) to form a partnership for the management of the facility and the distribution of the power produced. Public awareness increased when the projects were announced, and it culminated when cooperation between AER and TVA was made public. Much of the public awareness has now evolved into opposition. Embroiled are private individuals, landowners, public officials, county commissions, federal and state agencies, private industries, and public interest groups.

It can be said that there are more competing special interests involved in the outcome of this controversy than there are participants. However, through a selective (macroscopic) analysis, participants' in-

Research for this project was conducted from February 1997 through May 1997.

terests can be grouped into several broad goal categories. These include: (1) inviolability of private property rights; (2) access to the public decision process; (3) accountability of public agencies; and (4) indulgences in terms of wealth, power, and skill. For the concerned citizens who oppose construction of the power plants, the policy problem is that they feel excluded from the public decision process legally set forth, and many people feel that the process is being subverted by the public agencies charged with upholding the fairness and openness of the process and ensuring its inclusiveness.

Clarification of Standpoint

I have an academic, professional, and personal interest in this controversy. I have chosen to address this problem primarily to fulfill an obligation for a course project. My professional role stems from the fact that I am enrolled in a professional graduate program at Yale University to develop knowledge and skills that will advance my career in the fields of natural resources. While my specific field is forestry, I take an interest in all issues that affect public policy and the use of natural resources. It is in both my academic and professional interest to acquaint myself with the systematic, analytic tools of the policy sciences.

It should be noted that this is my first attempt at sizing up a problem using the functional, contextual framework of the policy sciences. Given the time allotted and the circumstances in which it is written, this paper cannot present all the pertinent information, issues, and trends, nor should its conclusions be taken as a comprehensive solution to the controversy.

My personal interest in this controversy is piqued because the proposed project sites are near my hometown. More particularly, the sites are in a valley which is culturally, aesthetically, and historically valuable to me. I have long felt that any major development in the valley would be detrimental to the values I place on the area.

Admittedly, because of my personal interest in this controversy, some degree of bias is present in my examination. I am opposed to the projects and therefore align myself with citizen opposition. Although I originally became involved in this controversy from a fairly neutral standpoint (by attending a conference to learn about the federal Environmental Impact Statement process), I was introduced to the opposition's viewpoint before I had heard the statements of the projects' proponents. My stance remains the same, having heard the proponents' position.

Through my introduction to the most organized group within the opposition—Save Our Sequatchie—I learned of other participants from the sos perspective. Most of the information that I gained from interviews I conducted while researching this paper was from people who oppose the projects, and that information has influenced my perception of the controversy. I should also note that the people who oppose the two plants were much easier to contact by phone and were much more willing to discuss the issues involved and clarify their personal standpoint to me. It is my opinion that, in this situation, the interests of sos are more representative of the "common good" than are those of AER or the TVA. As the goal of the policy sciences is to promote the common interest, I feel that it is appropriate to approach this policy problem from the point of view of sos.

My standpoint in this research involved interviewing participants and collecting and integrating information. Analysis of the policy problem was derived from primary research (interviews with participants), previous knowledge of the area and situation, and secondary research (publications and agency reports and documents). I also hope that my analysis and conclusions will help to resolve the controversy.

Policy Problem

Provisional Statement of Problem and Goal Clarification

As stated earlier, the policy problem in this controversy is exclusion from the decision process. This problem is coupled with suspicion that the TVA will not conduct the environmental assessment in an objective manner. Citizens opposed to the proposed projects (associated into the public interest group sos) have expressed a collection of goals that relate to problem resolution.

The principal goal of sos is authentic inclusion in the decision process. This means that all viewpoints expressed in the public commenting process will be given fair appraisal. The opposition hopes that achievement of this goal will naturally lead to the second goal—cancellation of the projects. Third, the group wants the Water Power Act of 1920 and the 1992 Energy Policy Act changed so that independent power producers (private industries) will not have access to the privilege of eminent domain, and, consequently, private property rights will be acknowledged (*United States Statutes at Large* 1920 and 1992). The group's fourth goal is greater accountability and neutrality of the public agencies that oversee the public decision process and environmental impact assessments for such projects.

Problem Context and Detail of the Social Process

In the social process each participant acts from unique perspectives in various situations to manage his or her base values to achieve certain outcomes and to enjoy them. I identified over one hundred participants and multiple overlapping organizational jurisdictions in this controversy. I aggregated participants into categories: government agencies at the federal, state, county, city, and town levels; community groups; public officials; journalists; private businesses and industries; public interest groups; and private individuals and landowners. For analytic purposes, I was selective, focusing on key participants: one federal agency (TVA), one private industry (AER), one public interest group (SOS), and several other significant, though peripheral, participants.

Key Participants

The Tennessee Valley Authority is the federal governmental agency charged with the management of the water and electrical resources of the Tennessee Valley (*United States Statutes at Large* 1933). This agency has been directed by the Energy Policy Act of 1992 to accommodate establishment of independent power producers in the region and allow these independent wholesale generators access to the TVA's power distribution grid (*United States Statutes at Large* 1922). As a federal agency, it is required by law (the TVA Act and the National Environmental Protection Act) to review any development that affects the Tennessee River or its tributaries and to create an Environmental Impact Statement (EIS) to assess the impacts of those developments. At the completion of the EIS, the TVA will choose an alternative and file this selection in the Federal Register as a Record of Decision (*United States Statutes at Large* 1971).

The TVA's current situation is conditioned by increasing public distrust and organized opposition to the agency. Operating at a financial deficit for some time, the agency may fear being decommissioned. Because of its current situation, the agency may be at risk of being deprived of certain values—power, wealth, skill, respect, and enlightenment. To address the public's growing disquiet, the TVA's current strategy is to conduct public relations campaigns to promote a better image within its region.

One approach the TVA has taken to address its financial insolvency is to sell some waterfront properties for private development. Another approach, which may be the case in this particular situation with AER, is to devise plans to sell power outside its designated jurisdiction. The TVA area is bounded by a "fence," outside of which the agency is not allowed to sell power. The authority may stand to gain operation and maintenance contracts for the proposed power generation facilities, and the sale of this "private" power may open a sizable hole in the fence through which other TVA power could be sold. The TVA's desired outcomes are to recoup its respect, become financially (wealth) and politically (power) stable, and to attain security for the agency's future.

Armstrong Engineering Resources is a private industry based in Butler, Pennsylvania. The company is planning a $2 billion pumped-storage hydroelectric power project—located at Reynolds Creek and Laurel Branch, along the rim of the Sequatchie Valley (Booth, personal communication 1997). AER, which has never built a power generation plant, is a spin-off investor enterprise of the Armstrong Group, also of Pennsylvania. The enterprise is interested in increasing its profits and dividends for its investors. Having already committed itself financially to the proposed projects, it must receive the necessary permits and funding to realize a return on its investments. The company does not communicate openly or directly with the public, which adds to public opposition.

The company seeks indulgences in terms of power, wealth, and skill. It is at risk of being deprived of these values, as well as respect. Strategies the company has used to affect preferred outcomes are aimed at local political entities (diplomatic, ideological), and they focus on the promotion of economic growth and development (diplomatic, ideological, economic). AER has billed the projects as memorials to a patriarch of the Armstrong family who promoted hydroelectric facilities as an environmentally benign source of power production. The company canceled one of its proposed projects, the Reynolds Creek facility, in December 1996, in the face of growing opposition. This may have been designed as an act of concession.

Outcomes the company prefers are receipt of permits and funding, an end to public opposition, and the successful completion of the project. If the Laurel Branch facility is built and run well, the company will profit from the sale of electricity. The company may even experience financial gain if neither project is completed. It is possible that AER could develop or sell the land acquired through eminent domain or profit by defaulting on the loans and declaring bankruptcy.

Save Our Sequatchie is an ad hoc public interest group composed of diverse business owners, landowners, and concerned individuals. SOS seeks power, enlightenment, respect, wealth, and skill. The group is opposed to the proposed hydroelectric projects for many reasons, including the potential for a decrease in water quantity and degradation of environmental quality in the Sequatchie Valley, the use of eminent domain by private industry, the TVA exceeding its jurisdictional limits, and a breach of public trust by government agencies (Patten, personal communication 1997).

sos is constrained by limited access to the decision process, the limited available time and influence of its members, and the size and power of its adversaries. Members of the group draw on various educational backgrounds and experiences. Members have had to shuffle their normal schedules to make time for sos and to learn laws and the form of the decision process. In trying to protect their interests, some sos members have been accused publicly of putting their personal interests ahead of those of the Valley community and obstructing economic development of the area.

Strategies of sos are propaganda—grass-roots organizing and promotion of the doctrine of private property rights and opposition to the use of eminent domain. The group's desired outcome is to halt both AER projects.

Other Participants

The other major participants are the U.S. Army Corps of Engineers (ACE), the Federal Energy Regulatory Commission (FERC), the Office of Surface Mining (OSM), the Environmental Protection Agency (EPA), the U.S. Fish and Wildlife Service (FWS), the Tennessee Department of Environment and Conservation (TDEC), public officials, county commissions, journalists, and other environmental groups. These participants are significant because they may be able to exert additional influence in the decision process. Relative to these participants' perspectives and desired outcomes, sos may want to lean on them to strengthen its own strategy.

The U.S. Army Corps of Engineers has responsibility for the nation's navigable rivers. The Sequatchie River meets the criterion of "navigable" and, as a tributary of the Tennessee River, thus falls under the overlapping jurisdiction of the ACE and TVA. The corps will have to file, in the Federal Register, a Record of Decision (ROD) concerning the chosen alternative of the EIS. This ROD is similar to the one the TVA will also file. The corps' mandated responsibilities for rivers are generally to ensure navigable quality and protection of associated wetlands. In this sense, their perspective on the situation will be to balance the water use needs of the independent power producer and the local communities against the volumetric requirements of the stream channel for maintenance of navigability and riverine ecosystem function.

The Federal Energy Regulatory Commission oversees the permission and licensing process for power producers. This commission has granted the preliminary permits for AER to begin its site review. The final permit and license, if granted, will allow AER to acquire funding for the project and, in conjunction with approval of the EIS, to begin construction. In the past, the FERC has required applicants to conduct an environmental assessment and file a report on the findings. In this situation, because a federal agency (the TVA) is involved, a federal EIS is required. In an effort to streamline the process in which two federal agencies are operating, the commission has agreed to acknowledge that the findings of the EIS overseen by the TVA will satisfy the environmental assessment part of the application. This information will be used by the FERC in deciding whether or not to grant AER a permit to construct the facility.

The Office of Surface Mining is expertly familiar with the geology and hydrology of this region. The agency is responsible for reviewing surface mining plans and deciding whether to permit such operations in the context of possible environmental impacts. The OSM has reviewed nearby sites previously, and has deemed the effects of disruption and exposure of several coal and shale seams to water to be too hazardous to allow surface mining. The agency will have joint jurisdiction with the TVA over the EIS portion of this project if more than 250 tons of coal will be removed in the excavation of reservoirs and tunnels. Under the National Environmental Protection Act (NEPA), the agency may be an official cooperator in the EIS if requested by TVA, regardless of the tonnage of coal removed.

The Environmental Protection Agency, a federal agency, and the TDEC, a state agency, are both charged with protecting environmental quality. The EPA has general oversight of federal EIS. It reviews these reports at their draft and final stages and may be responsible for administering mitigation if environmental damage occurs. The TDEC is much more focused on environmental issues at the state level.

The priority of the TDEC's Division of Water Pollution Control is to protect the waters of the state. This responsibility gives the organization shared jurisdiction with the TVA and the ACE over the Sequatchie River and the ground- and surface waters that empty into it. Furthermore, the mining section of this division is responsible for overseeing and issuing National Pollutant Discharge Elimination System permits for mining operations that remove less than 250 tons of coal, and thus do not meet the criteria of OSM jurisdiction. Other priorities of the organization include protecting public health, conserving habitat for wildlife, protecting threatened and endangered species, and safeguarding significant cultural and archaeological sites. All these priorities are scoping issues the EPA and TDEC may recommend for inclusion in the EIS.

The U.S. Fish and Wildlife Service of the Department of the Interior is another federal agency that may have jurisdiction over the proposed, remaining, project site. This organization is charged with implementing the Endangered Species Act to protect populations and habitats of federally listed species. If a listed species is found on the site, the service could halt the project in the name of species preservation. Additionally, the EIS of any federal project that would affect lands under Interior's jurisdiction is reviewed by the Office of the Secretary, USDI. If an endangered species is found on the site, the FWS and the Office of the Secretary of Interior are to conduct two additional reviews of the project's EIS.

Public officials and county commissions have an important say in the developments that take place in their districts. Public officials ultimately answer to their constituents—the citizens of the local region. The voices of these officials represent substantial power, respect, and enlightenment, and, ideally, these voices should reflect the opinion of the majority of the district's citizens. Depending on many factors, including who supports the officials, political voices could speak for or against the project. County commissions have the ability to offer or deny benefits and incentives to development projects.

Journalists have power, skill, and enlightenment to greatly influence public opinion. In this role, they often serve as watchdogs, raising awareness about contentious issues. Journalists have the opportunity to flag contentious issues or to limit public debate through the media. In the current situation, both approaches have been taken by journalists. Some papers in the region have refused to print articles or letters to the editors about the issue. In papers that do address the controversy, articles and editorials are often either one-sided or very limited in the information they convey.

Environmental groups, such as Save Our Cumberland Mountains, Sierra Club, The Nature Conservancy, Izaac Walton League, Environmental Defense Fund, Southern Environmental Law Center, and Natural Resource Defense Council, have substantial power, wealth, enlightenment, respect, and skill that could be employed in this cause. The involvement of these groups could increase chances of bringing the issue to national attention. If the opposition challenges the TVA in court, the legal and financial resources of several of these groups could prove extremely beneficial.

Identification of the Decision Function to Be Emphasized

The decision process comprises seven functions: intelligence, promotion, prescription, invocation, application, termination, and appraisal. This process is often disproportionately controlled by one or more participants. In this particular case, the decision process is formally established by federal law. It consists of the review process for applications for permits, which is overseen by the Federal Energy Regulatory Commission, and the public commenting periods prescribed in the EIS procedure, which is managed by the TVA. According to federal law, the decision process should remain open and inclusive of all participants. The problem is, though, that this process has been an exclusive one—many participants feel that their comments are not being given fair consideration.

In this situation, the goals of SOS appear more congruent with the common good than those of the TVA or AER. I approach the decision process from the point of view of SOS. The promotion function of the decision process is the function that SOS is most able to influence. Hence, it is emphasized here.

Status of Save Our Sequatchie Relative to the Decision Process

Save Our Sequatchie acquires intelligence from the collective backgrounds of its diverse membership. For example, one member of the group has extensive flow and volume records of the Sequatchie River and another has studied the local cave and spring network to assess the relationship of the reservoirs and tunnels to the karst hydrogeologic system. The group actively exchanges information about legal issues and raises community awareness through public demonstrations and letters to local newspapers. The group has also focused public attention on the fact that an earlier site review and assessment has been removed from public access by the TVA on the grounds that it is now proprietary information of the TVA and AER.

SOS does not have authority for prescription, but it influences what will be prescribed by participating in public hearings and by other legal means. The group feels that its goals should be adopted, invoked, and applied by public agencies; if not, the group will continue promotion to change laws until the policy is suitable. Termination comes for the group when the controversy passes and the group's interests have been protected—many members of the group state that even if the permit for the remaining project is denied, they will continue to press for legal change so that a similar situation will not occur in the future. SOS practices ongoing appraisal.

Analysis

Relevant Historic Trends and Factors That Shaped These Trends

The contextual description of historic trends starts with the creation of the Tennessee Valley Authority. The TVA was commissioned as part of President Franklin D. Roosevelt's New Deal to reverse the nation's economic troubles of the Great Depression. In the South, economic depression had persisted since its defeat in the War Between the States. Much southern capital had been invested and lost in the war, agricultural markets were depressed, and the economies of the southern states were in dire conditions. The TVA was seen as a way to get the rural South back on its feet through industrialization by producing and distributing electricity throughout the region and addressing the problems of flooding and navigability of the Tennessee River (TVA 1996c).

Yet, despite their lack of cash, the southern states had rich reserves of natural resources. As the nation continued to grow after the Civil War and World War I, as the economy rebounded after the Great Depression, and with the advent of World War II, much of the southern natural resources base was exploited by northern investors. Prime markets were developed for southern timber, coal, and laborers. During this same era, the South experienced increasing federal involvement through other channels. In addition to the lands taken for the TVA, other lands were condemned or purchased by the U.S. Department of the Interior, National Park Service, and the Forest Service and Rural Development Agency, both of the U.S. Department of Agriculture.

Factors Relevant to the Current Situation and Problem

The most striking factor in the decision process concerning this policy problem is the quantity and quality of public opposition to the projects. Opponents express themselves in public demonstrations, letters to the editors of area newspapers, bumper stickers, symbolic red carnations, and an active word-of-mouth network. SOS opposes the northern company, assisted by a federal agency, taking and developing southern land. An urban and rural conflict is also evident. Two other uncertainties associated with the proposed projects are prospects for regional economic development and the potential perturbations of environmental quality.

Proponents of the hydroelectric projects consistently wave the flag of regional economic development to rally support. However, several issues concerning employment remain unclear at this time: What number of jobs would be available to the local people? At what skill level are the jobs? How long will the jobs last? Also, future economic development made possible by the additional supply of electrical power and infrastructure are not necessarily in the long-term interest of the local communities of the Valley. Industrial and residential development, and the additional people that may migrate into the area, will further strain the present municipal infrastructure of roads, schools, sewers, and water supply. It is also difficult to predict how the positive and negative externalities of the proposed projects and future developments will be distributed across political boundaries. In the case of the Laurel Branch project, the facility is located adjacent to the county line. It is possible that one county (Bledsoe) will reap the economic rewards (tax income) of the facility while the other county (Sequatchie), downstream, will deal with the potential environmental degradation.

Another uncertainty is the future of local environmental quality. Concern for ground- and surface water quality is of utmost importance because of dependence of local communities on these resources. There are numerous concerns over whether the bedrock geology is suitable for the proposed projects because of toxic shale and coal seams, karst conditions, and joint fractures in the strata. If the Laurel Branch project proceeds and reservoirs are established, certain homes and properties will then be located in a newly formed floodplain. This result has human safety, environmental, and economic implications. Property owners may find the value of their properties and mortgages immediately diminished. To continue living in some areas beneath a reservoir may compromise residents' safety in the event of dam failure.

Future Trends and Projections

Based on the current situation, a number of future trends are possible. The projection of each trend is evaluated in terms of how its outcome meets sos's goals. In this analysis, trends are isolated; it should be noted, though, that in actuality one or more of these projections may be realized.

First is the prospect that the permit for the Laurel Branch facility will be denied. This possibility safeguards sos's values at stake in this controversy and may increase the respect the organization has for public agencies carrying out the decision process. Second is the denial of the privilege of eminent domain to an independent power producer and other private industries. For this to happen, legal changes to the 1920 Federal Power Act and the Energy Policy Act of 1992 are required. sos might lead or join a campaign to modify the acts. The organization stands to increase its power, skill, and enlightenment while furthering its goal of protecting private property rights.

Third, if the proposed project is approved and completed, economic growth in the region could be stimulated. This possibility would increase jobs in the area, in turn increasing the wealth and skill of some citizens. Other businesses and developments would be encouraged to take root in the area. This prospect could have a "ripple effect" as economic growth increases the tax base and supports additional services in the area, which increases the value of well-being. The goals of sos are not achieved in this outcome, but some individual members may find themselves better off.

Fourth, the completed project could lead to numerous environmental problems, and even disasters. Projections of this sort include groundwater contamination, the flooding of significant parts of the Valley, and the acceleration of plateau bedrock instability. Contamination of groundwater diminishes well-being for numerous participants. This is an especially important concern because a large portion of the population depends on wells. Catastrophic flooding threatens the wealth and well-being of a wide range of participants also. Altering hydrogeology may expedite formation of dissolution cavities in the bedrock. Should instability of the plateau bedrock pass a critical point, damages could be untold. This projection neither meets the goals of sos nor leaves any of its members better off. The group may find fulfillment in the loss respect for AER if the company is blamed for any disasters.

Fifth, if the project is approved, and AER obtains its funding, the company could scheme to default on its loans. This action could prove profitable to the company in two ways. First, it could declare bankruptcy and thus not repay loans. The company would benefit from money it received. Second, AER could arrange to develop and sell the properties it gained through eminent domain. While the company indulges its wealth, it deprives citizens of many values—wealth, respect, and power. If any of these possibilities come to pass, sos would certainly not have achieved its goal, but it might be seen locally as the hero of the "lost cause." As such a hero, the group might enjoy indulgences in terms of respect and affection as other community members realize how they have been deprived and recognize sos for trying to aid them.

Finally, if the project is completed, TVA might use the facility to launder power to sell outside its jurisdiction. This projection witnesses the TVA increasing its power and wealth values in a manner contradictory to its authorization. This projection is wholly contrary to the goals of sos and may amplify the effects of other trends.

Recommendations

Alternatives for Problem Resolution

Given the problem's context and possible future trends, several alternatives could address the problem. First is cessation of sos's opposition to the remaining project. In this alternative, sos commits no further resources toward problem resolution. Second, the organization continues along its present course. This alternative implies that sos maintains its current approach and continues to confront the problem in the frame of reference set by its adversaries. Third is for sos to direct opposition along a new course. This approach shifts the frame of reference and places the organization on the offensive. Fourth, sos allows construction of the Laurel Branch project to begin and, through monkey-wrenching, sabotages or destroys the facility. The next section evaluates each alternative.

Evaluation of Alternatives

Alternative one releases sos from the struggle but makes no contribution toward the achievement of any of its goals. This alternative does not address the crux of the policy problem as seen by sos—exclusion from the decision process—and offers no solution to future problems of a similar kind.

Alternative two notes that sos has gained significant ground following their present course. In fact, sos's approach may be sufficient in achieving most, if not all, of its goals. However, this approach may not allocate sos's resources efficiently. Alternative three guides sos along a course of action that enables the group to conceive of the problem more contextually and realistically. This could further empower sos and broaden its coalition by enlisting the influence of other participants. The approach also reveals which decision functions sos is most able to influence and how.

Alternative four may achieve some of sos's goals, but at significant cost to sos's respect and rectitude. This militaristic strategy prevents AER from capitalizing on the project and could prevent the company from having resources to engage in future projects. The approach also contributes little toward the achievement of sos's goals.

Selection and Justification of an Alternative

Alternative three is selected because it recommends two courses of action that increase the potential for sos to achieve its goals. The two concerted courses of action are (1) creating a broader-base coalition of opposition and (2) offensively shifting the frame of reference to enable sos to work more efficiently toward problem resolution and achieve its goals by concentrating its efforts where the group has the most influence.

The first course of action recommends that sos construct a broader-based coalition of opposition. This coalition should include more private individuals and landowners and the other major participants. Public opposition can be increased by tapping into larger reservoirs of discontent through the utilization of popular myths. sos could advance the view of an "out of control" federal government by emphasizing the TVA's debt, appetite for private land, and inability to compete in the recently deregulated environment of the electrical power industry. It could also stimulate the public conscience to recall the historic trends of regional exploitation by northern companies and federal agencies. The group can transform public discontent into effective opposition by posing poignant questions in public forums. Questions should be developed to expose the TVA's standpoint and the possibility of a conflict of interest over its involvement in the project and the EIS, and to investigate the feasibility of additional jobs and the potential effects of future regional development and economic growth.

Enlisting other participants in the social process is an essential component in a strategy to establish a new coalition of opposition. These groups must be involved, and their support is crucial. A more comprehensive examination of the environmental issues than the one carried out to date may be necessary to reveal how the proposed projects and their potential impact might affect each of these participants.

The second course of action could be carried out with the first to shift the frame of reference for problem discussion and resolution to an offensive stance. This strategy should take advantage of the instability of the TVA. The following suggestion considers all seven functions of the decision process. sos should thoroughly engage intelligence activities. The group must take advantage of the availability of journalists in controversies and public affairs. Media outlets should be utilized to question the TVA regarding the decision process in an open forum and to attract national attention to the environmental and legal issues that transcend regional boundaries. Notable issues are government corruption, the use of eminent domain by a private industry, obstruction of the NEPA process, environmental/legal injustice, and habitat loss associated with the perturbation of a significant watershed.

sos can best influence policy prior to prescription through promotion of its goals. The organization must explain publicly how the Laurel

Branch project falls within the jurisdiction of the other major participants and how these participants can act to stop the project. The inclusion of more participants expands the scope of factors considered in the
decision process. Additionally, the influence the TVA has in choosing
from among the alternatives proposed in the EIS is diluted by other participants whose desired outcomes are more congruent with those of SOS.

Invocation and application of any prescription that does not fully
address the goals of SOS must be challenged by the group. Critical appraisal of the decision process must be an ongoing function for the
group. An initial point for appraisal calls into question the suitability of
the TVA to make the EIS. If the agency stands to gain significant benefits
from permitting the Laurel Branch project, it probably cannot be re-
lied upon to conduct an objective analysis of the environmental impacts. This is a point that may need to be decided in the courts. Similarly, SOS must appraise the inclusion or exclusion of other federal
agencies that could act as official cooperators in the study.

NEPA regulations provide for a comment period after each phase
(draft and final) of the EIS process. SOS must be prepared to appraise
the EIS in terms of its methodology, findings, and proposed alternatives. If, after submitting comments in both periods, the organization
is dissatisfied with the alternative chosen by TVA based on the final EIS,
it can challenge that decision in federal court.

SOS must commit to continue its opposition until all its goals are
achieved. The group must not be satisfied with winning one battle (e.g.,
permit denial)—its goals are not achieved until the war is won (the decision process is inclusive; laws are changed; and public agencies are
objective and accountable). Tenacity will increase the chances for success and the prevention of similar problems in the future.

Conclusions

Restatement of Problem and Goals

The overriding policy problem in this controversy is that concerned
citizens feel excluded from the public decision process as it has been
legally set forth. Furthermore, many people feel that the process is
being subverted by the very public agencies charged with upholding a
fair process and ensuring its inclusiveness. The outcome of this process will be a determination of whether or not a permit will be granted
for the Laurel Branch project and, indirectly, whether a private industry will have use of eminent domain.

Many citizens (associated into the public interest group SOS) have
expressed opposition to the project. SOS is principally seeking inclu-

sion in the decision process. The group feels that if its positions and concerns are appraised fairly by the government agencies conducting the decision process, permission to construct the hydroelectric facility will be denied to AER. SOS also seeks to change the 1920 Federal Power Act and 1992 Energy Policy Act so that independent power producers will not have access to the privilege of eminent domain. Consequently, private property rights will be acknowledged. The organization's fourth goal is greater accountability and neutrality of the public agencies that oversee the public decision process and environmental impact assessment for such projects.

Alternative three guides the opposition along a course of action directed by a policy sciences analysis. Significantly, this approach enables the group to conceive of the problem functionally. The functional approach empowers SOS to analyze the social process and encourages the group to broaden its coalition of opposition and enlist the influence of other major participants. Finally, the approach reveals which functions of the decision process the group is most able to influence and strengthens the group's strategy relative to goal achievement.

LITERATURE CITED

Editorial staff. 1997a. "Eminent Domain for Profit," *The Chattanooga Times*, April 25, p. A10.
Editorial staff. 1997b. "Will AER Drain the River?" *The Chattanooga Times*, March 10, p. A6.
Energy Policy Act of 1992 (PL 102–486, 24 October 1992), *United States Statutes at Large* 106:2776–3133.
Flessner, D. 1997. "Armstrong Plan Fuels Debate over Land Use," *The Chattanooga Times*, April 12, p. B3.
Hodge, M. 1997. National precedent (Letter). *The Chattanooga Times*, June 3, p. A4.
National Environmental Policy Act of 1969 (PL 91–190, 1 January 1971), *United States Statutes at Large* 83:852.
Tennessee Valley Authority. 1995. "Energy Vision 2020." Brochure. 400 West Summit Hill Drive. Knoxville, TN 37902.
———. 1996a. "Building a Competitive Future." Brochure. 400 West Summit Hill Drive. Knoxville, TN 37902.
———. 1996b. "The Energy to Lead." Brochure. 400 West Summit Hill Drive. Knoxville, TN 37902.
———. 1996c. "Generation to Generation: A Short History of TVA." Brochure. 400 West Summit Hill Drive. Knoxville, TN 37902.
———. 1996d. "The Power to Lead: Annual Report." 400 West Summit Hill Drive. Knoxville, TN 37902.
———. "Executive Summary—Integrated Resource Plan and Environmental Impact Statement." 400 West Summit Hill Drive. Knoxville, TN 37902.
Tennessee Valley Authority Act of 1933 (18 May 1933), *United States Statutes at Large* 48:58–72.
Water Power Act (10 June 1920), *United States Statutes at Large* 41:1063–1077.

6 Use of Turtle Excluder Devices to Save Sea Turtles Around the World

David Kaczka

Around the world, most populations of sea turtles are declining (International Union for the Conservation of Nature 1995, hereafter "IUCN"). Human-induced mortality exceeds their natural reproductive capacity. The endangered status of sea turtles is widely known: all seven species are listed in Appendix I of the Convention on International Trade in Endangered Species of Wild Fauna and Flora (CITES). Many nations protect sea turtle nests and important nesting beaches. Unfortunately, gains from this protection have been more than offset by increased indirect mortality from human activities, particularly incidental capture in fisheries (IUCN 1995). Globally, shrimp trawling takes many more sea turtles than do other fisheries (Hillestad et al. 1995). To conserve sea turtles, it is necessary to control the impacts of shrimp trawlers; reasonable alternatives, especially turtle excluder devices (TEDs), are available that efficiently protect turtles from trawls. Accordingly, to improve the status of sea turtles worldwide, a central goal should be to reduce trawlers' impacts by increasing TED usage, or by fostering the development and distribution of other inexpensive and effective technologies. A variety of strategies can be used to accomplish this goal, including but not limited to traditional diplomacy, negotiation of multi-

Research for this project was conducted from February to April 1996. The paper was updated in October/November 1997.

lateral treaties and agreements, and application of trade sanctions on shrimp harvested with methods that have an impact on sea turtles.

This paper has three primary purposes. First, it describes, contextually and in detail, the threat to sea turtles from shrimp trawlers, and current policies to control these impacts outside of the United States. The primary prescription is U.S. Public Law 101-162. Additionally, the Inter-American Convention for the Protection and Conservation of Sea Turtles (hereafter, the "Salvador Convention") and the Marine Turtle Conservation Strategy and Action Plan for the Western Indian Ocean (hereafter, the "WIO Action Plan") are discussed. Second, the policy sciences are used to analyze current policy defects. The focus of analysis is primarily upon US PL101-162; the Salvador Convention is not yet in force, and the WIO Action Plan was developed only recently. Finally, I recommend alternative measures to increase TED use and improve sea turtle conservation.

At present, trade sanctions on certain shrimp imports to the United States are the primary mechanism used to encourage international use of TEDs. I was drawn to study this conservation problem because I believe that the current approach to the problem of sea turtle conservation is inadequate and damaging to U.S. trade and environmental objectives. The record of US PL101-162 Section 609, along with conflict, extended court battles, and small conservation gains, supports my conviction. Generally, I believe that trade liberalization can improve social welfare globally; however, it must be balanced with appropriate policies to ensure it does not exacerbate social concerns like child labor or environmental degradation. Open and equal competition is the objective of free trade, but there are simply some levels below which competition must not sink. The willingness to abuse an endangered species should not be the foundation of a competitive position. While the intent of the U.S. policy is to prevent such abuse, I believe that the policy's application limits its effectiveness and draws the sincerity of the U.S. effort into question. I am convinced that there are better strategies to achieve the conservation goal.

I have studied this problem as both a scientist and an advocate. Conservation strategies must be developed with respect to certain scientific facts. The goals of a strategy must be appropriately targeted to remedy the problem. I have attempted to gather enough scientific information to clearly understand the nature and scope of the problem, which is that shrimp trawlers around the world kill too many sea turtles. Additionally, I have reviewed the feasibility of potential control alternatives. My analysis of the problem, however, is also shaped by my view of trade and environmental linkages, and by my belief that a stronger institutional framework is needed to resolve international environmental problems.

Problem Description

Trawlers Take Too Many Turtles

Before the application of TED regulations, ten times more sea turtles were killed by U.S. shrimp trawlers than by all other human-caused sources of mortality combined.[1] In total, U.S. trawlers operating without TEDs were estimated to take 11,000 to 55,000 sea turtles each year. Mortality rates rose dramatically with the opening of shrimp fishing season and would fall just as drastically when the season closed. In areas where shrimp fishing was intense, nesting populations were declining, whereas populations were stable in areas where trawling intensity was low (National Academy of Sciences 1990, hereafter, "NAS").

The problem, however, is a global one. Sea turtles live in tropical and warm temperate waters worldwide, and often aggregate in shallow waters near nesting beaches and feeding areas. Shrimp trawling occurs around the world in shallow coastal waters, and many intensively trawled areas are near sea turtle nesting beaches and feeding areas. Trawlers may have an impact on all seven species of marine turtles, although the most frequently captured are loggerheads, greens, flatbacks, and Kemp's and olive ridleys (Hillestad et al. 1995).

Commercial trawl vessels are typically more than twenty-five feet long and pull as many as four trawls that are each forty feet wide and fifteen feet high (Center for Marine Conservation 1995, hereafter "CMC"). The threat they pose to sea turtles is related to the size of their nets and the duration of their tow times. Larger trawls are more likely to capture turtles. When overtaken by a trawl, turtles become entangled in the netting and are unable to surface to breathe. The longer the turtle is held underwater, the greater the chance it will drown. When trawl tow times are under forty minutes, few turtles are harmed. However, when tow times exceed ninety minutes, 70% of the captured turtles are comatose or dead when brought aboard the vessel (NAS 1990). Of the remainder, some may be revived and returned to the water. The effectiveness of such revival is uncertain. In general, vessels which use small trawls and short tow times appear to pose much less of a threat to sea turtles.

Although trawl-related sea turtle mortality is well documented in the United States, data on trawler impacts in other parts of the world is relatively scant. The studies that are available suggest significant impacts. By extrapolating from the U.S. shrimp fleet's take rate prior to the application of TED regulations, the Sea Turtle Restoration Project (STRP) of the Earth Island Institute (EII) has estimated an annual global take of 124,000 sea turtles (Court of International Trade 1995, hereafter "CIT"). The U.S. National Marine Fisheries Service (NMFS) has estimated the combined annual take of Brazilian, Chinese, Indian, and Filipino trawlers to be approximately 11,000 sea turtles (CIT 1995). A more recent study by Indian biologists reports that 5,282 sea turtles killed by Indian trawlers washed up on a 480-kilometer stretch of beaches in six months (Pandav et al. 1997). Including the nations discussed above, the trawlers of more than eighty nations may have an impact on sea turtle populations.

The high levels of mortality need to be understood in the context of sea turtle population size and the age range of individuals most likely to be taken. Most populations of sea turtles are seriously depleted and cannot tolerate mortality rates that are substantially higher than natural levels (IUCN 1995). For depleted populations, even low levels of human-induced mortality can prevent the recovery of the population and increase its vulnerability to extinction. Also, turtles in some age ranges may be more susceptible to capture and entanglement in trawls. According to age range, turtles have different conservation values based on their reproductive potential. Biological models show that the growth rates of sea turtle populations are most sensitive to the survival of large subadult and adult turtles and suggest that the loss of only a few hundred such females each year can lead to the extinction of a population (Heppell et al. 1996). Records of turtles that wash ashore dead on beaches in the United States suggest that trawlers have an impact primarily on these large subadult and adult sea turtles (Crowder et al. 1994). Therefore, the true impact of trawlers on sea turtle populations may be greater than indicated by the estimates of mortality rates presented in the preceding paragraph.

Shrimp trawlers are a serious global threat to sea turtles. Many scientists believe trawl-related mortality is the single greatest force driving sea turtles toward extinction (IUCN 1993). In areas where the ranges of sea turtles and trawlers overlap, any turtle conservation strategy must control the trawlers' impacts.

Controlling Trawlers' Impacts

Currently, there are three primary strategies that can be used to protect sea turtles from incidental capture in trawls. The first is to prohibit trawling in waters near known nesting beaches and feeding areas. This solution may be the most effective means to prevent trawler take of turtles. However, shrimp fishing is a valuable commercial activity. In fact, many nations subsidize their trawl fleets and have invested heavily to build their fleets and marine infrastructure. Few nations are likely to close areas to trawling in order to protect sea turtles. Given this condition, to protect sea turtles it is necessary to speed the release of turtles that enter trawls, either with TEDs or by reducing trawl tow times and releasing any turtles that are captured.

TEDs were developed in the United States in order to achieve the goals of saving sea turtles and allowing trawling. In the United States, since 1987 shrimp trawlers have been required to use TEDs in accordance with the provisions of the Endangered Species Act. The devices are relatively inexpensive and efficient and therefore may offer the best strategy for attaining environmental and economic objectives.[2] Easily manufactured and installed, TEDs are metal grates or mesh screens mounted at an angle in the neck of a trawl. A TED is essentially a filter that deflects large objects like turtles, tires, logs and debris through an opening and out of the trawl. Smaller objects like shrimp pass through the device and are collected in the net. Although TED use may slightly decrease the shrimp catch, it improves catch quality and purity. Furthermore, TED use can improve the management of other fisheries. In the United States, trawlers kill twelve pounds of finfish that would be commercially valuable at maturity per pound of shrimp harvested. The 12:1 by-catch ratio makes trawling one of the world's most wasteful and destructive fishing methods (WWF 1997). TEDs can reduce finfish by-catch by 50% or more.

Some degree of protection can also be obtained by requiring fishermen to limit their tow times and to release, unharmed, all turtles that are captured. However, it is important to note that commercial trawlers are designed specifically to allow the use of several large trawls and to facilitate longer tow times in order to obtain higher yields. Shorter tow times are likely to cut fishermen's productivity if they comply. When the United States applied TED regulations, it initially allowed fishermen to limit tow times as an alternative to TED use. However, this alternative was removed because it was difficult to enforce, and Coast Guard studies showed compliance was quite low.

Presently, sea turtles can be protected from trawlers by area closures, TED use, and tow time restrictions. It is possible that other alternatives may become available in the future. While many nations will find closing fishing areas to protect sea turtles too costly, the threat to turtles is serious enough that reasonable steps should be taken to prevent further harm. Currently, requiring the use of TEDs offers the best approach to meeting the objectives of commercial shrimp trawling and sea turtle protection. If trawl fisheries worldwide use TEDs effectively, sea turtle protection will be dramatically enhanced.

Initiatives that Promote the Use of Turtle Excluder Devices Internationally

Three initiatives currently aim to increase international use of TEDs to improve sea turtle conservation, including US PL101-162 Section 609 (hereafter, §609), the Salvador Convention, and the WIO Action Plan (IUCN 1996). Although this section briefly discusses all three, the focus of the subsequent analysis is on §609. This law has been in effect longer than the other initiatives, and has a significant performance record.

PL101-162 Section 609

Section 609 uses several means to encourage foreign trawlers to use TEDs but relies primarily on trade sanctions. It bans shrimp imports harvested in ways that have an impact on sea turtles. The law was enacted on November 21, 1989, and its trade control has been applied since May 1, 1991. Initially, the law was applied to fourteen nations in the wider Caribbean.[3] The limited area of application resulted in a series of court cases between the EII and the State Department. The culmination of these cases, *Earth Island Institute v. Christopher*, forced the global application of §609 on May 1, 1996. Since then, India, Malaysia, Pakistan, and Thailand have challenged the U.S. law through the World Trade Organization (WTO) dispute resolution process.

Section 609 has two subsections that use cooperative and coercive measures to promote TED use. Subsection 609(a) directs the executive branch to work with other governments to identify areas where shrimp trawling operations threaten sea turtles, and to improve protection of such sea turtles. The Department of State and the Department of Commerce are required to negotiate conservation agreements and amend existing international conservation agreements to include TED requirements. §609(b) bans imports of shrimp harvested in a manner that poses a threat to sea turtles. Imports from certified nations are unaffected; to be certified, a harvesting nation must adopt regulations that are comparable to U.S. TED regulations. Alternatively, nations may be certified if their fishing environment does not pose a threat to sea turtles. In practice, the State Department certifies a nation if its regulations are comparable to those of the United States and it has a credible enforcement program, or if trawlers fish in waters where they are unlikely to encounter sea turtles. The ban does not affect all shrimp imports from uncertified nations: shrimp harvested from aquaculture facilities or by artisanal means are not banned (*Federal Register* 1991).

Salvador Convention

The world's only treaty that specifically aims to protect sea turtles is the Salvador Convention (CMC 1996). Negotiation of the treaty concluded in September 1996, and it is not yet in force. Therefore, the Salvador Convention is not discussed in the Analysis section of this paper. However, features of the convention are relevant to future alternatives and are discussed in the Recommendations section. Parties to the treaty are required to reduce to the greatest extent practicable the incidental mortality of sea turtles, prohibit the capture and killing of sea turtles, protect important habitats, promote scientific research and environmental education, and promote efforts to enhance sea turtle populations. The Salvador Convention specifically requires the use of TEDs as a conservation measure. Although the treaty covers the inter-American region, it calls on nations in other regions to commit to protocols and agreements that are consistent with the objectives of the convention. The convention emphasizes national and regional management and the use of appropriate scientific information.

Western Indian Ocean Action Plan

The WIO Action Plan is the result of collaboration between the World Conservation Union (IUCN) Species Survival Commission, the World Wide Fund for Nature (WWF), the United Nations Environment Programme (UNEP), the Center for Marine Conservation (CMC), the Secretariat of the Convention for the Conservation of Migratory Species of Wild Animals (the Bonn Convention), and the representatives of many African nations. It was developed during a workshop that took place between November 12 and 18, 1995, in Sodwana Bay, South Africa. The WIO Action Plan recommends actions to be taken at the national level but suggests that the actions of any one nation are insufficient to ensure the long-term survival of sea turtles. Therefore, it stresses regional coordination of efforts. The action plan is a consensus document and as such does not legally bind the participating countries. It is structured to promote TED use and reduce sea turtle mortality through improved research and monitoring of marine turtle populations, management focused on sustainability, community participation, public education, regional and international cooperation, and funding for conservation programs. Like the Salvador Convention, the WIO Action

Plan emphasizes consensus, coordination of action at the regional level, a scientific basis for action, and environmental education.

In addition to the policies discussed above that specifically recommend or require TED use, other international treaties, conventions, codes, and agreements attempt to clarify a nation's obligations to conserve endangered species. These include CITES, the Convention for the Conservation of Biological Diversity (the Biodiversity Convention), the Convention for the Conservation of Migratory Species of Wild Animals (the Bonn Convention), the United Nations Convention on the Law of the Sea (UNCLOS), the United Nations Food and Agriculture Organization's Code of Conduct for Responsible Fishing, the 1992 Rio Declaration on Environment and Development, the 1982 World Charter for Nature, and the 1972 Stockholm Declaration on the Human Environment. These instruments promote conventional and customary norms intended to govern endangered species conservation, including conservation of marine turtles. Generally, they have not contributed to an increase in global TED use. The potential contribution of these instruments to the resolution of the shrimp–sea turtle problem is discussed in the Recommendations section.

Policy Process Overview

The policy processes that have led to the formulation of the initiatives discussed above occur nationally and internationally. Activities within and beyond the United States have resulted in the development and application of US PL101-162 §609. The other sea turtle conservation strategies discussed above, as well as the broad conventions for the protection of endangered species, have been developed internationally.

From the United States, participants in the policy process have included turtle biologists, environmental and animal rights advocates, shrimp fishermen, a fisheries trade organization, politicians, and federal civil servants from the Department of State and the Department of Commerce's NMFs. The record of a House subcommittee hearing on May 1, 1990, a year before §609 was applied, captures the general positions and objectives of many of these participants (House Fisheries Subcommittee 1990).

A representative from the Center for Marine Conservation was present at the hearing; the center is active in both the national and international arenas. Their objective is to promote sea turtle conservation in the United States and abroad. To do so, they aim to promote research, education, conservation of sea turtles in the wild, and passage of conservation legislation. With respect to §609, the CMC has placed greater emphasis on improving the compliance with TED regulations in the United States than on applying similar regulations to foreign producers (House Fisheries Subcommittee 1990). The CMC did support the use of targeted sanctions on Mexico and Japan under the Pelly amendment to stop trade in Hawksbill sea turtle shells in violation of CITES (House Fisheries Subcommittee 1990). Internationally, the CMC participated in the development of the WIO Action Plan, and provided technical support to the State Department during the negotiation of the Salvador Convention (CMC 1996).

In addition to the CMC's representative, a single representative of the Environmental Defense Fund (EDF) presented testimony on behalf of the fund, the National Wildlife Federation, the National Audubon Society, and Greenpeace. In essence, his testimony stressed the same points as those of the CMC representative: that U.S. TED regulations should be consistently applied and enforced, and that specific sanctions should be levied on Mexico and Japan to curb actions that diminish the effectiveness of CITES (House Fisheries Subcommittee 1990).

Several other environmental and animal rights organizations have had an important influence on the policy process, but they were not present at the House subcommittee hearing. These include the Sea Turtle Restoration Project (STRP) of the Earth Island Institute, the Sierra Club, the American Society for the Prevention of Cruelty to Animals, and the Humane Society of the United States. These organizations, in particular the STRP, made passage of §609 a priority, and later sued the Department of State and the Department of Commerce for failure to apply the law correctly. These groups address shrimp trawling and sea turtle conservation as an ethical problem. The STRP strongly favors confrontation and the application of trade measures to promote the use of TEDs, rather than international conventions or initiatives based on consensus.

Several associations represented the concerns of U.S. shrimp fishermen, who were chiefly concerned with competition from foreign shrimp producers. In regard to TED use, the organizations' positions ranged from belief that if U.S. shrimpers must use TEDs so should their competitors, to the belief that shrimp trawling was not the problem and that TED regulations should be softened or repealed. The underlying objective of the fishermen was to ensure that they could obtain a decent living wage from their labor and investments.

The National Fisheries Institute (NFI) is a trade organization that represents U.S. fisheries' producers, importers and exporters, distributors, and sellers. The NFI is based near Washington, DC, and it acts to influence legislative processes and to educate its members. The institute was concerned that §609 would be costly for its members that imported and sold foreign shrimp and shrimp products. The NFI sought to keep shrimp supplies and prices stable to protect the interests of its members.

In §609, elected officials attempted to craft a prescription to expeditiously fulfill the demands of their constituents. Although §609 does not relax the U.S. TEDs laws that were earlier enacted under the Endangered Species Act, it does require foreign producers to use TEDs in order to sell in the U.S. market. Therefore, it is designed to "level the playing field" and reduce competitive pressures on U.S. fishermen. By encouraging TED use in most areas of the world where trawlers may have an impact on sea turtles, §609 also strives toward a key conservation objective.[4] Furthermore, because §609 allowed other nations only eighteen months to comply, it is structured to provide a quick and visible solution to the problem. The politicians that drafted §609 sought to win and maintain the support of their constituents.

The State Department and the NMFS were to be responsible for applying §609—for determining which nations it covered, and how key phrases in it were to be interpreted and applied. In general, they were opposed to a broad ban on shrimp imports, and therefore were opposed to application of the law as written. Both, however, expressed support for the need to improve protection of sea turtles beyond U.S. waters.

According to the State Department representative at the May 1, 1990, hearing, if nations in the Wider Caribbean region could not comply with the provisions of §609, the resulting embargoes would have cost Latin and Central American nations several hundred million dollars (House Fisheries Subcommittee 1990). In 1989–91, the political and economic stability of the Central American region was beginning to improve, after resolution of the conflicts that plagued the region through the early and mid-1980s. The relationship between the United States and these nations was considered to be at a critical juncture. So, too, was the relationship between the United States and Mexico; negotiation of the North American Free Trade Agreement (NAFTA) was initiated during this time period. For the United States to inflict approximately $560 million in trade-related costs on Mexico and Central America to save sea turtles would have jeopardized other important foreign relations goals. Therefore, the State Department's objective was to minimize damage to the broad policy objectives of the Bush administration while providing protection to sea turtles.

Internationally, many of the same groups and organizations are participants in the policy process. Additional participants include the

UNEP, WWF, IUCN, foreign shrimp fishermen and exporters affected by §609, and the governments of nations covered by §609.

The UNEP, WWF, and IUCN collaborated with the CMC in the development of the WIO Action Plan. Their objectives are similar to those of the CMC: to contribute to the development of a comprehensive conservation strategy that respects regional and local needs and concerns, to ensure that such strategies are founded on a strong scientific basis, and to further research that will improve our knowledge of sea turtle biology.

The foreign fishermen and exporters seek stable markets for their harvest and products, and sufficient returns from their labor. The objectives of the State Department, the NFI, and the environmental groups that seek to build international consensus respect the demands of these fishermen and exporters. However, they are not direct participants in the U.S. policy process, and their interests may be represented only indirectly in domestic U.S. arenas.

Governments of nations covered under §609 are also participants in the policy process. Like their fishermen and producers, their value demands may not be strongly represented in the United States. These governments can be expected to have many objectives. Although protecting sea turtles from extinction may be one of them, it is probably not their only or their chief objective. In democratic societies, the most important objective is probably to satisfy the value demands of constituents. Another key objective of these societies is to affirm and defend their claims in the international community. A principal concern is the claim to govern their nationals without undue interference from other governments. An additional objective is to balance actions and policies that have consequences on strategically significant relationships. International trade is an important route to achieving economic objectives; therefore, trade relationships with key markets are strategically significant.

Conservation Strategy Goal Clarification

A strategy to conserve sea turtles should aim for several objectives. Above all, it must reduce the mortality rate so that population levels stabilize. Eventually, the populations may grow larger. To stabilize sea turtle populations, the impacts of shrimp trawling must be controlled. Increasing use of TEDs wherever trawlers and sea turtles occupy the same waters is a reasonable and appropriate way to reduce the number of turtles taken incidentally. Therefore, fostering the use of TEDs wherever they are needed must be a central objective of an optimal conservation policy.

In addition to trawl-related take, many other human activities affect turtle populations. These include dredging, beach filling, marine pollution, disease, collision with boats, recreational fishing, and removal of oil platforms (NAS 1990). The relative impact of each threat may vary around the world. There is insufficient data to determine the magnitude of these impacts on different populations. In addition to uncertainty about the importance of other factors affecting sea turtle populations, much of the life cycle of sea turtles remains poorly understood. It has been learned recently that immature loggerhead sea turtles in the surf near Baja California originate from nesting populations in Japan and Australia, 10,000 kilometers across the Pacific (Bowen 1995). This discovery demonstrates that even sea turtle biologists' knowledge of turtle migrations remains incomplete (IUCN 1995). Improved scientific information will improve our ability to identify threats to sea turtles, thereby aiding the development of more effective conservation strategies. Therefore, reducing scientific uncertainty should also be an important objective of an ideal sea turtle conservation strategy.

The shrimp–sea turtle conservation problem holds a unique distinction among international environmental concerns: it is one of the few problems for which an inexpensive and effective remedy already exists (Menotti 1994). As such, the problem also offers a unique opportunity to resolve an international environmental problem credibly, positively, and constructively. An appropriate and effective policy response may serve as a guide to addressing other international problems and achieving more sustainable practices and production methods. Therefore, the third goal of the shrimp–sea turtle conservation effort should be to develop a framework that is applicable to other international environmental problems.

To achieve these policy goals, it is essential to describe how the seven decision functions of the policy process (Lasswell 1971) should be structured. The functions of the decision process are intelligence, promotion, prescription, invocation, application, appraisal, and termination. Additionally, these goals, when clearly specified, can be used as an analytical benchmark against which to evaluate existing policies. Recommendations for addressing identified problems should be targeted based on the results of the analyses.

Intelligence

This function is the gathering, processing, and dissemination of information. For sea turtle conservation, information should be gathered at the local and national levels. Processing and integration of data should occur nationally and internationally. Enough information must be collected to support determinations of noncompliance with whatever prescriptions are established to protect sea turtles (see the section on invocation, below). Additionally, information should be sufficient to estimate regional mortality and population trends, and to identify populations at risk and causal factors. Such information should be openly available to decision makers, scientists, and others at the local, national, regional, and international levels. If the intelligence function is structured appropriately, it can ensure that areas where TEDs should be used are identified and that appropriate incentives are identified and offered to ensure that TEDs are in fact used, and it can help to reduce scientific uncertainty.

Promotion

Activities to raise concern about the impact of trawlers on sea turtles and the endangered status of sea turtles should be targeted to several audiences. Foreign shrimp trawlers should understand their role in the problem and the actions they can take to minimize that damage. Consumers of shrimp should realize that their choices could contribute to the extinction of sea turtles, and that shrimp harvested by less damaging processes can be purchased. Residents of coastal communities should also understand the problem and how they can be involved in its solution. Promotion can increase participation and concern, and contribute to the realization of all three policy goals.

Prescription

This function is intended to generate an accepted norm that prescribes the use of TEDs whenever shrimp trawlers may have an impact on sea turtles. Prescription should be regional and international, and it should be inclusive to incorporate the values and demands of the participants. It should allow some time for program application and should set penalties that are appropriate to the level and intent of deviations from the norm. Further, it should be flexible to change as necessary. If the prescription function is effective, it will support increased use of TEDs, as well as the creation of a policy that can be applied to other international environmental problems.

Invocation and Application

The invocation function must be able to determine whether or not fishermen are in compliance. Nationally, governments should be able to discern whether their fishermen use TEDs where necessary, and to punish noncompliance. Application is the administration of the prescription. Application should be conducted nationally and internationally. Effective invocation and application functions are needed to increase TED use.

Appraisal and Termination

Appraisal assesses the successes and failures of the TED policy to reveal if it is effective or should be altered. This function is crucial for ensuring the flexibility of the program to evolve along with conditions affecting sea turtle conservation. In particular, the actions and decisions of governments should be reviewed and evaluated to determine if they are sufficient to achieve policy objectives. For the policy to function effectively, the appraisal function should be strongly linked to the intelligence and invocation functions. Termination should provide the ability to radically revise or terminate the TED program if appraisal indicates it is unsuccessful.

In summary, a policy that aims to conserve sea turtles internationally should be measured by its ability to accomplish three goals: to reduce the incidental take of turtles by shrimp trawlers; to improve knowledge of sea turtle populations around the world; and to create a framework that can be applied to other international environmental problems. These goals can be characterized and analyzed according to the seven decision functions identified by Lasswell (1971).

Problem Analysis

Of the three initiatives that aim to encourage the use of TEDs, PL101-162 §609 has been in effect the longest; the WIO Action Plan was developed in late 1995, and the Salvador Convention is not yet in force. Because there is a substantial amount of information about the effectiveness of §609, and little information documents the effectiveness of the other initiatives, the focus of analysis here is §609. Features of the other initiatives are discussed in the subsequent recommendations.

Trends in Decision

In the eight years since its passage, §609 has been applied weakly and has not achieved widespread use of TEDs outside the United States. As written, the prescription clearly mandates the Department of State and the Department of Commerce to press for TED use wherever shrimp trawling might have an adverse impact on sea turtles. The primary mechanisms to encourage TED use were broad embargoes. The Department of State assessed the potential impact of the embargoes, and the secretary of state reported the department's findings and concerns to Congress in November 1990:

> While the Department of State supports fully the goals of Section 609 and the intent of the Congress in enacting this legislation, the specific language of the law presents problems and difficult choices. Because four of the five species protected under the U.S. regulation are known to occur worldwide, this law, if given the broadest possible interpretation, could affect shrimp imports from more than 80 countries totaling as much as $1.8 billion—more than 75 percent (by value) of all shrimp consumed in this country. The impact of the resulting embargoes would be unprecedented both internationally and domestically.
>
> In implementing the law, the Administration has proceeded on the assumption that Congress intended to take reasonable steps internationally to protect sea turtles but did not intend to force a situation that would create enormous market disruptions in the United States and major foreign policy problems with many countries. (CIT 1995)

Congress allowed the department to limit the geographical scope of §609 through its application. Rather than applying §609 to eighty nations worldwide, the State Department regulations covered the nations that border the Gulf of Mexico, the Caribbean, and the western Atlantic Ocean. Together, these nations are called the Wider Caribbean Region. Although Congress acquiesced to the State Department's concerns, it did not modify §609.

In addition to limiting the domain of §609, the Department of State relaxed the conditions for certification to make it easier for affected nations to comply. The section required nations to adopt, by May 1, 1991, a regulatory program governing incidental take of sea turtles that was comparable to the U.S. program. Instead, affected nations were given until May 1, 1994, to require all their trawlers to use TEDs. The nations covered by §609 do not appear to have had difficulty complying with its provisions, and on May 1, 1996, twelve of the fourteen nations in the Wider Caribbean Region were certified under §609. Wisely, the State Department deemphasized one of the conditions for certification. The section requires that the harvesting nation's fleet average incidental take rate be comparable to that of the United States. Such incidental take rates are exceedingly difficult to monitor. In addition, owing to factors like the frequency and type of turtles in the water trawled, foreign fishermen in some areas feasibly could exert greater effort to protect sea turtles than do their U.S. counterparts, yet find certification impossible. If certification is impossible to obtain, the foreign fishermen have no incentive to adopt conservation measures. Therefore, the Department of State has assumed that if foreign trawlers use TEDs, their fleet average take rates are comparable to U.S. rates.

The limited application of §609 provoked several U.S. environmental and animal rights organizations, notably the EII, to sue the Departments of State and Commerce. EII first challenged the executive branch's interpretation of §609 in February 1992, in the federal district court in San Francisco. That court, and later the U.S. Court of Appeals for the Ninth Circuit, held that the proper forum for the case was the U.S. Court of International Trade (CIT).[5] On December 29, 1995, the CIT ruled that the intent of §609 is clear on its face, and that the law is not limited in its geographical scope. Declaring that §609 is unambiguously global, Judge Thomas Aquilino ordered the State Department to apply §609 to all nations that export shrimp to the United States by May 1, 1996 (CIT 1995).

After a State Department attempt to obtain a stay of decision was denied in late April 1996, the department complied with the court's mandate and banned certain shrimp imports from uncertified nations on May 1. At this time, however, the State Department created a new exemption for exports from uncertified nations: if the shrimp were harvested by trawlers using TEDs, they were permitted (*Federal Register* 1996b). Such shrimp imports were required to be accompanied by a form that certified the shrimp were harvested by trawlers that used TEDs. A government official of the harvesting nation was required to sign the form (*Federal Register* 1996). The EII claimed the new exemption undermined the effectiveness of §609 by reducing the incentive for harvesting nations to pass TED regulations that would cover their entire fleets (CIT 1996). Instead, they claimed, the exemption encouraged only those vessels that exported their catch to the United States to use TEDs. On this basis, the EII once again challenged the application of §609 before the CIT. On October 8, 1996, the CIT ordered the Department of State to eliminate the exemption. The section is now being enforced as Congress initially intended (CIT 1996).

Unilateral actions taken to rectify international environmental problems are rarely well received when they are backed by trade controls. No nations came forward to support the unilateral attempt by the United States to protect dolphins by embargoing tuna imports when the measure was challenged under the General Agreement on Tariffs and Trade (GATT) dispute resolution procedure (Charnovitz 1994). The U.S. shrimp embargoes were quickly challenged by Malaysia, Thailand, Pakistan, and India. By December 1997 the World Trade Organization (WTO) Dispute Resolution Panel that is hearing the case will issue its ruling. If either side disagrees with the ruling, the decision will be appealed, and the WTO appellate body's report should be issued late in spring 1998. Therefore, the State Department's fears have been realized, and wide application of §609 has created trade friction as expected.

In summary, §609 has not spurred global TED use. The clearest reason for its failure is the State Department's reluctance to apply it in full. The agency has reduced the geographical scope of §609 and delayed its full application. While it has produced only small gains for conservation, §609 has generated significant national and international conflict. In relation to the goals of an optimal sea turtle conservation policy, the section falls short: it has not controlled sea turtle mortality or reduced scientific uncertainty, and the conflict it has generated makes it a poor model to apply to other international environmental problems.

Conditions that Have Shaped the Trends in Decision

The underapplication and conflict which characterize §609 are the result of functional problems in the decision process. The most conspicuous reason that §609 has failed to improve sea turtle conservation worldwide is its consistent underapplication. Congress assigned application of the law to the Department of State and the Department of Commerce, but the State Department effectively controls the application of §609. By controlling this function, the agency has monopolized the decision process (Clark 1995).[6] As prescribed by Congress, §609 had the potential to cause extensive foreign policy damage. The State Department's mission is to maintain and improve U.S. foreign relations and to serve the current administration. Full application of §609 may have had an impact on both these organizational objectives. Instead of global sea turtle conservation, the agency sought to minimize foreign policy damage. This goal inversion has blocked the section's effectiveness as a conservation tool.[7]

To further minimize the potential for damage to foreign relations, the Department of State created a relatively weak invocation function. The agency determines compliance based on documentary evidence submitted by the harvesting nation's government (*Federal Register* 1991). Acceptable documentation includes copies of laws, regulations, and voluntary arrangements between the government and the fishing industry. Additionally, to be compliant, all regulatory programs and voluntary arrangements must include reasonable enforcement programs. There is no provision for the agency to monitor foreign shrimp fishermen to determine whether they are using TEDs, and using them effectively. Therefore, the invocation function is unable to differentiate between cosmetic compliance, wherein a government's official position requires TED use but fishermen rarely use them, from true compliance, wherein all trawlers are required to, and do, use TEDs. For example, Thailand is certified to export shrimp to the United States under §609, but its trawlers rarely use the devices (WWF 1997). By relying on documentary evidence alone, the State Department determines compliance in the least intrusive manner. However, it is also the least reliable method to ensure that sea turtle mortality is reduced.

The underapplication of §609 and the structure of the invocation function debase the intelligence and promotion functions. When the State Department limited the area of application of §609 to the Wider Caribbean Region, it no longer needed to consider trawl-related take of sea turtles as a global problem. Therefore, it did not need to investigate fishing practices of most nations' trawler fleets, and did not need to raise global awareness and concern about the problem. Additionally, very little information is collected to support the invocation function. Although the policy aims to improve sea turtle conservation, it does not collect any information on rates of mortality or trends in sea turtle population sizes in each nation. As it is currently structured, the intelligence function cannot identify populations at risk or measure the effectiveness of conservation programs. Furthermore, it does not effectively gauge whether or not foreign shrimp fishermen actually use TEDs. As a result, the intelligence and promotion functions do not generate levels of effort or concern that are related to the magnitude of the threats to turtles, and they do not gather information that could contribute to human understanding of sea turtles.

The process of prescribing §609 was exclusive, and many of the groups that the law affects were not involved in its development. The result is a law which satisfies the value demands of the active participants in the policy process while shifting its costs to individuals and organizations that did not participate. Some of these costs include the "overwhelming trade disruptions" mentioned in the secretary of state's comment, above. Through goal inversion, the Department of State sought to reduce the cost of omitting important participants —foreign governments, foreign fishermen, and U.S. consumers—from the policy process.

If §609 had included formal appraisal and termination functions that allowed wide participation in the policy process, conflicts might have been diffused before they escalated into court battles. In the absence of a formal appraisal that allows participants to consistently target policy toward an agreed objective, participants must rely on other methods to steer policy and meet their value demands. There is no formal appraisal process under §609. Dissatisfied with the agency's application of the law, the EII challenged the State Department in court to break its decision monopolization. Their victory did that, and forced the worldwide application of §609 despite its political and economic costs.

Malaysia, Pakistan, Thailand, and India quickly challenged on the ground that §609 violates U.S. obligations under the GATT. The governments of these nations were not represented in the policy process but are strongly affected by its outcome. Acting through the exclusive policy process, U.S. interests promoted a policy that could achieve their objectives—without regard for the impact of the policy on widely held international values and objectives. As a result, §609 inflates the objectives of the organizations in the United States to a level at which they collide with other nations' demands for sovereignty and their claim to govern their citizens without undue interference from other nations. At this level, the actions of embargoed nations have little connection to their concern for sea turtle conservation. Instead, their actions are likely to be influenced by their demands concerning how power and respect are shaped and shared among nations. The result of this collision is further conflict and little additional sea turtle protection. By challenging §609 in the WTO dispute process, these nations are attempting to terminate the U.S. policy.

Projections

The WTO case over §609 is currently active, and a decision is expected by December 1997. If an appeal is filed, the appellate body report should be issued in late spring/early summer 1998. In the case, the complainants argue that §609 is a barrier to trade that violates several fundamental GATT principles. The United States claims that §609 is necessary to protect sea turtles and that it is entitled to an exemption from GATT obligations under Article XX Sections b and g.

There are several notable precedents to this case, particularly the two tuna-dolphin disputes and the U.S. reformulated gasoline case. The tuna-dolphin disputes, decided in 1991 and 1994, concerned the application of the U.S. Marine Mammal Protection Act (MMPA). The text of §609 that invokes embargoes was taken almost verbatim from the MMPA (House Fisheries Subcommittee 1990). In both cases, the GATT panels ruled against the MMPA (GATT 1991, 1994). In fact, all the prescriptions challenged in prior GATT/WTO trade and environment cases were found to violate the general agreement (Kaczka 1997). However, the appellate body report of the recent U.S. reformulated gasoline case opens GATT/WTO jurisprudence to the body of international law (WTO 1996). It can be credibly argued that the U.S. position is valid under international environmental law (WWF 1997). Although there is a potential for the panel to rule in favor of §609, most scholars who have written about the dispute believe it will be found to violate the GATT (Kaczka 1997; McLaughlin 1994; McDorman 1991; Stanton-Kibel 1997).[8]

If the panel exempts the U.S. policy, it will contribute strongly to the resolution of the trade and environment debate and international efforts to conserve endangered species. A decision that favors §609 would recognize and strengthen the validity of the customary and conventional norms that establish international obligations to conserve endangered species. If it is upheld, it is likely that many nations will pass TED legislation and will be certified under §609. Environmentalists in the United States will probably push to tighten the invocation function to bar cosmetic compliance from certification. The United States and other nations will continue to use unilateral action, including trade sanctions, to protect international resources and the global environment, and these actions will continue to be challenged on a case-by-case basis before the WTO. However, these actions would be judged on their merits in light of the full body of international environmental law.

A WTO decision against §609 will cause further conflict and dispute. In the United States, the decision will be perceived by many as one made by faceless bureaucrats withholding dearly needed protection for endangered sea turtles. Both Democrats and Republicans can draw strong symbols from the case: global bureaucrats reviewing U.S. laws, and endangered species suffering the consequences of free trade. Environmentalists in the United States will react strongly to the decision. They will claim that the trade regime allows other nations to gut U.S. environmental laws, and they will pressure legislators not to modify §609 in light of the WTO decision. Shrimp fishermen will also react strongly. They feel they are harmed by free trade. In competitive U.S. shrimp markets, they have not been able to pass the cost of using TEDs on to consumers. Instead, they have internalized the costs of sea turtle protection, while their competitors drive sea turtles to extinction and reap a competitive advantage from their poor environmental performance. New TED regulations may be applied in U.S. waters in spring 1998 (*Federal Register* 1996a). In all likelihood, shrimp fishermen in the United States would use the WTO decision as a tool to fight against stricter U.S. regulations.

A decision against §609 is also likely to jeopardize multilateral efforts to conserve sea turtles. Knowing that their actions cannot be bound by unilateral trade instruments, nations might be less willing to participate in and/or accept the results of multilateral efforts to establish real and enforceable conservation commitments (Chang 1995). The decision could also erode the validity of the norms that protect endangered species and international environmental resources. Beyond the shrimp–sea turtle debate, this ruling could weaken international efforts to protect other environmental resources.

In summary, a WTO panel ruling in favor of §609 could improve sea turtle conservation and help to resolve the trade and environment dispute, but it is generally believed that the WTO panel will find §609 to violate U.S. obligations under the GATT. This ruling will fuel the concerns of individuals who believe that trade fosters environmental degradation. Ironically, the U.S. effort to promote international use of TEDs may weaken resolve to protect sea turtles in U.S. waters. Environmentalists and shrimpers in the United States will pressure Congress not to modify the U.S. law. These factors will inflame further conflict, thereby making improved sea turtle protection unlikely. Therefore, at present, §609 is unlikely to achieve the goals of an optimal conservation strategy.

Recommendations

Identification of Alternatives

There are numerous policy instruments that can be applied to increase TED use and improve sea turtle conservation. These include: (1) a modified §609; (2) negotiation of multilateral treaties and agreements; (3) differential tariffs or taxes that could internalize the cost of sea turtle conservation; (4) consumer/producer education programs; (5) eco-labeling schemes; and (6) financial assistance. These alternatives should be selected strategically to address defects in the existing policy process.

Evaluation of Alternatives

Although §609 does not achieve the objectives of an ideal conservation strategy, it can contribute to the realization of policy goals. The law has enhanced sea turtle conservation where it has been applied since 1991. A recent study of Central American shrimp fisheries by the STRP indicates that TEDs are widely used in the Caribbean, and that annual regional take is approximately 520 sea turtles. In contrast, the Pacific fleets of the same nations, which rarely if ever use TEDs, take approximately 60,000 sea turtles each year (STRP 1996). The law has also provided an economic incentive for nations in the region to negotiate and participate in the Salvador Convention. Yet, the law does have several problems that should be remedied. In order to minimize the conflict between §609 and other nations' demands for sovereignty, the law should target the actions of individual producers rather than national policies. The need to conserve sea turtles would become an economic consideration that is integrated into each fisherman's day-to-day operations. If fishermen did not use TEDs, their harvest would be barred from the U.S. market. A modified §609 could provide a powerful incentive for fishermen to use TEDs without forcing their government to pass a law. Slight modification of §609 offers two other benefits: it defends the right of the United States to address harms caused by its own consumption, and it maintains the incentive for other nations to participate in multilateral treaties or conventions.

The second alternative is to establish multilateral conservation agreements that require TED use. An international framework convention with regional subagreements might be appropriate. This structure would allow delineation of conservation objectives and strategies through an inclusive decision process, and the flexibility to evolve in response to unique regional concerns and changing conditions.

There are several potential concerns with regard to the negotiation of a treaty: some important nations may not participate; negotiation may be slow; and the resultant treaty may be weak and ineffective. These concerns can be addressed in ways that correct existing defects in the policy process. First, a modified rather than weakened §609 will help bring all concerned parties to the table, and to accelerate negotiation. Second, there are already a number of international conventions that seek to govern the conservation of endangered species. The provisions of existing treaties like the Biodiversity Convention and the Bonn Convention could be modified to require TED use. This avoids potentially lengthy delays that may occur during the ratification process of a new treaty.

If a new treaty is developed, the intelligence, promotion, invocation, and appraisal functions should be inclusive, and structured to support decisions that are based on available scientific information. For instance, the invocation and appraisal functions could entail biannual conferences of the parties to the regional subagreements. At these gatherings, nations could present information on their conservation programs, the status of sea turtle populations in their waters, mortality counts, and population trends. All information should be open to review by other nations, nongovernmental organizations (NGOs), scientists, and other participants. Where problems are identified, action plans should be developed consensually and applied. At subsequent conferences, nations should report on progress under these action plans. Structuring these decision functions in this way yields several benefits. It is important to note that nations are generally reluctant to release power to international governmental organizations. The structure discussed above keeps power at the national level but makes the governments more accountable to other governments in their region, and to their own citizens. Provision of a formal appraisal process that is inclusive should help to regularly target conservation programs to achieve the conservation goals. Basing actions on scientific information rather than officially stated government policies should reconnect the intelligence and promotion functions with the conservation problem and should help direct efforts based on the scope and scale of regional problems. A well-structured and binding multilateral treaty should be able to satisfy all three goals of an optimal sea turtle conservation strategy.

Third, the United States could develop and apply differential tariffs or taxes. However, such taxes and tariffs could be inordinately complex and might not accurately reflect the nature of the threat to sea turtles in the different regions. Thus they might not provide incentives for protection which are consistent with the scope and scale of regional threats to sea turtles. Additionally, their development and application would take more time than would modification of §609, and they would likely be challenged under the WTO dispute resolution process. Whereas shrimp fishermen in the United States might support a differential tax that is sufficient to offset their competitive disadvantage, environmentalists in the United States would perceive such taxes as a slow route to uncertain levels of protection, and as the evisceration of an important conservation initiative. Therefore, a differential tariff would be unlikely to satisfy the value demands of environmentalists in the United States and is less desirable than simply modifying §609.

A fourth option for improving sea turtle conservation is consumer/producer education programs. By increasing consumers' and producers' understanding of the harms caused by shrimp trawling without TEDs and the advantages of trawling with TEDs, educational programs may help to reduce shrimp consumption and could provide corresponding value indulgences for conservationists. However, the failure of the NMFS's voluntary TED program (NAS 1990) suggests that a strategy that relies primarily on educational programs would probably be insufficient to effect the desired outcomes.

Fifth, an eco-labeling system could be used to inform U.S. consumers of the harmful impacts of their shrimp purchases. Eco-labeling would display a turtle-friendly seal or stamp of approval to inform consumers of the connection between shrimp consumption and sea turtle decline, and to help consumers make educated choices when they purchase shrimp. Such a scheme would indulge consumer respect and enlightenment demands without causing significant market disruptions. Eco-labeling would also, if effective, provide an incentive for compliance for all foreign shrimpers, and would indulge the value demands of shrimpers in the United States, who already use TEDs and would be in compliance. It would also help leverage consumer preferences for conservation and indulge the value demands of conservationists. The success of the tuna eco-labeling program provides a clear example of the power of this technique. However, an eco-labeling program for shrimp may be less feasible than for tuna. Shrimp are packaged and sold differently from tuna, and they are more expensive; if an eco-labeling program raises shrimp prices beyond an optimum level, it might encourage consumption of shrimp harvested using methods that have an adverse impact on sea turtles.

Sixth, the United States could offer financial assistance to prevent or limit hardships to foreign shrimpers, which may thus help improve and maintain compliance. However, this assistance should be structured carefully to be politically feasible, and to be resilient to changes in U.S. fiscal allocations. Conversely, fluctuations in the amount and availability of aid might lead to loss of wealth and respect among nations expecting financial assistance, thereby discouraging TED compliance.

Selection of Alternatives

Although no single alternative discussed above may be ideal, a combination of alternatives may help to achieve optimal policy goals and may offer the greatest net value gains. I present here the combination of alternatives that I believe is most likely to achieve policy goals.

PL 101-162 Section 609 should be modified to focus more clearly on preventing the damage caused by U.S. shrimp consumption. Section 609 should allow fishermen who use TEDs and who protect sea turtles access to the U.S. market. Currently, it does not. However, the United States should not coerce governments to pass TED laws. The law could also be modified to encourage U.S. importers to ensure that products are harvested sustainably. These modifications could be accomplished through revisions to the Section 609 certification process.

Section 609 should not be stripped of its teeth in response to a WTO ruling or to placate the international community. The United States recognizes that it is a significant contributor to the global decline of sea turtles because of its appetite for shrimp, and it has the right and the responsibility to act to mitigate the problem. Shrimp harvested in a manner that harms sea turtles should be barred from the U.S. market. The United States must protect its right to close its market to products that are inordinately harmful to the environment during production or consumption. Section 609 can create a powerful incentive to improve sea turtle conservation. Indeed, without Section 609, it may be difficult to develop consensus and create a binding multilateral treaty that addresses the full scope of the sea turtle conservation problem.

A binding multilateral treaty should be negotiated to create an international norm, and it should include the primary nations that produce and consume shrimp. Promotional activities for the treaty should be focused primarily on the European Union, Japan, and the United States—the world's primary shrimp consumers. Nonparties could not trade shrimp or shrimp products with parties to the treaty, in order to encourage all parties involved in all aspects of the international shrimp trade to become parties to the treaty and to prevent trade leakage and diversions that might undermine compliance. The primary strategy of the treaty system would be transfer of TED technology for sea turtle conservation. It would also include consumer and producer education programs and would provide financial assistance to support these programs through tariffs or taxes on shrimp harvested without the use of TEDs. I would recommend further study to determine the costs and feasibility of a shrimp eco-labeling program as a method of consumer education.

Education and training would be provided to relevant representatives of national governments and shrimpers by members of national, regional, and international NGOs and international governmental organizations like the United Nations Development Program. Members of national NGOs could be utilized to provide compliance assessments and monitoring. The results of this monitoring could be made accessible to conservationists and scientists, thereby contributing to the advancement and integration of scientific knowledge of sea turtles.

This combination of alternatives would address the functional deficiencies that have caused the failures in the current policy. Therefore, they should be better able to achieve the large conventional goals that I have outlined for the policy program in my policy goal clarification.

The functional problems that the recommended solution needs to alleviate are goal inversion, decision monopolization, the exclusive decision process, and participant value deprivation. The selection of alternatives that I have recommended will confront the problem of goal inversion by aligning the objectives and responsibilities assigned to the institutions with the conventional missions and functional value demands of those institutions. The nature of the prescriptions that would emerge from the treaty negotiation is uncertain, but the potential for goal inversion during this phase can be minimized if participants have a clear understanding of policy goals.

The inclusive nature of the treaty prescription process and the division of responsibilities that I have proposed will minimize the potential for decision monopolization, and will address the value deprivations that have resulted from the current exclusive decision process. Although the party/nonparty system may appear to institutionalize exclusivity on a larger scale, I believe that in practice it will provide a powerful economic incentive to encourage the widest participation. Such economic incentives can be crucial in international arenas where countries' individual and joint motivations are complex. They may also be crucial in order to help align the value demands of participants with the policy goal sought in the process (Brack 1996). Furthermore, in the party/nonparty system, affected parties can best influence policy development from within the process, providing an additional incentive for participation. The result, if structured effectively, should be a broadly inclusive policy process.

The recommended alternatives also address and should remedy the value deprivation that results from the existing policy and may result if the projections are accurate. If the proposed treaty results in the use of TEDs by shrimp trawlers on a global basis, it will fulfill the demands of conservationists worldwide and shrimpers in the United States. The combination of technology transfer, economic incentives, financial assistance, and educational programs should result in net value gains and some value indulgences for foreign producers and domestic consumers.

I believe that my recommendations are sufficient to redirect the policy process to the primary goal, which is to protect sea turtles through global use of TEDs by shrimp trawlers which may otherwise capture and drown them. Through design, the recommended course of action may also contribute to the realization of additional goals: it may strengthen international norms for addressing global environmental problems through multilateral action; strengthen existing international environmental institutions; be flexible and able to evolve as necessary; integrate with and contribute to other sea turtle conservation efforts; help further scientific understanding of the life histories and global threats that sea turtles face; and enhance respect for shrimp consumers and producers. Thus, I believe that this combination of alternatives can dramatically improve sea turtle conservation and TED usage.

Conclusions

Although sea turtles are endangered globally and there have been domestic and international conservation efforts, these efforts have not significantly alleviated the adverse impact of shrimp trawling on sea turtles. It is imperative that this cause of mortality be addressed effectively if sea turtle conservation programs are to succeed. Although an inexpensive and effective mechanism exists to minimize the worldwide incidental take of sea turtles by shrimp trawlers, the current policy to promote international TED use has been ineffective. The policy encourages goal inversion, decision monopolization that omits potentially affected parties and participants, and participant value deprivation. I believe that these problems can be resolved through a combination of measures. These measures should be structured through modification of §609 and negotiation of an inclusive multilateral treaty, and they should include economic incentives, educational programs, and technology transfer. Policy analyses suggest that these alternatives can contribute to the development of a policy process that yields the greatest net value gains for participants and therefore should achieve the optimal policy goals more effectively than does existing policy. On a higher level, the proposed program may facilitate the shift to sustainable production practices for fisheries and other resources, and to world order and improved relations among nations.

ACKNOWLEDGMENTS

I thank Tim Clark and Andy Willard for helping me to focus on and understand the central problem underlying the shrimp and sea turtle debate. Steve Charnovitz and Dan Esty contributed to my understanding of the potential GATT consistency of PL101-162 §609. Debbie Crouse, Marydele Donnelly, Todd Steiner, and William Gutting sharpened my understanding of the perspectives of participants in the shrimp and sea turtle debate. Any errors and omissions are solely the responsibility of the author.

ADDENDUM: PUBLIC LAW 101-162 SECTION 609

(a) The Secretary of State, in consultation with the Secretary of Commerce, shall, with respect to those species of sea turtles the conservation of which is the subject of regulations promulgated by the Secretary of Commerce on June 29, 1987—

(1) initiate negotiations as soon as possible for the development of bilateral or multilateral agreements with other nations for the protection and conservation of such species of sea turtles;

(2) initiate negotiations as soon as possible with all foreign governments which are engaged in, or which have persons or companies engaged in, commercial fishing operations which, as determined by the Secretary of Commerce, may affect adversely such species of sea turtles, for the purpose of entering into bilateral and multilateral treaties with such countries to protect such species of sea turtles;

(3) encourage such other agreements to promote the purposes of this section with other nations for the protection of specific ocean and land regions which are of special significance to the health and stability of such species of sea turtles;

(4) initiate the amendment of any existing international treaty for the protection and conservation of such species of sea turtles to which the United States is a party in order to make such treaty consistent with the purposes and policies of this section; and

(5) provide to the Congress by not later than one year after the date of enactment of this section—

   (A) a list of each nation which conducts commercial shrimp fishing operations within the geographic range of distribution of such sea turtles;

   (B) a list of each nation which conducts commercial shrimp fishing operations which may affect adversely such species of sea turtles; and

   (C) a full report on—

      (i) the results of his efforts under this section; and

      (ii) the status of measures taken by each nation listed pursuant to paragraph (A) or (B) to protect and conserve such sea turtles.

(b)(1) In General.—The importation of shrimp or products from shrimp which have been harvested with commercial fishing technology which may affect adversely such species of sea turtles shall be prohibited not later than May 1, 1991, except as provided in paragraph (2).

(2) Certification Procedure.—The ban on importation of shrimp or products from shrimp pursuant to paragraph (2) shall not apply if the President shall determine and certify to the Congress not later than May 1, 1991, and annually thereafter that—

   (A) the government of the harvesting nation has provided documentary evidence of the adoption of a regulatory program governing the incidental taking of such sea turtles in the course of such harvesting that is comparable to that of the United States; and

   (B) the average rate of the incidental taking by the vessels of the harvesting nation is comparable to the average rate of incidental taking of sea turtles by United States vessels in the course of such harvesting; or

(C) the particular fishing environment of the harvesting nation does not pose a threat of the incidental taking of such sea turtles in the course of such harvesting.

NOTES

1. The United States enacted TED regulations under the Endangered Species Act in 1987. PL101–162 §609 was enacted approximately two years later, in 1989. It encourages nations where shrimp fishermen may have an impact on sea turtles to enact TED laws that are comparable to those of the United States.

2. In the United States, several types of TEDs are available at costs of $50 to $400. In Malaysia, they are locally built at a cost of $26 (WWF 1997). Furthermore, in response to widespread application of §609, U.S. importers began to provide TEDs at no cost to their Asian suppliers.

3. The fourteen countries were Mexico, Belize, Nicaragua, Panama, Colombia, Venezuela, French Guiana, Guyana, Suriname, Trinidad and Tobago, Costa Rica, Guatemala, Honduras, and Brazil.

4. Section 609 may not promote conservation in all areas where trawlers may have an impact on sea turtles. If a nation does not export shrimp to the United States, it is not covered by the law.

5. For a complete analysis of *Earth Island Institute v. Christopher,* see Stanton-Kibel (1997).

6. Decision monopolization occurs when a powerful organization controls key functions of the decision process and steers the process to achieve its own organizational objectives.

7. Goal inversion occurs when the policy process targets objectives that differ from the openly acknowledged objectives (Clark 1995).

8. If the WTO panel rules against §609, the United States will be asked to change the law. If the United States refuses to change the law, the complainant nations will be able to embargo an equivalent value of imports from the United States. The panel does not have authority to strike down or otherwise modify U.S. laws.

LITERATURE CITED

Bowen, B. 1995. "Molecular Genetic Studies of Sea Turtles," in K. A. Bjorndahl, ed., *The Biology and Conservation of Sea Turtles,* p. 489. Washington, DC: Smithsonian Institution Press.

Brack, D. 1996. "International Trade and the Montreal Protocol," unpublished manuscript, The Royal Institute of International Affairs, London.

Center for Marine Conservation. 1995. *Delay and Denial, A Political History of Sea Turtles and Shrimp Fishing.* Washington, DC: Center for Marine Conservation.

———. 1996. "Countries Agree on World's First Sea Turtle Treaty," *Marine Conservation News* 8(4):1.

Chang, H. 1995. "An Economic Analysis of Trade Measures to Protect the Global Environment," *Georgetown Law Journal* 83:2131–2209.

Charnovitz, S. 1994. "Dolphins and Tuna: An Analysis of the Second GATT Panel Report," *Environmental Law Reporter* 24:10567.

Clark, T. W. 1995. "Making Endangered Species Policy: Learning from the Black-Footed Ferret Experience," in K. Kohm, ed., *On the Brink,* pp. 95–125. Washington, DC: Island Press.

Court of International Trade. 1995. *Earth Island Institute v Christopher*, 913 F. Supp. 559.

————. 1996. *Earth Island Institute v Christopher*, 942 F. Supp. 597.

Crowder, L., D. Crouse, S. Heppell, and T. Martin. 1994. "Predicting the Impact of Turtle Excluder Devices on Loggerhead Sea Turtle Populations," *Ecological Applications* 4(3):437–445.

Federal Register. 1991. "Turtle in Shrimp Trawl Fishing Operations Protection; Guidelines," 56 *Federal Register* 1051, Jan. 10, 1991.

————. 1993. "Revised Guidelines for Determining the Comparability of Foreign Programs for the Protection of Turtles in Shrimp Trawl Fishing Operations," 58 *Federal Register* 9,015, Feb. 18, 1993.

————. 1995. "Bureau of Oceans and International Environmental and Scientific Affairs, Certifications Pursuant to Section 609 of Public Law 101–162," 60 *Federal Register* 43,640, Aug. 22, 1995.

————. 1996a. "Department of Commerce, National Oceanic and Atmospheric Administration, Final Rule Amending Regulations to Protect Sea Turtles," 61 *Federal Register* 66933–66947, December 19, 1996.

————. 1996b. "Department of State, Bureau of Oceans and International Environmental and Scientific Affairs, Revised Notice of Guidelines for Determining Comparability of Foreign Programs for the Protection of Turtles in Shrimp Trawl Fishing Operations," 61 *Federal Register* 17342–17344, April 19, 1996.

GATT Dispute Settlement Panel. 1991. *United States—Restrictions on Imports of Tuna* (unadopted), International Legal Materials 30:1594.

————. 1994. *United States—Restrictions on Imports of Tuna* (unadopted), International Legal Materials 33:839.

Heppell, S., C. Limpus, D. Crouse, N. Frazer, L. Crowder. 1996. "Population Model Analysis for the Loggerhead Sea Turtle, *Caretta caretta*, in Queensland," *Wildlife Research* No. 23.

Hillestad, H., J. Richardson, C. McVea, and J. Watson. 1995. "Worldwide Incidental Capture of Sea Turtles," in K. A. Bjorndahl, ed., *The Biology and Conservation of Sea Turtles*, p. 489. Washington, DC: Smithsonian Institution Press.

House Fisheries Subcommittee. 1990. Transcripts of the House of Representatives Subcommittee on Fisheries and Wildlife Conservation and the Environment, Committee on Merchant Marine and Fisheries, Hearing on Sea Turtle Conservation and the Shrimp Industry, May 1, 1990.

International Union for the Conservation of Nature. 1993. Marine Turtles, an Action Plan for Their Conservation, Part 1:1993–1998. IUCN Marine Turtle Specialist Group, International Union for the Conservation of Nature. Printed from the IUCN Homepage.

————. 1995. A Global Strategy for the Conservation of Marine Turtles, IUCN Species Survival Commission, International Union for the Conservation of Nature. Printed from the IUCN Homepage.

————. 1996. A Marine Turtle Conservation Strategy and Action Plan for the Western Indian Ocean, International Union for the Conservation of Nature. Printed from the IUCN Homepage.

Kaczka, D. 1997. "A Primer on the Shrimp–Sea Turtle Controversy," *Review of European Community and International Environmental Law* 6,2:171–180.

Lasswell, H. D. 1971. *A Pre-View of the Policy Sciences*. New York: American Elsevier.

McDorman, T. 1991. "The GATT Consistency of US Fish Import Embargoes to Stop Driftnet Fishing, and Save Whales, Tuna, and Dolphins," *George Washington Journal of International Law and Economics* 24:477–525.

McLaughlin, R. 1994. "UNCLOS and the Demise of the United States' Use of Trade

Sanctions to Protect Dolphins, Sea Turtles, Whales, and Other International Marine Living Resources," *Environmental Law Quarterly* 21:1–78.

Menotti, V. 1994. "Trade Measures as Part of the Solution: How Compensatory Tariffs Can Help Save Endangered Sea Turtles," *Journal of Environment & Development* 3(2):97–106.

National Academy of Sciences. 1990. *Decline of the Sea Turtles*. Washington, DC: National Academy of Sciences Press.

Pandav, B., Choudhury, B., and C. Kar. 1997. "Mortality of Olive Ridley Turtles, *Lepidochelys olivacea*, Due to Incidental Capture in Fishing Nets Along the Coast of Orissa, India," *Oryx* 31(1):32–36.

Sea Turtle Restoration Project, Earth Island Institute. 1996. New Study Shows 60,000 Sea Turtles Caught Each Year in Central American Shrimp Nets, San Francisco, CA.

Stanton-Kibel, P. 1997. "Justice for the Sea Turtle: Marine Conservation and the Court of International Trade," UCLA *Journal of Environmental Law and Policy* 15:57–81.

World Trade Organization Appellate Body Report. 1996. *United States—Standards for Reformulated and Conventional Gasoline*, WT/DS2/AB/R.

World Wide Fund for Nature. 1997. WWF Amicus Brief to WTO Shrimp-Turtle Dispute. Gland, Switzerland.

7 How Everything Becomes Bigger in Texas

The Barton Springs Salamander Controversy

Katherine Lieberknecht

The revamping of the 1973 Endangered Species Act (ESA) has been a hot topic in public and private arenas lately. For example, the five-year-long struggle over the listing of the rare Barton Springs salamander has been attributed to the unpopularity of the ESA. I argue in this paper that the problem of unsuccessful recovery of this species can be attributed not simply to the failure of one federal law, but instead to complex interactive social and decision processes taking place at local, state, and federal levels. Use of Harold Lasswell's policy sciences framework allows for analytic refinement of the problem, its context, and recommendations (see Lasswell 1970, 1971; Lasswell and Kaplan 1950; Lasswell and McDougal 1992).

This paper describes the social and decision process constituting current management of the salamander. A summary of the former conservation plan, the Barton Springs Salamander Conservation Agreement and Strategy (BSSCAS) is given (Texas Parks and Wildlife Department et al., 1995). Recommendations for how the species should be successfully managed are presented and critiqued.

My standpoint includes my position as a conservation biologist and my Austin roots. I have a personal and academic interest in seeing the salamander preserved. I believe that biodiversity should be conserved,

Research for this project was conducted in 1997.

for the value of species in their own right as well as the aesthetic, economic, and ecological values associated with them. Austin is my hometown, so I have a personal interest in seeing that biodiversity and the environmental qualities associated with salamander habitat, such as clean water, are conserved there. In many ways I am an "insider" to the situation, as an Austin resident and as someone who swims in the Barton Springs Pool. In some ways, however, I am also an "outsider." I am going to school "out east" at Yale University, and my family does not live in the part of Austin undergoing current development.

My role is that of an appraiser of past decision making and an advisor to future decision making. I am biased, as are all humans. As a conservation biologist, I have an interest in seeing that biodiversity is preserved. I also believe that humans, including Austinites, should have access to good quality water, which will not be the water coming out of the Edwards Aquifer if proposed development there proves unsustainable. From the perspective of area developers, the policy problem is that Austin's environmentalists refuse to allow them to make a living. From my perspective, the problem is that the salamander is not being successfully managed in the common interest. However, I also feel that the salamander should be managed in the most effective way possible, and if this means no ESA listing, and only regional action, so be it.

To orient myself to the policy problem, I familiarized myself with the participants, their perspectives, arenas, and other key variables. I acted as a participant observer and interviewed citizens, government agency staff, environmental groups, private property groups, and developers by phone. In addition, I was a spectator to the social process. While in Austin on a recent visit, I observed reactions to the controversy while spending time at Barton Springs Pool, by listening to the radio and by scanning the salamander website chat line. As a collector and researcher, I reviewed newspaper articles, press releases, journal articles, biological and hydrological reports of the salamander and its habitat, and the BSSCAS. I also read *A Review of the Status of Current Critical Biological and Ecological Information on the Eurycea Salamanders Located in Travis County, Texas* (Bowles 1995), the scientific basis of the BSSCAS management plan.

Background on the Barton Springs Salamander Case

The Barton Springs salamander is a rare species found only in four hydrologically connected springs, which make up the Barton Springs area of the Edwards Aquifer in Austin, Texas. Spring outlets are in Zilker Park, which is owned and operated by the city of Austin. Prior to

April 1997, salamanders were found in only three springs, which had been previously lined with concrete (U.S. Fish and Wildlife Service 1995). Salamanders were discovered in a fourth spring in early April 1997; this spring flows only during wet periods (Haurwitz 1997). The main spring forms a large pool named Barton Springs Pool. This 200-meter-long pool is the centerpiece of the Austin park system and is used by swimmers year-round. Because, unlike most salamanders, the Barton Springs salamander retains external gills throughout its life, it is dependent on a continuous flow of clean, clear, and cool water. Because of this characteristic, the salamander is seen as a modern-day canary in the coal mine. It is thought to be an indicator of the "health" of the Edwards Aquifer.

Austin, Texas, is experiencing the growing pains of a formerly small city. There are plans for extensive residential and highway development in the southwestern part of town, which is also the area overlying the Edwards Aquifer. The Edwards Aquifer is a "karst" aquifer made of permeable limestone and characterized by caves, sinkholes, and other conduits (Marytyn-Baker et al. 1992). Scientists consider it the most environmentally sensitive aquifer in the state. Because of these characteristics, Barton Springs is thought to be heavily influenced by the quality and quantity of runoff. It is thought that urban development will lead to increases in pollutants in the water that flows through the salamander's habitat, therefore endangering the population (Chippendale et al. 1993).

In 1992 the U.S. Fish and Wildlife Service (USFWS) received a petition from a University of Texas scientist, Dr. Mark Kirtpatrick, and an Austin environmental group, the Save Our Springs Alliance (SOS), to list the Barton Springs salamander under the ESA (Haurwitz 1995b). The USFWS determined that the petition presented environmental threats indicating that the listing might be justified. The agency continued its status review of the species and solicited information. On February 17, 1994, it proposed that the species be listed as endangered. However, all listing actions were suspended by a congressional moratorium on new listings that began on April 10, 1995, and ended on April 26, 1996 (Save Our Springs Alliance 1996). The USFWS reopened the comment period on June 24, 1996, as a result of the potential for new information on proposed regulatory protection under state authorities and disagreement over existing regulatory mechanisms. Department of the Interior Secretary Bruce Babbitt encouraged the Texas state government to create an agreement for state control of the species' conservation. On August 13, the USFWS, Texas Parks and Wildlife Department (TPWD), Texas Natural Resource Conservation Commission, and Texas Department of Transportation created a document entitled the BSSCAS, committed to the development and implementation of measures needed to conserve the species and Barton Springs. On September 4, the USFWS withdrew the proposed listing of the salamander, based on a belief that the BSSCAS would prove sufficient (U.S. Fish and Wildlife Service 1996; Haurwitz 1996f). The SOS and Dr. Kirtpatrick immediately sued Secretary Babbitt, stating that he had violated the ESA when he decided to withdraw the proposed endangered listing of the Barton Springs salamander. On March 26, 1997, the federal court ruled that Babbitt had violated the ESA, stating that he had missed ESA deadlines and failed to follow proper procedures concerning information used to make the decision. On April 22, 1997, the USFWS and Secretary Babbitt agreed to list the Barton Springs salamander as an endangered species.

The Policy Problem

The policy problem is that, despite five years of management effort, short- and long-term threats to the salamander's survival continue to exist. The salamander's population dynamics show a fluctuating population. When first discovered in 1946, the salamander was reportedly "abundant." In 1992, at least 50 salamanders were present in the main spring. After a fish kill resulting from chlorine application in the pool in late 1992, only 10 salamanders were observed (Chippendale et al. 1993). Permanent transects were put into place in June 1993, and survey counts have ranged from 1 to 27 individuals from 1993 until April 1995. After a 1994 flood, only 1 to 6 were found over several months. From April 1995 until April 1996 3 to 45 individuals were counted (Haurwitz 1996b). However, in April 1997, a record 188 salamanders were found within the three springs (Haurwitz 1996a, 1997).

Short-term threats to the salamander population occur in the form of pool maintenance procedures and nearby road development. Although only the Barton Springs Pool is still maintained for swimmers, all four spring habitats are connected hydrologically, so pool management affects the whole population. Chlorine and high-pressure hoses have been used in the past to clean the pool for human use (USFWS 1994). All the recorded salamander deaths so far have been attributed to pool maintenance procedures, and the city park department has discontinued them. Although the known pool threats have been eliminated, another short-term hazard remains. As more roads are built in southwest Austin, heavy truck travel has increased over the aquifer. A massive oil or chemical spill from an overturned tanker truck could wipe out the entire salamander population.

The long-term threat of increased development and the subsequent chronic degradation of water quality is the other major concern. Pollutants such as fertilizers, sediment from erosion, pesticides, and auto oil and grease run off the land and taint the aquifer feeding Barton Springs. As development grows, so does the threat. At present, neither the toxic spill threat nor chronic water degradation has been properly addressed.

Why Haven't the Threats Been Resolved?

A provisional statement of the problem can be made. The policy and management problem is that there exists a rare species with a limited habitat range and present and future threats to its habitat. Over the past five years, there has been much controversy over its listing as an endangered species, and a management plan did not go into effect until August 1996. The population has continued to fluctuate throughout the past few months, and most of the threats to survival have not been removed.

Goals can be clarified in relation to this problem. The overall goal is to restore and maintain a viable salamander population. This population should be managed to ensure sustainability, and management should focus on cultivating public awareness of and involvement in the recovery of the species.

I used three policy sciences methods in order to clarify the problem. First, I looked at the social process and mapped the perspectives of key participants. Next, I examined the decision process and the seven decision functions. I also took an in-depth look at the written plan, the Barton Springs Salamander Conservation Agreement and Strategy, using the policy sciences analytic framework. When the salamander was listed under the ESA, the BSSCAS was nullified, so I present only a summary of the analysis here.

Analysis of the Social Process Behind the Barton Springs Salamander Case

Conflict in decision making can be related to many factors, including the perspectives of participants (Clark and Brunner 1996). Perspectives can be described in terms of identities, expectations, and demands. Although there are many formal and informal participants within this case, six main groups have had the most influence in the decision process: the state government, the federal government, scientists, developers, environmentalists, and citizens. Most of these groups comprise one or two main power-holding organizations. By examining the identities, expectations, and demands of these main participants, we can clarify the policy problem.

State Government

The Texas Parks and Wildlife Department (TPWD) had led the BSSCAS, the former recovery plan for the species. Many critics have accused the agency of being influenced significantly by development and private property interests. During the heat of the salamander controversy in summer 1996, three of the TPWD's endangered species biologists resigned. It has been suggested that the departures are indicative of deeper problems at the agency, such as bureaucratic harassment of employees who do not support private property rights over conservation (Bryce 1996). The resignation of the three biologists came after months of friction between TPWD managers and endangered species biologists. In summer 1995, the TPWD reprimanded Andy Price, its leading aquatic biologist, for working on a report that listed threats to the region's salamander population (Bryce 1996). The department also recently terminated the Texas Natural Heritage Program, which had been targeted previously by private property advocates because it monitored the status of endangered species (Haurwitz 1996c).

As an agency responsible for species management, the TPWD is expected by the public to be concerned with conservation of the salamander. As a state agency, the TPWD believes that successful management is done at the state level. The department expected that the federal government would not list the species as endangered and would continue to allow the state to manage its own biological resources. They stated, "Texas, not Washington, can best manage Texas' natural resources" (Haurwitz 1996c:25).

Scientists

The group of University of Texas scientists who named the salamander as a separate species in 1992 has continued to conduct research on the salamander's habitat requirements. The scientists are operating under the view that "good science" is what policy making is or should be based on and that scientific knowledge is nonbiased. These scientists originally proposed the endangered species listing to the USFWS and demanded that the species be listed for three reasons: it has a small and fluctuating population; it has a limited and interconnected range; and there are present and future threats to its survival. Additionally, one scientist, Dr. Mark Kirtpatrick, has worked with the SOS on lawsuits concerning the listing.

Developers

The major developer, FM Properties, is from the Austin area. FM Properties and other developers believe in the free market. Austin is a boom town, there is a market for new development, so development should be provided. They contend that water quality is adequate and evidence of endangered status for the BSS is lacking. They expect that the ESA listing of the species will be a large constraint on their business. With a listing, it becomes illegal to harm the salamanders or their habitat. As a result, many large development projects over the aquifer may be prohibited. Therefore, the developers demanded that the species not be listed and that state and local governments continue to let them develop almost without restraint. This is evidenced by the development community's contributions to state politicians' campaign funds (Haurwitz 1996d). Now that the species is listed, developers have demanded that any sanctions be weak and that any regulations be only mildly enforced.

Environmentalists

Like state and federal officials, Austin's environmental groups also state that they are committed to conservation of the salamander. The salamander is a symbol of clean water and environmental health. The environmentalists' expectation is that if the salamander population declines, so will the quality of the water used by forty thousand Austinites. They demand that the species regain a viable population, along with associated controls on land development above the aquifer. The SOS is the most influential environmental organization, and it had an additional demand that the salamander be listed as endangered and managed by the federal government, not the state (Arnold 1997). Their expectation is that now that the species is listed under the ESA and a new recovery plan will be written, the salamander will survive and water quality will not degrade. To ensure that their demand was carried out, the SOS twice sued Secretary of the Interior Bruce Babbitt over his actions with respect to the process of determining whether the salamander should be listed as endangered.

Federal Government

The major federal participants in the salamander controversy are the USFWS and Secretary of the Interior Bruce Babbitt. Under the ESA, the USFWS is responsible for the prevention of species extinction. It expected that the salamander could be managed successfully under the former recovery plan (BSSCAS) led by Texas agencies. However, Texas is a conservative state and much of the decision process took place in late summer 1996, an election year. Some observers think that the hope of winning electoral votes in Texas swayed Secretary Babbitt not to list the salamander and instead to allow state management of the species. In the March 1997 court decision in which Secretary Babbitt was found to have violated the ESA, the judge wrote that the court had found that "strong political pressure was applied to the Secretary to withdraw the proposed listing of the salamander" and that the record suggested "that the political lobbyists for the development community worked with the political appointees of the Secretary" (SOS 1997:1).

Citizens

Texas is a conservative state, and even though many Texans consider Austin liberal, many local citizens value private property rights and autonomy. However, many citizens are also in favor of conservation. A popular T-shirt reads "The Barton Springs Salamander—not just any amphibian, an Austin original," and it has a drawing of the salamander complete with a diving mask and flippers. Even if not all citizens are fond of the salamander itself, most are in favor of the clean water associated with the salamander's habitat. And no one is concerned that the Barton Springs Pool will be closed to swimmers because of threats to the salamander. The USFWS has stated that humans and salamanders can share the pool without threats to either population (Haurwitz 1996b). Most citizens want the species to survive, although there is much controversy over the endangered species listing. Also, most citizens feel very removed from the decision process concerning the salamander.

A subgroup of Austin residents is the private property advocates. This group expects that the ESA listing of the species as endangered will threaten their property rights. Unlike most citizens, including the members of the SOS, they do have power in the decision process. They have a lobby called "Take Back Texas" and are allied with development interests. Prior to the ESA listing, both groups made large campaign donations to state politicians (Haurwitz 1996c).

Analysis of the Barton Springs Salamander Decision Process

Examination of the decision process clarifies the policy problem and suggests solutions. I examine the decision process by analyzing the seven decision functions as described by Lasswell (1971): intelligence, promotion, prescription, invocation, application, termination, and appraisal.

Appraisal

The decision process has been characterized as weak. The USFWS took authority and control away from the state agencies who wrote the BSSCAS. It did so when it deemed that the salamander was not being managed successfully, even though there was no specification of how "success" was to be measured. The BSSCAS had an appraisal schedule of two semiannual meetings written into it, but this did not ensure that a thorough, independent appraisal would ever occur.

Appraisal of management for the salamander also takes place in informal settings. The media have served as appraisers but have no authority to enact any changes if problems are found with decision activities. Likewise, citizens have participated by writing editorials to the newspapers and letters to formal power holders. However, the only real appraisal achieved by citizens occurs when they join organized groups, such as the SOS. The SOS lawsuits have served as outside appraisal factors, but overall, a comprehensive and independent appraisal function is lacking to date (see Haurwitz 1996e).

Intelligence

The gathering, processing, and dissemination of information has so far been concentrated on positivistic scientific knowledge. Despite the 1995 assessment of the ecological knowledge of the species by scientists at the University of Texas, little is known about the biology of the salamander. The viable population size and habitat requirements are unknown. This makes it difficult to create a clear and precise plan. In addition to the urgent need for better biological understanding, the need for information on other aspects of successful management has been ignored. For example, intelligence in the form of knowledge about effective recovery plan creation and the policy sciences is totally lacking.

Promotion

Promotional efforts have dominated the decision process. Although its goal is to save the salamander, the state government seems to be motivated more by keeping management (i.e., prescription, invocation, and application) authority and control (i.e., power) within the state. Likewise, the main environmental group, the SOS, has worked harder at making sure the species is listed than at promoting solid conservation recommendations, educating the public about the situation, or other actions that will ensure survival. The federal government (USFWS) also has a stated goal of conserving the salamander along with other species, yet it has put all its efforts into promoting the presumed benefits of state control of conservation, despite a lack of proof that this is the best conservation strategy. The TPWD has stated that it did not want the species listed because the three state agencies understand more about the salamander and the situation than does the federal government (McKinney, personal communication 1997). However, there is no reason why the state cannot play a major role in the planning and invocation of the recovery plan now that the salamander is listed. Overall, there has been a sharp and dominant focus on the promotion of power, not salamander conservation, by the TPWD, USFWS, and SOS.

Prescription

The former plan, the BSSCAS, had admirable goals, but they were vague and lacked explicit methods for achievement. The only comprehensive rules it had were in the water quality protection program, which had been written before survival of the salamander population had become a concern. These were also the only rules with any authority and sanction behind them.

Invocation

Invocation has been successful in the sense that all the agencies assigned duties in the BSSCAS agreed to conform to them. But because no new regulations were authoritatively imposed by the BSSCAS, public agreement did not require much effort. For example, the TPWD was assigned responsibility for working with the city of Austin to create more specific rules about Barton Springs Pool maintenance. Together they made a pool management plan, but it lacked specificity about allowable procedures. Later, more die-offs resulted from a draining procedure not addressed within the new plan.

Application

Although the application of new pool maintenance procedures has been partially successful, no other management actions have been taken specifically to reduce short- and long-term risks to the salamander. If any other regulations besides the previously enacted water quality rules had been generated by the BSSCAS, compliance would have been voluntary. The current ESA listing provides an opportunity for better invocation and application.

Termination

On March 26, 1997, the federal court found that Secretary Babbitt had violated the ESA and the Administrative Procedures Act when he decided to withdraw the proposed endangered listing of the salamander (SOS 1997). The court held that it was improper for the secretary to base his decision to withdraw the proposed endangered listing on the current existence of the BSSCAS. The court maintained that the BSSCAS had failed to address threats identified by Secretary Babbitt in his 1994 proposal to list the salamander. It also held that Babbitt had considered factors other than those contemplated by the ESA. The secretary was ordered to make a new decision based on the salamander listing within thirty days of the decision. On April 22, 1997, Babbitt decided to list the species as endangered. However, federal listing is not a guarantee that threats to the salamander will be reduced or eliminated.

Problem Refinement

After my first look at the situation, I defined the policy problem: only a few concrete steps had been taken toward reducing short- and long-term threats to the survival of the salamander. By mapping the social process and analyzing the decision functions, I was able to refine the initial policy problem statement. A "goal substitution" by key participants, the influence of developer and private property interests on the state government, and an absence of citizen power in the decision process are key factors in the social process of this policy problem. In the decision process, a lack of knowledge about the salamander's biology (intelligence), missing comprehensive and independent appraisal, a misdirected promotion function, and a weak plan (a prescription leading to ineffectual invocation and application) have all contributed to the policy problem. Owing to these factors, the policy problem can be redefined: short- and long-term threats to the survival of the Barton Springs salamander have not been reduced because of weaknesses in the social and decision processes. The goal substitution by participants in the social process is directly or indirectly linked to all the other process failures. This diagnosis is similar to the ones made for other endangered species cases (see Clark et al. 1994; Clark 1997).

The root of the policy problem is found in goal substitution in the social process. Participants, such as the federal government in the form of the USFWS, the state government in the form of the TPWD, and environmentalist groups, including the SOS, have substituted the pursuit of power for their stated goal of salamander conservation. This goal substitution is clear in all decision functions, especially promotion. Both the federal and the state governments promoted the best option as state control through the BSSCAS, while the SOS promoted the ESA listing as the only viable option for survival. Two other examples are the USFWS agreement to allow the TPWD to write and enact a recovery plan (i.e., BSSCAS) that lacked specificity and effectiveness, and the SOS's concentration on efforts to sue Secretary Babbitt over his decision to not list the species, while refusing to make specific suggestions on how to improve management of the species. This goal substitution has complicated the conservation problem and has contributed to the long time—five years—it has taken to make a decision about the salamander's legal status. It has also contributed to the creation of a weak plan, because the underlying goal of the BSSCAS writers was not to conserve the species but to maintain state control—a power issue!

This substituted goal can be explained by the maximization postulate, which says: "living forms are predisposed to complete acts in ways that are perceived to leave the actor better off than if he had completed them differently" (Brunner 1987:10). Texas state government agencies believed they would benefit more by pleasing the development and private property interests who donate large campaign funds than they would by ensuring the survival of the salamander population. Development and private property interest groups have affected TPWD behavior in other arenas, such as the firing of the three conservation biologists and the termination of the Texas Natural Heritage Program. This influence has been felt beyond the state agencies. The recent court decision against Secretary Babbitt stated that there was evidence that political lobbyists for the development community had worked with Babbitt's political appointees (SOS 1997).

An in-depth analysis of the BSSCAS shows that although the goals were comprehensive and admirable, methods for achieving them lacked specificity. Because the plan was weak, invocation and application could not be effectively carried out. The Texas Natural Resource Conservation Commission was the only agency with real power to carry out the BSSCAS, but its water quality regulations were created prior to the BSSCAS and did not consider recovery of the salamander. The general and diffuse threats of accelerating urbanization to the salamander population were not specifically addressed. The conservation commission also has a history of inconsistently enforcing its own rules and of cutting fines in half a few weeks after enforcement (Haurwitz 1996c). Although the urbanization water quality measures were weak and were not created for the salamander, they were the only part of the plan that included contingencies and sanctions. The other two threats, pool maintenance problems and catastrophic spills, were never addressed with specific measures within the BSSCAS. Even if measures had been created, they could have been met only on a voluntary basis. Overall, the BSSCAS was weak and did not constitute a realistic management plan, ensure the survival of the salamander, or serve as a good substitute for endangered species listing. Although the April 22 decision to list the species nullified the BSSCAS, its faults should be kept in mind as a new recovery plan is written.

The problems with promotion and prescription are directly linked with the goal substitution, but intelligence and appraisal are also indirectly connected to the goal substitution. There is still a lack of knowledge about the biological characteristics of the salamander and the hydrological and chemical characteristics of its habitat. The BSSCAS listed research as a goal, but like its other goals, no real steps were taken toward achieving it. The appraisal function is the ongoing evaluation of all the decision functions. If appraisal of the problem had been carried out completely, then weaknesses in the other functions may have been identified and possibly modified. However, the efforts of appraisals so far have been either coercive and motivated by the desire of the SOS for an endangered species listing, or token, self-serving efforts in the form of semiannual meetings attended by the groups in control of the BSSCAS.

The problem can therefore be further refined: short- and long-term threats to the survival of the Barton Springs salamander have not been reduced because of weaknesses in the social and decision processes. The overall goal is still to restore and maintain a viable salamander population while focusing on public awareness of and involvement in recovery. The goal substitution by participants in social process is directly or indirectly linked to all other process failures, which include a lack of knowledge about the salamander's biology (intelligence), a lack of comprehensive and independent appraisal, a misguided promotion function, and a weak prescription leading to ineffectual invocation and application.

What Set the Stage for Management Failure?

Trends Behind the Ineffective Management of the Barton Springs Salamander

Five trends have been discussed throughout the paper and are summarized here:

1. The decision process has been slow. It took five years from the original petition for listing to the recent court decision requiring Secretary Babbitt to make a final decision.
2. Within this time period, although a recovery plan was created, it was not successfully applied.
3. The salamander population has continued to fluctuate.
4. Meanwhile, urbanization of the land over the aquifer that feeds the springs has continued, and threats to the salamander presented by increased urbanization have not been addressed.
5. Known threats to the salamander population caused by pool maintenance methods have been resolved, but the threat of a catastrophic highway spill has not been realistically addressed.

Factors Shaping These Trends

These five trends have all been shaped by several conditioning factors, including: a lack of biological knowledge of the salamander and its habitat requirements; a goal substitution of power for conservation by the major participants in the decision process; a weak plan; the Texan desire for autonomy; the influence of development and private property interests in the state government; the federal government's wish for Texan support; and the federal government's reluctance to impose the ESA because of its unpopularity.

The slow pace of the decision process resulted primarily from two factors. These are the goal substitution of power for conservation, which kept the decision process in promotion, and a weak prescription that made invocation and application difficult because it lacked specificity. The goal substitution was the result of many conditioning factors. Texas has always considered itself an autonomous state, and residents and lawmakers alike resent when the federal government tries to "interfere" with management decisions. The federal listing of an endangered species is considered a threat by developers and private property advocates, both of whom have considerable influence in the state political arena. Therefore, the state government demanded state control of all management, which the federal government conceded in the form of the BSSCAS. The conditioning factors behind this decision were the present unpopularity of ESA and the federal wish for the state's political support in national elections and other issues.

Before the Barton Springs salamander was listed as an endangered species, successful recovery was not going to be achieved because the recovery plan (i.e., the BSSCAS) was weak. The BSSCAS was ineffectual because there exists a lack of knowledge about the species and because a powerful plan could have restrained influential developers and private property advocates. The plan was voluntarily written by a group of state agencies. Although their participation is described as voluntary, they acted with the threat of an endangered species listing. They chose to write the plan themselves rather than have a federally directed one. The official reason for this was that Texans "know their natural resources best," so the state should be responsible for management. The unofficial reason for this position was the influence that development and private property advocates have on the state's agencies. When there is a threat of regulation, people and groups agree to comply voluntarily only if the costs of volunteering are less than the costs of regulation. Therefore, the state agencies agreed to write the plan and to take on responsibility knowing that the resulting plan would put less constraint on development and private property than would a federal listing. Although the BSSCAS has been nullified, however, a successful replacement recovery plan has yet to be designed and adopted.

The very small salamander population has continued to fluctuate because of pool maintenance procedures and because of a lack of knowledge about what the species requires. Basic threats continue, and future threats loom because urbanization has continued. The possibility of a catastrophic spill has not been addressed because of developer and private property power and the vague and powerless BSSCAS.

Projections of Salamander Survival Based on These Factors

If the current conditioning factors had continued, I would have projected the following: the decision process would remain ineffective. The BSSCAS would continue in place, so a successful plan would not be crafted. The salamander population would be likely to continue to fluctuate or even go extinct because the BSSCAS does not address known threats adequately, and much is still unknown about the biology of the species. Rapid urbanization would continue, and future threats to the salamander would persist.

One of these conditioning factors changed, however. On April 22, 1997, the USFWS and Secretary Babbitt decided to list the salamander as endangered. This action represents a power indulgence for the SOS and a power deprivation for the Texas state agencies, because the state government may lose control of management. However, despite what the sos appears to believe, listing of the salamander under the ESA does not guarantee successful management or USFWS leadership. USFWS management decisions could still be affected by development, private property, and state interests in a large way. None of the other conditioning factors are likely to change either. The promotion of substituted goals will likely be transferred into the process of writing a new plan, and its application, if it occurs, might also be infected.

Ideally, a comprehensive, contextual recovery plan should be written and successfully applied; the salamander population should be made viable; and urbanization in southwest Austin should be managed in a way that does not jeopardize the salamander population and the aquifer.

Recommendations for Successful Conservation

The goal of the decision process is to restore or maintain a viable salamander population. Analysis of the problem in this paper emphasizes facets of the social and decision process involved. If the goal is to be achieved, the effect of participant goal substitution must be lessened significantly. If this is accomplished, it may be possible to take steps toward solving the other dimensions of the problem, which include a lack of knowledge about the salamander's biology, inability to prescribe and apply an adequate recovery plan, a lack of comprehensive and independent appraisal, influence of developer and private property interests on the state government, and an absence of citizen power in the decision process. Therefore, alternatives to lessen the effect of the power struggle in the social and decision process must be addressed, along with alternatives for an appraisal system, for more scientific research, and for public participation in the recovery process.

Goal Substitution Alternatives and Evaluations

The goal substitution of the pursuit of power for conservation of a viable salamander population within the state and federal governments and the sos has made the decision process highly conflictual and ineffective and has resulted in a weak management plan. It is unrealistic to believe that one can change these groups' desires for power, but it is possible to reduce its effect on the decision process. One way to do this is by creating an independent and comprehensive appraisal clause in the new plan (see Backhouse et al. 1996). This should include a timetable of semiannual appraisal meetings in which the performance of each of the seven decision functions is systematically and thoroughly addressed. Either the appraisal exercise should include participants other than those people who are directly part of the decision-making process or a separate, outside appraisal team should be formed to provide independent appraisal. One consideration is the identification of *independent* participants. I can suggest two people to lead the search for independent participants. Lady Bird Johnson, the widow of former President Lyndon Baines Johnson, is the epitome of a well-loved public person who is involved with natural resource management. Mrs. Johnson has earned the respect of environmentalists, citizens, and politicians throughout Texas by creating a wildflower restoration project and opening the National Wildflower Research Center in the 1980s. She is perceived as someone who loves and values Texas' natural resources but who is little influenced by "extremists." My second suggestion is based on the fact that Austin is intimately associated with the University of Texas (UT). Dr. Mark Kirtpatrick, a professor at the university, has worked with the SOS in the past, but other members of the faculty can be considered as independent with respect to a salamander appraisal committee. I recommend Dr. Clark Hubbs, an emeritus professor of zoology. Dr. Hubbs is familiar with salamander biology and is well liked and respected. He is also recognized as an independent person who lacks political or extremist environmental ties. Together, Mrs. Johnson and Dr. Hubbs could put together a board of independent and informed citizens and outside, knowledgeable persons, perhaps some being national figures, to advise the plan, appraise the process of writing a new plan, as well as monitor the new decision process that is eventually established.

A second way to reduce the influence of the goal substitution by participants is to limit the amount of campaign donations the development community and private property advocates can make to state political campaigns, thus reducing their influence on state politics. Given the popularity of national campaign reform, this might not be as difficult to achieve as might otherwise be expected, even in Texas.

A third way is to break the control of state government over federal government agencies. This has been partially achieved with the recent decision to list the species as endangered under federal law. No new evidence was produced in support of a listing. In this process, Secretary Babbitt and the USFWS received a lot of negative press after the March court decision to list. The media can be used to create and direct public attention, which can in turn be used to bring pressure on the federal government to restore the species. The media have done a fair job of drawing attention to development and private property influence on state government, but a series of articles could be written tracing political campaign funds from development and private property interests to the state government, and the state's pressure on the USFWS and Secretary Babbitt. It would also help if the issue received more national public attention. Additionally, the public could get involved in the decision process by writing more editorials and letters to Secretary Babbitt and Texas Governor Bush.

It should be noted that these three alternatives concentrate primarily on how to reduce the power and goal inversion of the state and federal governments. The fact that the SOS concentrated on listing the salamander and that it subsequently neglected planning for the successful management of that population is not as large or as immediate a concern now because the species has been listed as endangered. Although the species did not have to be listed to be successfully managed, the listing may encourage its survival as long as a new decision process can be made to work for salamander conservation.

Plan Alternatives and Evaluations

Even if state control of the decision process is lessened, this is not a guarantee that the new plan will be comprehensive and successful in reducing threats to or ensuring survival of a viable salamander population. The following are three approaches to achieving the overall conservation goal.

One option is to ignore the long-term threats and return to management prior to the BSSCAS. Under previous management, the city of Austin regulated the Barton Springs Pool cleaning techniques. This approach ignores degradation of water resulting from urbanization and potential catastrophic spills that threaten the salamander. In addition, it ignores key elements in the complex social and decision process underlying the salamander problem. With this option, nothing would be done to increase public participation or to increase scientific knowledge about the salamander. I project that the salamander will go extinct eventually under this approach.

A second approach is to create a plan similar to the BSSCAS and to apply it. My projection here is also for species extinction, because present and future threats are not adequately addressed.

A third approach is to write a new plan. To be successful, this new plan would have to address all the decision process and content problems listed above. Unlike the BSSCAS, the new plan would explain specifically how, when, to what participants, and through what arenas the planning actions as described below should be invoked and applied.

These measures would have authority and control, including sanctions. The amount of funding needed would be explained. A time line would establish when research goals would be achieved, and a list would specify which groups would do the work. The viable salamander population and what actually constitutes threats to this population still need to be fully clarified. If the hydrology of the aquifer were better understood, perhaps other conservation alternatives could be created. For instance, if most of the aquifer's water converges in one place before forming the four springs, perhaps a sediment and pollution filtration system could be put in place to reduce threats to the salamander and its habitat. In addition, provisions for citizen participation throughout the decision process would be made. For instance, citizens could organize themselves and restore the two cemented springs that are not used for human recreation. Finally, a clause for an independent appraisal team would be written into the plan (see Clark 1996).

Furthermore, salamander conservation might be achieved by: a plan concentrating on traditional regulation of threats; a plan using land credits; a habitat conservation plan; creation of a new habitat; and a breeding refugium and release program.

First is the regulation of threats to salamander survival. This plan would mimic the goals and strategies of the BSSCAS but make them specific and enforceable. It would focus attention on urbanization over the aquifer, the risk of catastrophic highway spills, and pool maintenance by regulating the agencies responsible. Economic sanctions such as fines and coercive threats in the form of jail sentences might be used.

Second is the use of land credits, distributed according to a "habitat transaction method." Todd Olson, an environmental planning specialist, was hired by a county north of Austin to prepare a proposal describing how a "habitat transaction method" could be used to prevent ESA listing (Haurwitz 1995a). The habitat transaction method is a free-market approach to conservation that makes it profitable for landowners to conserve land, design development projects, and manage their land in ways that achieve specified environmental objectives. The method calls for properties in a target area to be assigned conservation credits based on their biological value—the more crucial the property is to the preservation of a species, the more credits are issued. Credits may be freely bought and sold so that a landowner can conserve habitat and sell the resulting credits to landowners who desire to develop elsewhere. The plan has drawn support from the USFWS, but it has not received final approval by that agency. This approach may not be effective in Austin because the salamander lives in a very small range, only on public land, and is affected by all activities that could potentially influence the quality of the Edwards Aquifer, which is thirty-three square meters in size. Therefore, credits would have to be used for all the property owners whose land-use practices might affect the section of the aquifer that feeds Barton Springs, and even if this plan proved workable, many other activities that might adversely affect the aquifer's quality would not be addressed.

Third is use of a habitat conservation plan procedure. Under this option, the Barton Springs salamander would be included in the list of species protected by the existing Balcones Canyonlands Conservation Plan, a $160 million project that will preserve more than 12,140 hectares of land in and around Austin. The conservation plan is centered around two endangered bird species and six endangered cave insects. A key feature of the plan allows landowners to bulldoze or pave over lesser-quality habitat in return for fees that the city and county would use to buy high-quality habitat. The plan generally has been locally and nationally well received. Austin citizens have supported it, as evidenced by their enacting a $22 million bond package to buy preserve land. Developers like it because it gives them a definite answer about where development is permissible. Private property owners begrudgingly admit that it seems like the best option. However, a habitat conservation plan is limited by the salamander's dependence on the entire Edwards Aquifer, not just a few acres of good habitat. Because degradation or a spill on any part of the land may affect the species, it would be difficult to prioritize the land. Thus, if the salamander were to be managed like the other endangered species in the Balcones Canyonlands Conservation Plan, the entire thirty-three square meters would have to be bought by the government with fees collected from areas outside the aquifer's watershed.

Fourth is habitat creation, but this option has not yet been considered by the community. Both physical and social factors are involved in making a new habitat. The Barton Springs complex maintains a constant 68°F temperature throughout the year, and salamander kills have occurred when the temperature has risen as a result of pool maintenance (Haurwitz 1996b). An artificial or new natural habitat would also have to maintain these constant, cool temperatures throughout Austin's hot summers. However, little else is known about the species' requirements besides a correlation with certain vascular plants, so design of an artificial habitat or the selection of a substitute natural one may be quite challenging. On the other hand, there would probably be much public support for habitat creation efforts. The biggest hurdle in this regard may be objections to spending tax money on research to determine habitat requirements and create the habitat. State officials will not object if state agencies lead this effort, developers will not object if the new habitat is not located in areas they want for development, and private property advocates will not object if the new habitat is not located on private property. Environmentalists might be satisfied only if the salamander's current habitat is maintained in addition to the created habitat. So, despite the challenges of designing a new habitat for the species, I project that this can be done with some funding and with public support.

The fifth option for conserving the salamander population is captive breeding and release. Captive breeding increases population size and maintains genetic diversity. Plans for a breeding program were included in the BSSCAS, but no progress was made toward establishing one. One limiting factor is money, and although this, like the habitat creation option, may be somewhat unpopular because it would use public funds, it would not seem to be too unpopular with the agencies and groups in the power struggle. Like the creation of habitat, it is not guaranteed to work, but it is a politically viable option because it is unlikely to create much opposition among the development community and private property advocates.

Selection of Conservation Alternatives

Analyzing the salamander conservation problem by looking at the social and decision processes involved shows that two strategies must be taken to meet the stated goal. First, a way must be found to deal effectively with the influence of development and private property interests on the state government. Second, a clear and effective plan must be written and successfully applied. To accomplish these goals, I recommend a complex of strategies. To cope with the influence of developers and private property interests on government management of the salamander, an independent appraisal committee should be created. Semiannual meetings could appraise plans and their invocation and application. If problems were found, results could be reported and strategies to resolve them developed and applied. If no problems were found, this could be reported and publicized as well. Successes and failures could be tracked honestly. Community members who are informed, respected, but independent of the SOS, the state government, and the federal government could organize and operate this committee. I nominate Mrs. Lady Bird Johnson and Dr. Clark Hubbs. Additionally, I would recommend more media attention and public involvement in the form of letter writing to bring the issue into the public arena, especially at a national level. This kind of public action contributed to Secretary Babbitt's decision to list the species. Trying to limit campaign donations by private property and development interests may also help.

I recommend that the new recovery plan address the problems identified in my analysis of the BSSCAS. This new plan could reduce threats to the salamander's survival. I recommend starting with a traditional enforcement system of fines and legal punishment for developers who violate standards set by the USFWS concerning acceptable levels of pollution and development. Todd Olson's model of a habitat transaction method is not a viable option. Inclusion of the salamander into the Balcones Canyonlands Conservation Plan should be examined. The conservation plan has wider public support than traditional methods of regulation, but it has not proved successful as yet. In addition, more research into the habitat requirements of the salamander should be made so an artificial habitat may be created if feasible. If successful, this would remove the salamander from the water and controversy of the Edwards Aquifer. Efforts to create a captive breeding program should be continued; like the creation of artificial habitat, captive breeding is not as politically volatile as restrictive regulations on commercial development.

Conclusions

The goal is to restore and maintain a viable salamander population while focusing on public awareness of and involvement in recovery of the species. However, a "goal substitution" of power for salamander conservation by key participants, the influence of development and private property interests on the state government, and an absence of citizen power in the decision process are factors in the overall policy problem. In the decision process, a lack of knowledge about the salamander's biology, missing comprehensive and independent appraisal, a misdirected promotion function, and a weak prescription leading to ineffectual invocation and application have all contributed to the policy problem. Almost all these problems can be traced back to the "goal substitution" of pursuit of power for salamander conservation by major participants. This substitution by the state and federal governments and the SOS has made the decision process ineffective, resulted in a weak management plan, and limited progress toward conserving the salamander. Owing to these factors, the policy problem can be restated: short- and long-term threats to the survival of the Barton Springs salamander have not been reduced because of failures in the social and decision processes.

I recommend two remedial strategies. First is to reduce the effect of "goal substitution" on the management of the salamander. This could be aided by creating an independent appraisal committee, by increasing media attention, and by encouraging more public involvement. Second is to write a new recovery plan which seeks to avoid all the problems in the former plan. The new plan should use traditional regulation of the watershed and Barton Springs Pool maintenance but should research the possibility of including the salamander in the existing Balcones Canyonlands Conservation Plan. In addition, a captive breeding and release effort should be considered, and the salamander's habitat requirements should be researched so that a new habitat might be created if feasible and desirable.

ACKNOWLEDGMENTS

Many thanks to Andy Willard, Tim Clark, Christina Cromley, the members of FES 891b, and Brian Roberts for their help with this paper and their grasp of Harold Lasswell's analytic framework.

LITERATURE CITED

Arnold, M. 1997. The SOS Movement. http://www.sosldf.org.

Backhouse, G. N., T. W. Clark, and R. L. Wallace. 1996. "Reviewing Recovery Programs for Endangered Species: Some Considerations and Recommendations," in S. Stephens and S. Maxwell, eds., *Back from the Brink: Refining the Threatened Species Recovery Process*, pp. 170–179. New South Wales: Surrey, Beatty, and Sons.

Bowles, D. E. 1995. "A Review of the Status of Current Critical Biological and Ecological Information on the Eurycea Salamanders Located in Travis County, Texas." Texas Parks and Wildlife Department, Austin.

Brunner, R. D. 1987. Conceptual tools for policy analysis. Center for Public Policy Research, University of Colorado, Boulder. Unpublished pamphlet.

Bryce, R. 1996. The TPWD Brain Drain. http://www.auschron.com.

Chippendale, P. T., A. H. Price, and D. M. Hillis. 1993. "A New Species of Perennibranchiate Salamander from Austin, Texas," *Herpetologica* 49:248–259.

Clark, T. W. 1996. "Appraising Threatened Species Recovery Efforts: Practical Recommendations," in S. Stephens and S. Maxwell, eds., *Back from the Brink: Refining the Threatened Species Recovery Process*, pp. 1–22. New South Wales: Surrey, Beatty, and Sons.

———. 1997. *Averting Extinction: Reconstructing Endangered Species Recovery*. New Haven: Yale University Press.

Clark, T. W. and R. D. Brunner. 1996. "Making Partnerships Work in Endangered Species Conservation: An Introduction to the Decision Process," *Endangered Species Update* 13(9):1–5.

Clark, T. W., R. P. Reading, and A. L. Clarke, eds. 1994. *Endangered Species Recovery: Finding the Lessons, Improving the Process*. Washington, DC: Island Press.

Haurwitz, R. K. M. 1995a. "Biological Barometer of the Environment," *Austin-American Statesman* June 6:D25.

————. 1995b. "Salamander Is Awash in Crucial Fight," *Austin-American States-man* August 24:D1.

————. 1996a. "Salamander Surprise: 188 At-Risk Amphibians Found Near Barton Pool," *Austin-American Statesman* February 2: D1.

————. 1996b. "Urban Habitat," *Austin-American Statesman* April 21:D20.

————. 1996c. "The Barton Springs Salamander: Presidential Election Issue?" *Austin-American Statesman* June 6:A25.

————. 1996d. "U.S. Won't List Barton Springs Salamander as Endangered," *Austin-American Statesman* August 8:A29.

————. 1996e. "State's Main Environmental Agency Chastised," *Austin-American Statesman* December 11:D12

————. 1996f. "Babbitt Rejected Advice on Salamander," *Austin-American States-man* December 12:D5.

————. 1997. "Salamanders Are Found in a New Spot," *Austin-American Statesman* April 4:D1.

Lasswell, H. D. 1970. "The Emerging Conception of the Policy Sciences," *Policy Sciences* 1:3–14.

————. 1971. *Pre-View of the Policy Sciences*. New York: Elsevier.

Lasswell, H. D., and A. Kaplan. 1950. *Power and Society: A Framework for Political Inquiry*. New Haven: Yale University Press.

Lasswell, H. D., and M. McDougal. 1992. *Jurisprudence for a Free Society: Studies in Law, Science, and Policy*. 2 vols. New Haven: New Haven Press.

Marytyn-Baker, J., R. Fieseler, and B. Smith, eds. 1992. "Hill Country Oasis: Barton Springs-Barton Creek-Edwards Aquifer," Austin Parks and Recreation Department, Edwards Aquifer Conservation District, and Save Barton Creek Association, Austin, Texas.

Save Our Springs Alliance. 1991. Securing a Safe Future for Barton Springs. http://www.sosldf.org.

————. 1996. Salamander Trapped in Collision Between Science and Politics. January 1. http://www.sosldf.org.

————. 1997. Federal Court Finds Improper Political Influence on Salamander Decision: Orders New Decision within 30 Days. March 26. http://www.sosldf.org.

Texas Parks and Wildlife Department, Texas Natural Resource Conservation Commission, Texas Department of Transportation, and U.S. Fish and Wildlife Service. 1995. Barton Springs Salamander Conservation Agreement and Strategy.

U.S. Fish and Wildlife Service. 1992. "Endangered and Threatened Wildlife and Plants: Notice of Finding on Petition to List Barton Springs Salamander," *Federal Register* 57(239):58779–58781.

————. 1994. "Endangered and Threatened Wildlife and Plants: Proposal to List the Barton Springs Salamander as Endangered," *Federal Register* 59(33):7968–7978.

————. 1995. "Known Distribution of the Barton Springs Salamander," U.S. Fish and Wildlife Service, Austin, Texas.

————. 1996. "Endangered and Threatened Wildlife and Plants: Withdrawal of Proposed Rule to List the Barton Springs Salamander as Endangered," *Federal Register* 50(17):1018-AC22.

8 The Killing of Grizzly Bear 209

Identifying Norms for Grizzly Bear Management

Christina M. Cromley

The U.S. Fish and Wildlife Service listed grizzly bears (*Ursus arctos horribilis*) as "threatened" in the continental United States in 1975 under the U.S. Endangered Species Act (16 U.S.C. §1533; 50 C.F.R. §17.11 at 74). According to the Grizzly Bear Recovery Plan, the goal of grizzly bear recovery has been to establish "viable populations in the areas where the bear existed or [was] thought to exist in 1975 and which are capable of supporting viable populations." However, people now occupy or use many of the areas suitable for grizzly recovery. People and bears on the same land leads to conflict over resources on that land. Although often discussed as "nuisance bear management," which implies a problem with managing bears, the policy problem in dealing with conflicts between humans and bears is really a problem of managing people's expectations about how resources shared by humans and bears are allocated and how conflicts over those resources are resolved.

One of the last remaining populations of grizzly bears lives in the Greater Yellowstone Ecosystem (GYE), approximately 7.6 million hectares of public and private land comprising two national parks, seven national forests, three national wildlife refuges, and three national wilderness areas. A series of incidents began in 1992 between grizzly bears and ranchers—and subsequently government agencies and the

Research for this project was conducted in 1997 and 1998.

public—in the community of Jackson Hole, Wyoming, which lies within the GYE. These incidents, which continue today, make it clear that many citizens, government officials, ranchers, and conservationists differ in their expectations about how people can and should handle themselves and bears to prevent, reduce, or deal with conflicts between bears and humans.

One of the most intense reactions to these management incidents involved the killing of Grizzly Bear 209, a nine-year-old male. Officials from the National Park Service, Wyoming Game and Fish (WYGF), and the U.S. Fish and Wildlife Service had agreed that 209 was of "nuisance status" (Bruscino, interview 1997; Cain, interview 1997; Moody, interview 1997). WYGF personnel captured 209 in Grand Teton National Park (GTNP) on August 4, 1996. They took Bear 209 to their office in Lander and lethally injected him (Moody, interview 1997).

The public responded to this decision immediately and intensely in the days and weeks following the incident, and the death of Bear 209 was still being referred to a year later. This reaction, in addition to numerous court cases over grizzly bear recovery, indicates that expectations about whose activities—grizzlies' or humans'—should take precedence in which cases and about who should resolve such conflicts remain unclear.

Standpoint and Methods

In this paper, I examine events leading to the decision to kill Bear 209 and the reaction that followed to understand better how participants —citizens unaffiliated with other organized interests, government officials, conservationists, ranchers, and researchers—think the decision process in the grizzly bear recovery program should be run. Whose views do they think it should include and under what circumstances, and how do they intend to include and decide among conflicting values or find areas of common ground to secure common interests? Based on this understanding, I recommend ways to improve the process of decision making so that management actions resulting from grizzly bear–human conflict will meet the expectations of those concerned with grizzly bear recovery and will minimize deprivations among participants.

The idea of investigating a particular incident, such as the death of Bear 209, and the reaction to it in order to clarify normative expectations comes from international law and policy. W. Michael Reisman and Andrew R. Willard, scholars at Yale Law School, have devised a method called "incident analysis" (Reisman and Willard 1988). Reisman argues that understanding normative expectations is important because "shared notions of what is right influence perception, reaction, and capacity for mobilization. These inferences about what other actors think is acceptable behavior . . . are almost entirely derived from the responses of key actors to a critical event" (Reisman and Willard 1988:5). In analyzing reactions to the killing of Bear 209, I have found that in their reactions to his death, participants have articulated their notions of what is right—of what they expect in a normative sense from grizzly bear recovery policies. Some key questions are: To what extent and how do different actors pay attention to the responses of other actors, and how does ignorance or misunderstanding of others' expectations affect grizzly bear recovery?

To understand the complex issues and interactions that led to the decision to kill Bear 209 and the subsequent reaction, I acted as a collector, spectator, interviewer, and participant observer. I began to understand the perspectives of residents toward grizzly bears by conducting an attitudinal survey in the summers of 1996 and 1997. I began studying the Bear 209 case by surveying books and articles on grizzly bear management written by academics and researchers. These provided me with analysis and a historical context. I then read government documents, such as the environmental assessments of grazing allotments and reports on grizzly bear–human conflicts, confrontations, and management actions, to understand better the written prescriptions. Also helpful in understanding reactions to the killing were newspaper and magazine articles and editorials about Bear 209 and grizzly bears in general, and records of personal communication among citizens, conservationists, and government officials. I then spent June to August 1997 in Wyoming interviewing ranchers, government officials, unaffiliated citizens, conservationists, and researchers. Finally, I attended and participated in several meetings regarding natural resource issues in the GYE.

Throughout the research and writing process, I was driven by differing motivations. As a student pursuing a doctorate in natural resources policy, I am interested in the role of myths, symbols, and institutions in shaping the perspectives of participants in policy processes and how to use the policy sciences analytic framework to improve natural resources policy. I am also dedicated to finding ways of improving public participation in policy processes. Grizzly bear management has provided an interesting case study in attempting to apply the policy sciences framework and to understand the ways in which myths shape our behavior, how we identify ourselves, the symbols we encounter on a daily basis, the institutions within which we operate, and our role in

making policy decisions or in influencing those decisions. In my recommendations, I discuss my ideas on creating decision-making processes, which include a range of participants and their perspectives in
gathering and disseminating information; finding creative and effective ways of working through recurring conflicts; prescribing, invoking, and applying these alternatives; assessing existing policies; and
finding ways to change policies that no longer work.

In these recommendations, it will become clear that I am biased in
the sense that I believe more inclusive processes will lead to better decisions, meaning that they will address the concerns of more participants and reflect democratic ideals. I also recognize that I have two
salient predispositions that have influenced my interpretations of my
research results: (1) I was drawn into the field of natural resources because of my concern for the environment (therefore, for keeping species like grizzly bears alive), and (2) I am interested in people; I was
drawn into policy because of my concern for including people in studies of the environment not as obstacles to overcome, but as part of
working, living, functioning, viable ecosystems. I hope stating explicitly my motivations and biases will allow readers to draw their own
conclusions about my analysis and data.

The Killing of Bear 209

On August 4, 1996, the Wyoming Department of Game and Fish captured Bear 209, a 9-year-old, 550-pound male grizzly bear, on the Elk
Ranch East (ERE) in GTNP. He was trapped in a leg snare in the park
and lethally injected. The capture and subsequent destruction of the
bear came in response to the bear depredating on eleven calves over a
three-week period in July 1996 (Gunther and Bruscino 1997). The bear's
long history of cattle depredation and earlier attempts at translocation
influenced the decision to destroy the bear.

Bear 209 had preyed on cattle previously and was captured by the
WYGF on July 14, 1993, on the Blackrock/Spread Creek (BRSC) grazing
allotment in the Bridger-Teton National Forest and on September 8,
1995, on the ERE allotment in GTNP in an area bordering the western
side of the BRSC allotment (Table 8.1). Both times, Bear 209 was moved
to the northern end of Yellowstone National Park, and both times he returned to the BRSC grazing allotment. He had also been captured on
the BRSC grazing allotment earlier, in 1994, but was released on site to
prevent jeopardizing a study of cattle-grizzly interactions being conducted on the BRSC and ERE allotments. Bear 209 was responsible for
a total of sixteen cattle depredations.

Table 8.1. History of Management Actions Involving Grizzly Bear 209

	July 14, 1993	Aug. 5, 1993	June 18, 1994	Sept. 8, 1995	Aug. 3, 1996
Activity	1st management capture	Lost radio collar	Research capture	2nd management capture	3rd management capture
Location	Baldy Mt., BTNF, Situation 2		Spread Creek Allotment Situation 2	ERE Allotment, GTNP Situation 2	ERE Allotment, GTNP Situation 2
Management action	Move to Yellowstone National Park		Released on site	Move to Yellowstone National Park	Euthanized

Note: This information is maintained by and was obtained from the Interagency Grizzly Bear Committee. Management actions involving grizzly bears are guided by a management situation (MS) zoning system. In MS 2 habitat, grizzly bear and human activities hold equal priority, and it is legal to remove bears from these zones. A more thorough consideration of this system is given in the text. BTNF = Bridger-Teton National Forest; ERE = Elk Ranch East; GTNP = Grand Teton National Park.

Widespread public reaction followed the death of Bear 209. Newspapers printed numerous articles, editorials, and letters to the editor;[1] correspondence and personal communication occurred between the superintendent of GTNP, conservation groups, and unaffiliated citizens (e.g., Lichtman 1997b; Neckels 1996; Taylor, interview 1997); letters were written to Secretary of the Interior Bruce Babbitt and to the head of the National Park System (Stratton 1996a); in 1996 a citizen organized a petition protesting the killing and collected 831 signatures from residents and tourists; conservation groups mentioned Bear 209 in their newsletters; and an epitaph for Bear 209 appeared on the World Wide Web (Landreth 1996; Stevens 1996). The office of public relations at GTNP responded to public outcry over the bear's death with a management summary of the death justifying the decision (National Park Service 1996). The death of Bear 209 is even being invoked in current policy debates over whether to allow grazing to continue in GTNP on leases that have expired under the original authorizing legislation but which have been extended by the park superintendent and a congressional bill (S. 308). Bear 209 is also being referred to in editorials about conservation biology and the need to improve conservation policy (Camenzind 1997a; Camenzind and Taylor 1996; Taylor 1997; Turner 1997).

Social Context

The widespread public reaction to the death of Bear 209 occurred in the context of an ongoing and heated debate about grizzly bear recovery. Although members of the WYGF were responsible for trapping and killing Bear 209, a number of other participants—operating under different perspectives, in different situations, with different base values, using various strategies, seeking often opposing outcomes—were involved in the decisions and events leading up to and following the death. These participants include the bears themselves, researchers, government officials, ranchers, conservationists, and unaffiliated citizens. Each individual differs in his or her perspectives; yet, the discussion that follows concerning the perspectives of these participants necessitates some generalization of participants in each category.

It is difficult to determine the perspectives and base values of the bears, other than in biological and well-being terms. Bears require large tracts of continuous habitat and large amounts of food to survive. Their situation is bleak in terms of these requirements—they have been reduced to 2% of their former range and depend on a decreasingly available natural food supply. Although bears are omnivorous, which means they eat foods as diverse as moths, ants, whitebark pine, biscuit root, and carcasses, bad natural food years tend to lead to increased mortality because as bears search for alternate sources of food, they often come into conflict with humans (Craighead et al. 1995). They also occasionally kill calves, be they elk or domestic cattle, and seem unable to distinguish between domesticated and wild calves (Bruscino, interview 1997; Moody, interview 1997). Grizzlies seem to avoid people—use of prime habitat decreases with increasing human recreational use, for example (Gunther 1996). This may be advantageous, because increased contact with humans leads to higher mortality rates. Bear biologist David Mattson says that grizzly mortality in the GYE—which is almost solely human-caused—is based on two factors: (1) frequency of contact with humans and (2) lethality of contact for the grizzly (Mattson, personal communication 1997; Mattson et al. 1995). Grizzlies also have slow reproductive rates, so the death of one or two bears could potentially affect the genetic and population viability of bears, although the impact of each death on the population given its current size is currently being debated.

The debate over the viability of the existing population of bears takes us to the next category of participants—researchers. This category overlaps with government agencies because much of the research used to inform and justify decisions comes from the Interagency Grizzly Bear

Study Team (IGBST), an effort to coordinate information gathering and processing among agencies involved in grizzly bear management. Although researchers, whether working for the government or not, typically have similar perspectives and base values, government researchers operate in different situations and experience different outcomes from researchers not affiliated with the government. For example, scientists on or working with the IGBST have more authority and control in the decision-making process than do those outside the IGBST.

Government researchers continue to conduct research using positivistic methods and to inform, make, or justify decisions in accordance with the data gathered and their interpretation of prescriptions. IGBST data focuses on biology, and ecology, and the effects of human activity on bear behavior and populations, with few if any studies on the influence of human values on decision making. The public has criticized many of these decisions, and senior government officials have often ignored the advice of lower-level officials. Not surprisingly, some researchers and lower-level officials have experienced feelings of disrespect (Bruscino, interview 1997; Moody, interview 1997).

The practices of researchers—governmental or not—seem to be shaped by a faith that research is value-neutral and that scientists can form nonsubjective views based on facts. Many of the researchers consider themselves objective and demand access to information, the ability to study bears in the parks, and the authority to influence decisions with the data they gather and analyze. They use enlightenment and skill in their attempt to attain a certain amount of power and respect. This strategy has often failed, as indicated by those outside the IGBST who express frustration over not having their data considered in management decisions. For example, according to Craig Pease, a University of Texas zoologist who presented data challenging population estimates of the IGBST, "'one side of the argument . . . has been cut out of affecting management policy'" (Neal 1995). John Craighead, a grizzly bear researcher whose work with Yellowstone National Park ended in conflict, reports that "despite being allowed some latitude in its research, the IGBST was constrained by park administration from full interagency cooperation and from using the data from, or even citing, the many publications documenting our study" (Craighead et al. 1995).

Government officials include not only those conducting research but also those making decisions. Government agencies involved include Grand Teton National Park (National Park Service), the WYGF, the U.S. Fish and Wildlife Service, and the Bridger-Teton National Forest (U.S. Forest Service). All agencies hold a certain amount of authority, but the WYGF apparently maintains control over situations involving bear management actions in GTNP and the surrounding area. The focus of attention, zone of activity, and level on which government officials in each agency operate seem to influence their base values, situations, strategies, and outcomes. Officials who have continual contact with certain members of the public seem to be driven by affection as much as enlightenment and skill. Continual contact with those affected by bear activity seems to lead to a sense of loyalty by officials toward those people with whom they work and from whom they hear complaints on a regular basis. Although these officials recognize, through such contact, the emotional and political side of bear management, they think of themselves as objective experts.

I was unable to speak with any of the highest officials in any government agency—either my calls went unreturned or I was rerouted to public relations specialists—so my analysis comes secondhand from those who have dealt with such officials. Many top officials seem to be most interested in power and respect and tend to focus their attention and energy on those who can enhance their power. Citizens have complained that top officials express annoyance when the public makes demands that oppose their decisions. Many of their strategies seem coercive, and although they have maintained power and gained respect from other powerful community members, they have lost the respect and loyalty of many conservationists, unaffiliated citizens, researchers, and even lower-level members in their own organizations.

I found a range of perspectives in the ranching community. Overall, they have expressed a sense of insecurity about the future of ranching. In magazines such as *Range* (e.g., McInnis 1997) and in my discussions with ranchers, many spoke of the "good old days," the services ranchers provide for the country, the role of ranching in maintaining open spaces, and the public's lack of appreciation for ranching. The identity as a rancher who is loyal to other ranchers or as a cowboy loyal to the cowboy culture seems strong even among many who no longer ranch as their main occupation or source of income. Ranchers are concerned about what the public outside their community thinks about allowing grazing in the park and on public land, and they have a great deal of respect for the government officials who make decisions that affect their ranches. For example, one member of the Elk Ranch East claimed he would have preferred that Bear 209 remain alive because he knew a public outcry would result, but that he respected WYGF officials' need to enforce their own policies as they see fit (Mead, interview 1997).

Independent citizens involved in responding to the killing of Bear

209 and conservation interest groups seem similar in their perspectives. Members in these groups include people as diverse as photographers, lawyers, outfitters, and advocates. Many seem driven by rectitude and feel compelled to act, to do something to change what they dislike about the policy process they observe and in which they participate. For example, the citizen who organized the petition claimed she wants to maintain good relations with both GTNP and Yellowstone National Park (YNP), but she feels strongly enough about preserving bears and wolves to question activities such as killing grizzly bears for eating cattle in GTNP (Stratton, interview 1997). These participants maintain expectations about the goals of national parks and the actions of officials in those parks. For example, a member of the Jackson Hole Alliance for Responsible Planning remarked that the alliance "mentioned this [in 1995]—they're moving bears in a national park in response to grazing. They continue to extend this grazing and make decisions based on that. This begs for a review" (Thuermer 1996a). The situations in which these groups operate include channels of personal communication, letters to the editor, and the courts. Thus, they use persuasive, assembling, and processing strategies. They often win court cases but may not feel any sense of power or respect in the long run.

Overall, participants expect that people and wildlife can get along. However, the formulas for doing so are subject to debate, with some people favoring more control of human activity and some favoring more control of wildlife activity. The same phenomena—the bears themselves, the incidents in which they become involved, the decisions made, the numbers generated by scientists—are used to support the differing perspectives. The debate, then, is shaped significantly by myths.[2] Expectations concerning how bears should be managed —the prescriptions and their invocation and application—emerge from the myths people hold, how they identify themselves in the context of those myths, and where they place their loyalties. When myths clash, the outcome is that some people feel indulged in seeking certain values, but most people feel deprived in some way. Trust seems eroded in the decision-making process. Those seeking respect feel disrespected; those seeking power feel powerless; those seeking affection feel rejected or torn between competing loyalties; those seeking rectitude feel slighted; those operating under enlightenment and skill feel unable to fulfill the quest for more knowledge or perceive their pursuit as futile. Although the clash of myths often manifests itself in debates over data and specific decisions, at a deeper level it is about these human values.

The Conditioning Decision Processes

Countless decision processes intersected in the decision to kill Bear 209. It is beyond the scope of this paper to discuss (and beyond my ability to identify) all the decisions made in grizzly bear recovery that led to the death and the reaction to it. However, those decisions about grizzly bear recovery that seem most influential in the Bear 209 case are discussed below. Such decisions include determinations to move other bears because of their depredation of cattle and determinations about grizzly bear research, grazing in GTNP, delisting the bear, and the allocation of authority and control in decision making about grizzly bears in general.

Trends and Conditions

Grizzly bears have lived in North America for at least 50,000 years (Servheen 1993). The downward trend of grizzly bear numbers and the increasing predominance of human activity over bear activity began almost two centuries ago. Grizzly populations rapidly decreased as European settlers moved westward and killed large numbers of bears (Botkin 1995). By the 1920s grizzlies inhabited only 5% of their former range, and between 1800 and 1975 the population dropped from esti-mates of over 50,000 to fewer than 1,000 (Mattson and Craighead 1994; Servheen 1993).

Bears first received protection in 1886 when the superintendent of Yellowstone National Park prohibited the killing of animals in the park, although hunting regulations for bears outside of YNP did not begin until the 1930s (Craighead et al. 1995; Primm 1993; Schullery 1986). When a new administration took over Yellowstone in 1916, it instituted a pol-icy of predator control. Bears, however, were not targeted because they had become a tourist attraction; tourists came to observe bears congre-gating at dumps to feed on human refuse (Craighead et al. 1995).

Despite a certain level of protection from hunting and predator control actions, statements as early as 1894 report shipping bears to zoos. Superintendent James B. Erwin in 1899 felt that "it might be necessary to 'kill some [bears] to prevent [property] destruction,'" and by 1907 bears were being killed in control measures (Craighead et al. 1995:32–36). Since the establishment of this control policy in 1907 in Yellowstone, there have been fluctuations in the numbers of bears killed in control actions. Overall, from 1907 to 1965 at least 159 griz-zlies were killed or shipped to zoos. Craighead and colleagues point out, however, the difficulty of interpreting such statistics; throughout the early history of management control actions, management control and reports of that control were sporadic: "No guidelines were established defining what constituted a personal injury or damage to property. District Rangers made independent judgments about what should be recorded. . . . [The number of incidents recorded] reflect the judgment of the times" (Craighead et al. 1995:38).

Ecological and biological research that dealt with human–grizzly bear conflicts was also fairly nonexistent. The first field research project on the bears of Yellowstone with management implications was conducted by Olaus Murie, who determined bears could survive without dumps, on natural foods (Schullery 1986).

The issue of dumps was to arise again with the researchers Dr. John Craighead and Dr. Frank Craighead, Jr., in 1970 (Mattson and Craighead 1994). The Craigheads began the first long-term study of bears in Yellowstone in 1959 (Weaver 1996). Throughout the Craighead study, bears continued to congregate at open-pit dumps containing human food waste. By 1967 the park instituted a "natural regulation" policy that called for an immediate closure of park dumps. The Craigheads claim they were prevented from continuing their research in YNP after they openly opposed the park's plan to close the dumps immediately, recommending instead that the park slowly phase out the dumps (Mattson et al. 1995). Their research with the park terminated in 1970.

Following the closure of the dumps, grizzlies began to wander into human developments in search of human garbage, the food to which they had become accustomed at the dumps. They were killed in record numbers during hunting season and in management actions. Thus, scientists, the public, and managers became concerned about the future of the Yellowstone grizzly bear population. However, three years lapsed before, in 1973, the Interagency Grizzly Bear Study Team was created to study the "new" bear population, which had no access to dumps as a food source (Schullery 1986).

The IGBST originally consisted of three members, one from the National Park Service, one from the U.S. Forest Service, and one from the U.S. Fish and Wildlife Service (Mattson and Craighead 1994; Schullery 1986). There are mixed perspectives about the IGBST, its research, and the information generated by its members that is meant to inform grizzly bear recovery policy.[3] According to YNP historian Paul Schullery, "the entire conservation community was waiting to see what the team would find, and was probably also waiting for the slightest opportunity to criticize those findings" (Schullery 1986:143). The Committee on the Yellowstone Grizzlies, a National Academy of Sciences (NAS) committee, convened to review research needs in 1974 and recommended that most research on grizzlies be conducted by nonagency scientists. Although the IGBST informally includes research conducted by the Wyoming Game and Fish Department and Idaho Cooperative Park Studies unit, bear researchers David Mattson and John Craighead claim the IGBST "does not reflect the original recommendations of the NAS committee that reviewed research needs in 1974 . . . that most research be conducted by non-agency scientists" (Mattson and Craighead 1994:105). They also point out that despite any original professed intentions to engage in interagency cooperation, the only two formal members of the IGBST are two National Park Service personnel. The result, Mattson and Craighead claim, is effective control over research by a single agency (the National Park Service) and in effect, less dependable, reliable, and comprehensive research.

Government agencies became the main actors on research for grizzly bears through the IGBST about the time grizzly bears were listed as "threatened" under the U.S. Endangered Species Act (ESA) in 1975. The ESA preempts any state law inconsistent with its mandates, places authority for grizzly bear management at the federal level, and prohibits federal agencies from executing or supporting projects that could harm a threatened species or its habitat (Miller 1992). A "Grizzly Bear Recovery Plan," required under the ESA (16 U.S.C. §1533 (f)), serves as the key document in recovery and was authored solely by the U.S. Fish and Wildlife Service (USFWS recovery coordinator) (Mattson and Craighead 1994).

The USFWS has authority over the recovery programs of species listed under the ESA, yet the first governing body with control over grizzly bear recovery was the Interagency Steering Committee, established in 1975. The committee was composed of researchers and mid-level managers from agencies with jurisdiction over land considered suitable for grizzly bear habitat and recovery. In addition to fulfilling other responsibilities, it "provided general review and direction for the IGBST research program" (Mattson and Craighead 1994:105). By 1982, the population of bears seemed to be declining and the committee terminated itself, recommending the formation of a committee of higher-level managers with more decision-making power (Mattson and Craighead 1994). Subsequently, an interagency memorandum of agreement led to the formation of the Interagency Grizzly Bear Committee (IGBC) (Gunther and Bruscino 1995).[4]

The mission of the IGBC is to "guide implementation of the Grizzly Bear Recovery Plan by promoting interagency coordination of policy, management, planning, information and education, and research" (IGBC 1996:2). The IGBC is composed of officials from the USFWS; the U.S. National Forest Service (USFS); the National Park Service (NPS); the Bureau of Land Management (BLM); state fish and game agencies of Montana, Wyoming, Idaho, and Washington; and management authorities from British Columbia and Alberta. Using the "Grizzly Bear Recovery Plan," written by the grizzly bear recovery coordinator for the USFWS, the IGBC created guidelines that serve as the main prescription in decisions concerning "nuisance bears" (Moody, interview 1997). A nuisance bear is defined as "any bear involved in a bear-human conflict situation," and episodes involving nuisance bears are defined as "incidents in which bears injured people, damaged property, killed livestock, damaged beehives, obtained anthropogenic foods, or obtained garden and orchard fruits and vegetables" (Gunther and Bruscino 1997:3, 5).

It is written into the revised 1993 version of the "Grizzly Bear Recovery Plan" that agencies should follow the "Interagency Grizzly Bear Guidelines" and its Management Situation (MS) zoning system in managing nuisance bears, and that "since the inception of the Guidelines, all agencies have worked to implement the policies stated in the Guidelines within and surrounding grizzly bear recovery zones" (Servheen 1993:10). The revised recovery plan states later that "all areas within the recovery zone will be managed as either MS I, II, or III under the Interagency Grizzly Bear Guidelines" (Servheen 1993:17). The recovery plan also states that "bears residing within the recovery zone are crucial to recovery goals and hence to delisting" (Servheen 1993:18), and that "the Guidelines detail protocol for nuisance bear management and also detail grizzly bear habitat management policies" (Servheen 1993:10).

The guidelines give specifications for the agencies which manage land containing grizzly bear habitat, including mandates for national parks that state: "The Park Service will identify, within Park boundaries, grizzly habitat requirement. . . . Active management programs, where necessary, will be carried out to perpetuate the national distribution and abundance of grizzlies and the ecosystems on which they depend, in accordance with the Fish and Wildlife Service" (Mealy 1986). As these mandates for parks indicate, the guidelines require each agency to classify habitat under its jurisdiction according to management situation zones to guide decisions when human and grizzly activities conflict (Mealy 1986).

MS 1 zones contain habitat considered critical for the recovery and long-term survival of grizzly bears, so resolution of grizzly bear–human conflicts in MS 1 zones requires alteration of human activity before bear activity. MS 2 zones contain areas "unnecessary for survival and recovery of the species, or the need has not yet been determined but habitat resources may be necessary" (IGBC 1986:3). In this ambiguous zone, bear and human activities are given equal priority, and if management decisions must be made in favor of bears, the area should be reclassified. In MS 3, grizzlies may inhabit the area but their presence is infrequent because of the presence of campgrounds, resorts, or other human use facilities. Grizzly bear presence is discouraged in these areas, and any bears found in these areas are controlled. MS 4 zones contain suitable habitat, but grizzlies do not currently live in such areas and grizzly bear–human conflict is not a management consideration. In MS 5, grizzlies are not present or are present only rarely and any bear in the area will be controlled (Mealy 1986).

Although both guidelines for managing bear activity in different MS zones and consultations with the USFWS under ESA Section 7 in theory address trans-jurisdictional boundaries, a congressional committee that appraised natural resources policies in the GYE recommended in 1986 that managers "scrap the Management Situation concept" because "the Situation Management concept currently used by the agencies is not a useful management tool for preventing deaths of grizzly bears" and does not reflect grizzly bear density or habitat quality (Congressional Research Service 1986:15, 144). Law professor Robert B. Keiter makes it clear that "despite clear language in the ESA authorizing the FWS to designate critical habitat for the grizzly bear (16 U.S.C. §1533(a), (b)), the government has refused to extend this additional level of protection to the bear, choosing instead to rely upon an administratively-constructed habitat zoning scheme that does not carry nearly as much legal weight" (Keiter 1991:251). The ESA requires that recovery zones—areas in which members of an endangered population are protected—be declared for endangered species. Keiter believes that MS zoning offers less protection than the ESA prescribes, pointing out that management zones were created without complete data on habitat, do not correspond to grizzly bear use of habitat, and "reflect a series of administrative compromises" (Keiter 1991:251).

Another problem with the MS zoning system is that, in practice, habitat quality and grizzly bear density often have not been given the prominence one would expect from reading the recovery plan or the guidelines in determinations of which parcels of land are designated as which MS zones. An area may be designated as MS 2, 3, 4, or 5 because it is near a road or a recreation area, but it may contain prime grizzly food or habitat conditions. For example, the tourist area known as Fishing Bridge in Yellowstone National Park is designated MS 2 because it is developed with stores, campgrounds, and other services.

However, it is also prime grizzly habitat. Of all injuries related to interaction between grizzlies and humans between 1968 and 1983, 62.5% occurred at Fishing Bridge, resulting in the death of nine female grizzlies (Schullery 1986). Despite park plans in 1974 to close this area because of grizzly habitat quality, commercial interests have taken precedence and Fishing Bridge remains in operation (Primm 1993).

The congressional committee that appraised natural resources policies in the GYE pointed out the discrepancies in maps of bear densities and the zoning of land under the MS zoning system. For example, much of the area in the Gallatin National Forest with the highest density of bears is zoned MS 2, which does not fit the criterion set by the IGBC for MS 2 that "grizzly presence is possible but infrequent" (Congressional Research Service 1986:54–55). In addition, many areas identified as "mortality black holes," in which high numbers of grizzly mortalities occur, lie in MS 1 and 2 habitats (Congressional Research Service 1986:55–56). Therefore, controversy often results over MS zoning designation because an agency may classify areas more on the basis of current human uses than on the basis of habitat quality. This is one more example of the pervasive perspective that bears should be managed, not people.

An additional jurisdictional factor that shaped the reaction to the killing of Bear 209 is the location of the depredation and capture within a national park. The National Park Service Organic Act of 1916 states that parks must "provide for the enjoyment of [the scenery and the natural and historic objects and the wildlife] in such a manner and by such means as will leave them unimpaired for the enjoyment of future generations" (Shanks 1984). This is usually interpreted to mean that extractive uses of park resources—including such activities as hunting, cattle grazing, and mining—are prohibited. However, when Grand Teton National Park expanded in 1950, park officials allowed cattle ranchers with existing leases—on land to be acquired by the park—to continue grazing their cattle in the park until the death of the leaseholder's children (PL 81-787). As I discuss below, the Bear 209 case clarified the differing expectations participants have regarding how decisions should be made when grazing mandates conflict with mandates to leave the park's resources unimpaired and mandates to protect endangered species.

Another complicating factor is that under the original park expansion legislation (PL 81-787), the two ranchers on the ERE, from where Bear 209 was removed, are grazing cattle on expired leases. Park superintendent Jack Neckels extended these leases, justifying the extension by claiming that ranching helps maintain open space in

Jackson Hole (Thuermer 1996a). A bill (S. 308) introduced to Congress by Wyoming senator Craig Thomas to study the relationship between ranches and open space in the West, approved by Congress in 1997, extended the leases for another three and a half years. However, I discuss below the conflicting claims regarding the legality of these extensions.

Precipitating Events that Influenced the Decision to Kill
Bear 209 and the Reaction to the Death

Delisting

Just nine months before WYGF decided to move Bear 209 from Grand Teton National Park in 1995, the IGBC voted in December 1994 to support a motion brought by the Wyoming Game and Fish director to delist the grizzly bear in the Greater Yellowstone Ecosystem. This attempt to delist the grizzly by the WYGF was blocked subsequently, in part by the outcome of a court case over the recovery plan in 1995. The court case had been initiated by nearly forty conservation organizations in 1993 because they felt the plan failed to measure adequately habitat degradation, threats from disease and livestock, or population size (Carlton 1995; Collins 1995; Shelton 1995a; Thuermer 1995b). In addition, a group of National Academy of Sciences researchers stated that the "Grizzly Bear Recovery Plan" was "deficient because it is not based on scientifically credible data or analysis" (Shaffer 1994). The plaintiffs won the court case, and the judge ordered the U.S. Fish and Wildlife Service to revise the recovery plan. This outcome had the effect of blunting the efforts to delist the grizzly.

The court case is over and conservation groups won that battle, yet debate about bear population numbers still rages. Criteria for delisting include reaching a minimum number of female grizzlies with cubs and not exceeding a maximum number of allowable deaths, a percentage based on current grizzly bear population estimates (consequently, the larger the population, the more allowable grizzly bear deaths). A large part of the deliberation over delisting thus centers on the difficulty of determining the size of the bear population and the number of females with cubs. Data generated by nongovernmental researchers continues to conflict with numbers generated by the IGBST. For example, Craig Pease, a University of Texas zoologist who studies bear populations, says that the IGBST overestimated the numbers of grizzlies and that data generated by the IGBST "are not obtained through rigorous scientific method" (Wilkinson 1995). Such a discrepancy over numbers, because of delisting criteria, affects delisting debates and management decisions regarding "nuisance bears"; each individual bear is more important to a population close to extinction than it is to a population close to or reaching recovered status.

In addition to debates about numbers, proper research methods to arrive at those numbers, habitat protection, and delisting criteria, there is suspicion about the motivations of various participants for promoting or attempting to prevent delisting. According to one news report, "the Wyoming Game and Fish Department has advocated delisting the grizzly because it wants greater flexibility to manage bears in Wyoming outside the boundaries of Yellowstone and Teton national parks" (Collins 1995). Some conservationists claim that government officials want to delist the bear because they need an endangered species success story. On the other hand, ranchers and government officials have accused members of the conservation community of opposing delisting because they are organized around grizzly bear conservation. Rancher Paul Walton commented in a meeting of the IGBC that "'I've never been in a room with so many people that made their living off bears. . . . I'm trying to make a living despite them'" (Thuermer 1995a).

Despite the grizzly's current status as "threatened," it seemed that talk of delisting, or at least of classifying the population as growing (although there continues to be debate about this claim), had preconditioned public officials to feel justified in first moving Bear 209 from Grand Teton National Park in 1995 and then killing him in 1996. One official at the forest service believes that it is time to turn attention away from individual bears and turn toward the population (Puchlerz 1997). Additionally, the WYGF bear management officer, referring to the fact that Bear 209 was a nine-year-old male, stated that "'totally biologically speaking, the removal of an adult male bear does nothing to affect long-term growth'" and that "'an important point here is that the population seems to have recovered sufficiently to switch from preservation of all individuals to conservation of the population'" (Adams 1996d). Grand Teton National Park superintendent Jack Neckels agreed with WYGF officials, saying in response to citizen concerns over the death of Bear 209, "I believe that the future of grizzly bears in the greater Yellowstone area is secure" (Neckels 1996). Ranchers also seem to agree that the bear can be managed with more discretion. Cliff Hansen, the rancher who leases land on the ERE allotment, said in reference to a proposal to move cattle from the BRSC allotment, "I am persuaded from what I hear from Game and Fish that the grizzly bear has successfully recovered and I think that Mr. Walton ought not to be required to move his cattle from that range" (Adams 1996b).

Thus, the classification of grizzlies as "threatened" remains the same as when grizzlies were first listed under the ESA in 1975, but the application of procedures to manage a threatened species seems to be changing as many decision makers are beginning to regard the grizzly as a species close to, if not already, recovered. Bear 209, according to the claims of many decision makers, was one adult male bear contributing little to the recovery of the species.

Controversy over Bear 203 and Moving Bear 209 in 1995

Cattle depredation has been a relatively recent public problem, but one that has received much attention (Moody, interview 1997). While it is difficult to determine the exact "beginning" of the conflict between bears, cattle, and people (see Botkin 1995; Schullery 1986), and the "beginning" of the current formulation of factors that led to the decision to kill Bear 209, it can be shown that the debate over bears and cattle in Jackson Hole began to become public and receive sustained attention around 1992.

About that time, ranchers on the Blackrock/Spread Creek (BRSC) allotment—the allotment adjacent to Elk Ranch East—in the Bridger-Teton National Forest began experiencing high cattle losses caused by grizzly bear depredation. Depredation on this allotment had been recorded and studied in 1948 (Murie 1948), but there seemed to be few interactions between cattle and bears in the intervening years (Claar et al. 1984). The important point is that interactions began to receive wider attention among current participants in 1992.

In response to the increase in depredations from 1992 to 1993 (from six in 1992 to twenty-five in 1993), the Wyoming Game and Fish Department began a study in 1994 in cooperation with a number of agencies to "identify grizzly bear–cattle interactions on [BRSC] and the adjacent Elk Ranch East cattle allotment (ERE) by monitoring bear activity relative to cattle distribution and documenting cattle depredation caused by grizzly bears" (Anderson and Moody 1997:13). It is not difficult to understand how events on one allotment affect the reaction to events on the adjacent allotment. They both occupy prime grizzly habitat and prime grazing land. The original study design allowed depredations to continue for two years to identify habitual depredators, then to remove those individuals the third year to assess the effectiveness of selective removal (Anderson and Moody 1997).

In 1995, the second year of the study, ranchers on the BRSC allotment began to experience high levels of depredation. The frustrated range manager for the allotment said that the state let the depredations get out of control. He was "tired of dealing with . . . the Game and Fish biologist in charge of the study" who refused to move the depredating bears and "went over [the biologist's] head and called the Game and Fish department director for relief" (Shelton 1995b). WYGF state biologists, who normally make decisions to move or not move bears, were also in charge of conducting the study. They opposed any such translocation because of its effect on their study design. However, because of the complaints from the range manager, the bear management officer was ordered by the director of the WYGF to move Bear 203, the bear responsible for depredating cattle on the BRSC allotment (Bruscino, interview 1997; Moody, interview 1997; Shelton 1995b).

The trap was set for the capture of Bear 203, but Bear 209, which had been depredating cattle on the ERE allotment on MS 2 habitat adjacent to the BRSC allotment, was captured in Grand Teton National Park. He was moved, according to the official with the authority to make such decisions, because the order to move Bear 203 would already compromise the study design. In the meantime, the Sierra Club Legal Defense Fund and other area conservation groups filed an intent to sue to halt the proposed translocation of Bear 203 because his depredations occurred on habitat designated as MS 1. (MS 1 habitat requires that human activity be altered before bear activity occurs.) The conservation groups invoked the ESA and "formally warned the U.S. Forest Service that it will be in violation of the Endangered Species Act if it allows grizzly bears to be trapped and moved from prime habitat in Wyoming without first considering and implementing alternative actions" (Staff 1995). So Bear 203 was scared away, and his removal was blocked by mandated protection under MS 1 habitat.[5]

According to the final WYGF study report, the intended study design of monitoring depredations for two years before taking management actions in the third year was changed "due to the contentiousness of the depredation issue" and "resulted in the early translocation of 1 bear [Bear 209] from MS 2 habitat" (Anderson and Moody 1997:13). MS 2 habitat allows bear activity or human activity to be altered to resolve conflicts between bears and humans, and it is significant that the study specifically states that the bear was removed from MS 2 habitat; the intent to sue stabilized expectations that bears receive priority on MS 1 habitat.

Despite the controversy surrounding the study, WYGF researchers suggested management strategies as a result of the study's findings. For example, they found that older males are the most common chronic depredators, but killing cattle is an individualistic trait in bears (Anderson and Moody 1997). Consequently, the researchers proposed to continue the policy of moving depredating bears and killing ones that return. The authors state, "It is not realistic or in the best interest of the grizzly bear population to force cattle operators to live with management burdens created by depredating grizzly bear, or to require them to abandon all allotments inhabited by grizzly bears" because it would increase the likelihood of ranchers "implementing their own management actions" (Anderson and Moody 1997:52). Other agencies involved in managing land on which bears kill cattle agree with such perceptions. For example, Dale Gomez, a wildlife biologist for the Bridger-Teton National Forest, said that "removing cows would end up getting bears killed because of the fragile acceptance of bears in rural areas. The tolerance of good bears would really drop" (Adams 1996a).

The study recommended no change in current management practices, which means grazing will continue on the BRSC and ERE allotments, cattle carcasses will continue to be removed to prevent bears from being attracted to the allotment, and aversive conditioning techniques will continue to be used on Bear 203 (Anderson and Moody 1997). There are no contingencies concerning what should be done if different bears, displaying behavior different from that of Bear 203, begin to depredate these allotments.

The series of events during the WYGF study stimulated criticisms of the grizzly bear recovery program. In response to the removal of Bear 209 in 1995, the Jackson Hole Alliance for Responsible Planning sent a letter to GTNP superintendent Jack Neckels that stated, "by condoning the removal of a grizzly bear off National park land, you have reneged on your word, you have wasted limited agency funds, and you have potentially seriously jeopardized an important study" (Garland and Lichtman 1995). An editorial in the *Jackson Hole News* mentioned that the removal of Bear 209 in 1995 "gummed up the scientific integrity of the study" (Anonymous 1995). Even those conducting the study expressed similar sentiments that because of orders from top WYGF officials, "the study fell apart at that time" (Moody, interview 1997).

The claims about the WYGF study, particularly charges that it had been changed to accommodate special interests, reiterate complaints made previously about intelligence in the recovery program. The effect of moving Bear 209 on the WYGF study illustrates an absence of creativity, reliability, and openness in the intelligence phase of the grizzly bear recovery program. Participants have expressed concern that the design was compromised for the interests of one rancher, and this decision led the public—and even those conducting the study—to call into question the dependability and openness of the decision-making process. One editorial claimed that "Before 203 could be caught, he slipped away. Instead, another cow-killer in nearby Grand Teton National Park, Bear 209, was trapped. With nine cattle kills to his credit this year, he appears to have been a good enough substitute, and so was shipped off to Yellowstone. . . . Now thousands of dollars, long hours of study and difficult work with dangerous animals has been undermined. And the move didn't even satisfy the complaints of the main protester—Terry Schramm—nor take care of the principal 'problem' —Bear 203" (Anonymous 1995). Thus, the WYGF study, as part of the larger trend described above, indicates the lack of public trust in the dependability of data collected that informs decision makers about how to proceed in grizzly bear recovery.

Interviews with agency officials, on the other hand, indicate that they feel deprived in terms of respect because of the lack of public trust in their decisions. They see themselves as skilled decision makers and scientists, and feel that they know better than anyone else the reasons for their decisions. While some participants object to the institutionalization of intelligence within the IGBST and its exclusionary nature, they do not seem to dispute the competence of individual researchers in that system. For example, one of the lessees says that he supports the decisions WYGF officials feel they must make (Mead 1997), and a member of the Jackson Hole Alliance for Responsible Planning expressed similar respect for the research abilities of those in WYGF (Camenzind, interview 1997).

The removal of Bear 209 from GTNP and his death in 1996 followed a long-standing controversy over how to handle grizzly bear-cattle interactions as well as how to collect data and how to proceed with the recovery program. Controversy had already increased with the removal of Bear 209 in 1995, with no change in the procedure for handling grizzly bear–cattle conflicts or public reaction to management decisions. With the death of Bear 209, however, the stakes increased. This was an endangered species in a national park eating cattle that, according to some participants, were in the park illegally.

Reactions to the Death of Bear 209

The location of Bear 209 on MS 2 habitat—but habitat that was in a national park—two of the three times he was removed (see Table 8.1) raised the issues of conflicting authority and conflicting prescriptions in grizzly bear management. Both the Blackrock/Spread Creek and Elk Ranch East allotments include primarily MS 1 (47%) and 2 (39%) zones, with some MS 3 and 5 zones (14%) located near roads (Anderson and Moody 1997). In MS 1, grizzly activity receives priority; in MS

2, bear and human activity receive equal priority; in MS 3 and 5, grizzly presence is discouraged and grizzlies are controlled (Mealey 1986).

Jay Lawson, wildlife division chief of the WYGF, justified the killing of Bear 209 because the bear was on MS 2 habitat (Barron 1996). Pam Lichtman of the Jackson Hole Alliance for Responsible Planning stated that although she disagreed with the decision, "they have every right to move this bear because it's in Situation 2" (Adams 1996d). Steve Thomas of the Greater Yellowstone Coalition also expressed concern that there were no grounds for legal action in this case because Bear 209 was in MS 2 habitat (Landreth 1996). However, an attorney for the Sierra Club Legal Defense Fund said "MS designations had nothing to do with habitat quality. [The land from which 209 was taken] was designated MS 2 because they already had grazing" (Angell, interview 1997). Some conservationists, however, are reluctant to push this point because they apparently feel that following the intent to sue over the planned removal of Bear 203, grizzlies at least receive protection in MS 1 habitat (Lichtman, interview 1997).

The management situation zoning system is one component of a larger interagency effort to coordinate management. In theory, MS zoning offers prescriptions for handling human–grizzly bear conflicts uniformly across agency jurisdictions. However, many participants referred to the conflicting mandates of protecting resources in national parks and following management guidelines established for different management zones. Jack Neckels, GTNP superintendent, remarked that "If the Greater Yellowstone area is to be managed as a whole, with all agencies involved cooperating for the good of the resources, especially recovering populations like the grizzly, then management actions must adhere to determined guidelines that work to protect resources and the human interests in the area. In the long run, such cooperation will do the most to protect this precious [open] space" (Olson 1996). Additionally, a GTNP management summary of the killing of Bear 209 states that "as full cooperators in the interagency grizzly bear management agreement represented by the Guidelines, GTNP may occasionally be required to take action against livestock-depredating grizzly bears" (National Park Service 1996). Thus, the park superintendent and other park officials justified in part the decision to kill Bear 209 by identifying GTNP as part of a larger area with resources crossing political boundaries. They claimed essentially that if they are going to cooperate with other agencies in managing resources, decisions like the one to kill Bear 209 are inevitable.

Another complicating factor in this case is the mandate to allow grazing in GTNP. Park officials argue that in addition to coordinating management efforts, they are also legislatively required to permit grazing, which burdens them with additional responsibilities in making management decisions: "The action [the killing of Bear 209] is atypical for the National Park Service and somewhat inconsistent with the values and management direction the public associates with the park. The uniqueness of the situation lies in the fact that, unlike almost all other national parks, GTNP has been mandated by legislation to permit livestock grazing within its boundaries" (National Park Service 1996). Thus, GTNP is aware that the killing of Bear 209 was "atypical" for activity within a national park. However, it justified this action on two grounds: (1) that it is part of an interagency effort to manage bears (and other resources), and (2) that it was necessary because of the unique grazing situation in the park. It may be significant that the language used in the management summary quoted above states that the inconsistency of Bear 209's death with park values and management direction emerges from what the "public associates with the park," and not necessarily with how park officials view those park values and management directions.

In contrast to the acceptance of the claim of legality under MS zoning (despite debate about the zoning itself), many participants do not see interagency coordination efforts or the special grazing situation in GTNP as legitimate reasons to kill a grizzly in a national park. A petition organized by a local citizen and signed by residents and tourists makes clear the expectation that resources should be protected in national parks, regardless of other uses: "We, the undersigned, protest the August 5 [4, *sic*], 1996 killing of grizzly bear #209 to protect cattle grazing interests within Grand Teton National Park, Wyoming. We believe that a threatened species such as the grizzly bear should receive priority protection on its own habitat, particularly when that habitat lies within the boundaries of a National Park" (residents petition 1996). Other participants expressed similar sentiments about the outrage they felt over their perception that interagency efforts and ranching interests took precedence over the life of an endangered species in a national park. The citizen who organized the petition argued that "this is a threatened species protected under the Endangered Species Act, surviving on marginal habitat on public land and within a National Park" (Stratton 1996). She "felt betrayed by a Park Service that [she] had believed was doing all it could to protect a threatened species within its borders" and cited the lack of response from a top park service official as well as frustrating encounters with other officials as evidence of citizen demands being ignored (Stratton, interview 1997). Jack Turner, writing about conservation biology, mentions Bear 209 as an example that "these animals live in one of the largest protected preserves in the U.S. It didn't save them from management policy and wildlife biologists" (Turner 1997). Conservationists argued that "national parks are supposed to be refuges for native flora and fauna and the processes that support them" (Lichtman, interview 1997) and that "in a national park bears should be protected, period, unless there is a human threat" (Camenzind, interview 1997). Steve Thomas of the Greater Yellowstone Coalition remarked that "'We're thinking about what to do with that issue right now, and we haven't come to any conclusion on that: the issue of taking bears in a National Park'" (Landreth 1996). Sierra Club Legal Defense Fund attorney Jim Angell observed, "It is bad when bears are taken even from the park" (Angell, interview 1997). Even some officials from the WYGF see the problem as "should cows be grazed in the Park" because "if cows weren't there, then bears don't have a problem" (anonymous interviewee 1997).

The WYGF officials who made the decision to euthanize the bear and carried out that decision, however, felt there were no other options. Dave Moody, large predator coordinator for the WYGF, said that "We were out of options in managing No. 209. There was nowhere in the Greater Yellowstone ecosystem area to relocate him that he would not return from; zoos did not want him; and other states will not willingly receive Wyoming's problem bears, especially a bear habituated to human contact or livestock" (Thuermer 1996b). WYGF officials thought that the location of cows within GTNP led to a situation in which they had to make a decision about killing a grizzly inhabiting a national park. Given this situation, they did not see any option other than killing the bear because he had been moved twice before and returned and no captivity sites were available.

However, this justification points not only to the issue of removing bears from parks but also to the effectiveness in general of moving and removing bears. Moving bears is one nonlethal option. However, many bears return to their "home ranges," or the areas from where they were removed (Bruscino, interview 1997).

Other participants claim that options other than moving or killing bears should be considered. A number of conservation groups have pointed out that moving bears does not work, an observation that is especially crucial when considering that the bears are being moved from national parks: "In 1948, a study by Adolph Murie was published in the *Journal of Wildlife Management* on livestock depredations by bears in this same area. In it, he remarked 'A number of grizzlies have been taken off the cattle range over a period of years but the predation has persisted.' It is time to learn from history that sim-ply moving grizzlies will not solve the problem" (Stevens 1996). Other conservationists have said that "there seem to be other things that have worked to reduce bear-cattle incidences, and there is no reason to be removing bears" (Adams 1997) and that "in a national park, bears should have as many strikes as they need if the strike is for killing livestock and not human safety" (Lichtman, interview 1997). One conservationist has stated her belief that bears are listed because in years past, the strategy was to remove "problem bears" (Taylor, in-terview 1997).

Even members of the forest service, in an Environmental Assess-ment of the Blackrock/Spread Creek allotment, stated that "there may come a point in time when moving/removing grizzlies is no longer po-litically nor biologically sound" (U.S. Forest Service 1997). The study conducted by the WYGF discussed above found, in reference to the ef-fectiveness of moving bears, that "the removal of bear 209 and the ab-sence of 203 did appear to reduce losses during 1996," but that "in cases involving adult, habitual depredators, such as we observed dur-ing this study, a reappearance of problem individuals is expected, and translocation efforts are not a long-term solution" (Anderson and Moody 1997:50). The study went on to state that attempts at translocations can help identify "problem bear, chronic situations or habitual offend-ers" (Anderson and Moody 1997:50). In the cases where these habitual offenders are identified and killed, however, "new bears moving into the areas could become chronic depredators, thereby creating a pop-ulation sink" (Anderson and Moody 1997:51).

The ranchers involved in the Bear 209 case from the time he was first moved in 1995 differ in their opinions on the effectiveness of mov-ing bears. One member of the family with an allotment in GTNP sees cattle losses to bear depredation as just one of many risks of running cattle on land on which grizzlies live, and seems satisfied with the com-pensation program currently in place that pays ranchers for livestock lost to grizzlies. However, he "respected that WYGF felt they had an obligation to enact the ["three-strike"] policy" (Mead, interview 1997). The other rancher on the Elk Ranch East allotment in GTNP seems satisfied with the "three-strike" policy in which "the rules are if a bear is caught three times, he is killed," but he does not think that ranchers should have to deal with repeated depredations (Gill, interview 1997). The rancher and range manager on the Blackrock/Spread Creek allot-ment on forest service land thinks that grizzlies may cause deaths not only directly through depredation but also by causing stress in animals grazing near the bears. The WYGF study addresses these allocations, but the findings were inconclusive (Anderson and Moody 1997). Those working on the BRSC allotment tend to favor the removal of depredating grizzlies and have pushed for such removals (Anonymous 1995; Bruscino, interview 1997; Moody, interview 1997; Primm, interview 1997). The range manager for this allotment appeared in the fall 1997 issue of *Range* magazine wearing a shirt that read "Screw the bears and wolves, save the cowboy" (Raab 1997).

The consensus seems to be that moving bears is not always effective, because bears often return to the areas from where they were taken. However, moving depredating bears seems to help managers identify the circumstances under which habitual depredators return. When those circumstances are identified, participants' expectations differ. The justifications for the death of Bear 209, together with statements in the WYGF study and from an environmental assessment done on forest service land, all indicate that many government officials see killing depredating grizzlies as the only viable option to create tolerance for bears. Other participants, including conservationists and citizens concerned with wildlife and national parks, think that other options may be more effective and meet the expectations of more of the public. Ranchers are more divided on the issue—some expect compensation, others expect bears to be killed.

Overall, participants do not view what happened in the Bear 209 case as an indication that this is a ranching versus wildlife issue. One citizens has said that "we are intelligent and creative enough to have ranching and wildlife coexist" (anonymous interviewee 1997). A letter from the Jackson Hole Alliance for Responsible Planning states that "cattle grazing and bear habitat are not inherently incompatible; they are made so through management decisions that favor cattle grazing over bears" (Lichtman 1997). When discussing alternatives to prevent outcomes similar to that in the case of Bear 209, participants have mentioned several specific strategies to minimize contact between grizzlies and cattle, such as removing dead cattle carcasses that may attract grizzlies to an allotment, using guard dogs to chase away bears, or moving cattle to different allotments. A frequently suggested alternative strategy is to encourage more dialogue among participants. Some conservationists are suggesting that "all interested parties move forward on finding a long term solution" (Staff 1995). A member of the Greater Yellowstone Coalition has said that "there can be cattle on public land, and there can be bears if we do things right" and that she preferred trying to work with ranchers and agencies rather than bringing all controversies to court (Taylor, interview 1997).

Some attempts at dialogue have been made in the past and have failed for various reasons (Primm 1994, interview 1997). One reason is the contentious and polarized political atmosphere surrounding grizzly bear recovery, which seems to intensify with time. For example, many participants have mentioned that they would be willing to work with groups that have differing perspectives on these issues, but that they expect other participants to "say no to everything" (Gill, interview 1997). Moreover, because some agreements have been violated—as in the WYGF study of grizzly bear–cattle interactions on the BRSC and ERE allotments (Anderson and Moody 1997)—some participants now have a hard time trusting other participants.

This does not mean that dialogue is impossible. A member of the Greater Yellowstone Coalition has pointed to the strategy of blowing up cattle carcasses that may attract bears as a successful compromise that has been reached among ranchers, conservationists, and agencies (Taylor, interview 1997). Additionally, in a workshop on problem solving held in Jackson, Wyoming, in September 1997, which examined ways of overcoming problems in the decision-making process, grizzly bear policy was discussed at length. I participated in this workshop and observed that although there was much disagreement, participants in the workshop—who represented a broad range of interests—generally agreed that the gathering was a step in the right direction and that more such workshops might help alleviate tensions. Every participant said they would recommend the workshop to other people, including managers, people involved in policy appraisals, group facilitators, journalists, people in the private sector, "people with power, in groups containing opposing views," agency employees, and academics (unpublished manuscript 1997). Although some people felt that certain groups participating in grizzly bear policy, such as government officials and ranchers, are criticized too harshly and used as scapegoats, the overall feeling was that progress was being made.

More specifically, participants in the workshop said that decision process appraisals helped them to "think through the whole decision process," to "keep social process in the conscious forefront," and to "direct attention to [a] comprehensive range of components" (unpublished manuscript 1997). They also mentioned that, overall, "the human element needs a lot more attention" (unpublished manuscript 1997:3). They discussed specific aspects of decision and social process appraisal. For example, they mentioned the need to "always be willing and open to assessment/appraisal" and that "this process forces us to rethink the 'paradigms' . . . by allowing us to discover the 'anomalies' in how we get involved in issues we're passionate about" (unpublished manuscript 1997:3–4). Thus, while some attempts at dialogue have met with limited success, many participants in this policy process seem enthusiastic about engaging in dialogue to solve problems.

Projections

As mentioned earlier, both houses of Congress passed a bill (105 S. 308) in 1997 that authorizes a study of the relationship between open space and ranching in the West and permits grazing in GTNP to continue for the duration of the three-year study. The bill awaits President Clinton's signature. Therefore, it is likely that grazing in GTNP will continue for at least three years. Conservation groups will most likely not sue over such an extension as long as wildlife is protected during that time. However, it is important to note that the Jackson Hole Alliance for Responsible Planning pushed for two amendments to the congressional bill, one mandating the priority of wildlife needs over the needs of cattle and another that limited the grazing extensions to the duration of the three-year study (Camenzind et al. 1997). Because the bill was passed without the amendments despite letters from conservation groups to Senator Thomas (who introduced the bill to Congress) (Camenzind 1997b) and repeated correspondence with park superintendent Jack Neckels (Camenzind and Taylor 1996), conservationists can expect that depredating bears will continue to be removed from GTNP land. In addition, park officials can expect that removal or killing of grizzlies in the park in response to depredation will trigger a reaction from conservationists similar to if not more dramatic than that stemming from the death of Bear 209. Conservationists have expressed their desire to maintain discussions about grazing in the park that are open, out of the media focus, and more on a one-on-one basis with concerned parties. They have also expressed the potential need, however, to call more attention in a public arena to the current grazing situation and original legislation that outlawed such grazing (Taylor, interview 1997, and *Jackson Hole News* 1997). Conservationists anticipate problems because "park grazing has other consequences" such as the death of Bear 209 for "killing cattle on the Grand Teton grazing allotment" (Thuermer 1997a), and they feel that "repeat situations are unacceptable" (Thuermer 1997b). To prevent future conflict similar to that surrounding Bear 209, it will be necessary to use strategies to stabilize expectations and meet participant demands.

In any case, it is also likely, given historical and recent trends of cattle depredation by grizzly bears on the BRSC and ERE allotments, that depredations will continue. This will be especially true if there is a bad natural food year. Bear 203 returned to the BRSC in 1997 and will most likely return again in upcoming years, and aversive conditioning may or may not continue to scare him away. However, it is also likely that new bears will move in to replace Bears 209 and 203. Controversy will probably continue, whether the depredations take place on park service or forest service land. Government officials are concerned with creating tolerance for "good bears," and they project that not removing "bad" bears will lead ranchers to kill all bears. Ranchers seem divided on this projection. Some ranchers may feel compelled to take matters into their own hands, while others seem content with receiving compensation, and others may push for the continuation of the three-strike policy or another policy to remove depredating bears.

While certain agencies have been pushing for delisting of the grizzly, it seems that this will not happen soon. Part of the criteria for delisting involves limited grizzly mortality. Seven bears were killed in fall 1997 alone by hunters who claimed self-defense. However, when it comes to depredation, bears will probably continue to be treated more as a recovered species, and bears that continually depredate will probably continue to be killed if aversive conditioning does not seem to work. Continuation of such action will probably lead to continued public outcry with little response by decision makers. It is likely that if such conflict and erosion of trust continues, there will be a court case in the future. Some ranchers have said that they will stop reporting depredations, which would allow them to avoid the public outcry.

Groups with diverse and apparently conflicting interests are beginning to organize around what appear to be less contentious issues, such as how to learn about decision-making processes. Many contending groups in the grizzly bear case have found common ground on other issues such as bison management. Thus, while the ability of participants to discuss grizzly bear issues has been shaky in the past, the current movement toward collaborative problem solving seems to be slowly eroding the layers of mistrust that have built up over the years.

Appraisal and Recommendations

Normative Issues

Expectations About When It Is Appropriate to Move Bears

A number of normative issues are at stake in this case. The first such issue, one that is very specific, is what to do when bears come into conflict with cattle. Almost all participants agree that this is natural behavior—bears cannot distinguish between elk and domestic calves. However, the WYGF, the "Interagency Grizzly Bear Guidelines," and the "Grizzly Bear Recovery Plan" consider depredating bears to be "nuisance" bears, which means management actions must be taken to reduce cattle–grizzly bear interactions. The guidance for determining such actions comes from the grizzly's status as "threatened" under the ESA and the "Interagency Grizzly Bear Guidelines" on how to handle nuisance bears within different management situation zones.

The first normative issue, then, is how to determine the circumstances under which it is acceptable or unacceptable to move or kill grizzly bears, especially those that depredate cattle. A range of situations exists. Depredations may take place on public lands such as forest service, park service, or Bureau of Land Management land, or on private lands. They may take place in various management situation zones. Depredators which display behavior such as Bear 209 are almost sure to return even if translocated, while ones such as Bear 203 may respond to aversive conditioning (scare tactics, such as gunshots) and not return for the grazing season.

Expectations about the Management Situation Zoning System

One strategy that decision makers have used to make determinations about how to handle nuisance bears was to establish and apply the management situation (MS) zoning system described above. The intent to sue that helped block the removal of Bear 203 reaffirmed the expectation that bears should not be captured in MS 1 habitat. However, bears have large ranges and typically do not conduct all their activity in one MS zone. Consequently, deciding upon the appropriate action against a depredating grizzly involves, in turn, determining whether the MS zoning system is to be applied in reference to where the bear is captured, where it lives primarily, where it depredates cattle (or commits another offense that renders it a "nuisance" bear), or a combination of all three factors. Currently, grizzly bears captured in MS 1 habitat cannot be killed or removed for livestock depredation, as the successful attempt to block Bear 203's removal has clarified. However, it is almost inevitable that a grizzly will be captured in MS 2 habitat even though it depredates primarily or even episodically in MS 1 habitat. This eventuality will stir controversy; talk of this possibility arising with regard to Bear 203 was common in summer 1997 and was discussed in an environmental assessment of the BRSC and ERE (U.S. Forest Service 1997).

In addition, it cannot be assumed that the MS zoning system is always the appropriate instrument or frame of reference for making decisions about depredating bears, as the reaction to the Bear 209 case exemplifies. For example, it remains unclear how to proceed when mandates according to the MS zoning system conflict with other agency mandates. Park and WYGF officials seemed to agree that trapping Bear 209 within the park and euthanizing him was permissible because of interagency cooperation established for grizzly bear management. However, concerned citizens and conservationists argued that interagency cooperation should not necessarily trump mandates of the park service to protect wildlife within its borders, especially endangered wildlife.

The second normative issue, then, involves clarifying how participants expect the MS zoning system to be applied to determine appropriate management action. Clarification requires a determination of which grizzly bear activity the zoning system applies to (bear range, conflict, or capture), how the zoning system relates to other agency mandates, and whether to apply it to determine the status of a bear as a nuisance on the basis of property damage on public land or in terms of human safety only. It also involves determining how the MS zoning system relates to other agency mandates, such as the National Park Organic Act of 1916.

Expectations about Allowing Grazing in the Park and the Relationship of Grazing to Wildlife

Participants have also made claims about the desirability and legality of allowing grazing to continue in GTNP. Ranchers claim that they cannot move their cattle to another allotment because it would cost too much in time, labor, and money and that the only way they can stay in business is to continue using parkland for grazing their cattle. They have claimed that inheritance taxes would require their children to subdivide land outside the park, rather than keep it as open space, if they cannot continue to operate within the park or find alternative solutions. Government officials have claimed that this issue requires further study; the bill mentioned above that is awaiting President Clinton's signature would authorize such a study and would allow grazing to continue for at least three more years. While concerned citizens and conservationists feel that any grazing in a national park is inappropriate, the demand they are making is that grazing and other commercial uses of GTNP should not take precedence over wildlife within the park's borders, regardless of what happens outside the park (Camenzind 1997a; Taylor, interview 1997).

This discussion leads to a third normative issue, which involves

GTNP's relationship with land surrounding the park. The issue in this case relates specifically to land outside the park owned by the same ranchers who graze their cattle within the park. How should park policy relate to wildlife and/or cattle within its borders when, as part of living, the wildlife and/or cattle also spend time beyond the park boundary?

Expectations about Interagency Decision Making and about Participation in Decision Making

The final normative issue I discuss, one that encompasses all the others, centers on who is involved in the decision-making process— who decides, for example, under what conditions removing grizzlies for cattle depredation is acceptable. Many people claim that the entire decision-making process is unreliable and that there is a lack of trust overall. They claim that the decision to kill Bear 209 in a national park for killing cattle grazing on expired allotments and the reaction that followed typified the reasons for the lack of trust on all sides, because no one seemed truly happy with the situation or the decision.

Nonetheless, the Bear 209 case clarified people's expectations about at least one grizzly bear–human conflict—cattle depredation in national parks. Many citizens, tourists, and conservationists demand that wildlife take precedence over cattle in national parks, regardless of management situation zoning. The WYGF does not want to have to make decisions about grizzly bears that depredate on cattle in national parks. Ranchers want to avoid public outcry. Given that the recent bill approved by both houses of Congress will extend grazing in the park for at least three more years, and that Bear 203 returned in summer 1997, cattle depredation is likely to occur again in the park. To prevent another public outcry, the decision-making process that responds to and shapes the demands of grazing interests and conservationists must be, at the very least, far more inclusive and transparent than in the past. This change would relieve Wyoming Game and Fish of the burden of having to make unilateral decisions about bears in national parks, and it would lessen the perceived public need to scrutinize ranchers.

In summary, then, a final normative issue emerges: Who should make or have input into decisions about grizzly bear policy, and how should those decisions and input be made?

Alternatives

The goal of grizzly bear recovery is to increase the population of bears, which involves reducing grizzly mortality. The grizzly bear recovery plan states that one of the "leading challenges" in recovery is the reduction of human-caused mortality (Servheen 1993:10). Dave Mattson has said that mortality is related primarily to the frequency of bear contact with humans and the lethality of that contact (Mattson and Craighead 1994). Reducing mortality therefore involves reducing the frequency and lethality of contact. Reaching these goals involves changing human practices so that, over time, such practices engage grizzly bears less often.

The Bear 209 case has clarified the expectations of participants on many normative issues involved in the attempt to reach grizzly bear recovery goals. In addressing the appraisal and recommendations that follow, I make specific recommendations wherever possible regarding how to stabilize expectations concerning a given normative issue. These issues include how to handle cattle depredation by bears in national parks and possible alternatives to create a more inclusive decision process. Regarding other normative issues, including how to apply and/or improve the MS zoning system and how interagency coordination relates to the application of this system and to the Endangered Species Act, I feel it is not yet possible to make specific recommendations. The wide range of expectations regarding these issues has been evidenced in the discussion of Bear 209, and few alternatives can be offered without further clarification of participant expectations and demands by the participants themselves. Creating a more inclusive decision-making process can help clarify such interests.

Clarifying Expectations and Demands about the Conditions under Which Moving or Killing Bears Is Acceptable and about Grazing in the Park

Many of the normative issues outlined above are related to the decision to permit grazing to continue in GTNP. Many participants feel that allowing cattle to take precedence over wildlife in the national park is unacceptable. Therefore, they do not feel that the MS zoning system is sufficient justification for making decisions such as the one to remove Bear 209 from GTNP or to kill him. Many participants mentioned that if Bear 209 had been captured outside the park, the decision would have been no more acceptable, because the depredations did occur in the park; therefore, it matters both where a bear depredates and where it is captured.

Some participants have suggested that the majority of parkland should have been designated as MS 1 habitat initially and might now be redesignated accordingly (Angell, interview 1997; Primm, interview

1997). Redesignating parkland as MS 1 habitat has biological and political benefits. Historical grizzly habitat and current grizzly range indicate that the area designated as MS 2 habitat in GTNP where the grazing allotments currently exist may be more appropriately designated
as MS 1 (Hoak et al. 1981). Vegetation maps of food sources may
provide information on habitat quality, and could provide a tool for
conservation groups who claim that the zoning of the BRSC and ERE allotments as MS 2 is inappropriate (Angell, interview 1997). Such a redesignation would also relieve the WYGF of having to make decisions
about human-bear-cattle conflicts on that land because expectations
have been stabilized about land designated as MS 1; grizzly bears cannot be touched on this land without first significantly altering human
activity. Compensation for depredations can continue to address deprivations experienced by ranchers.

However, although GTNP ultimately has authority to redesignate
land within its boundaries and could use biological data such as vegetation maps, habitat quality, and bear distribution on the land to
make such a decision (Mealey 1986:3), the political environment may
preclude such an option. Given the long-standing debate about the
effectiveness of the MS zoning system (Keiter 1991), efforts at redesignation may not occur in a timely manner and may exacerbate the ongoing debate over grizzly bear–cattle conflict in GTNP. For example, redesignation sets up a potential win-lose situation between ranchers
and conservationists and could draw more public attention to an issue
that is already contentious. Generating public attention may be a legitimate strategy for conservationists in the future, but it has been
mentioned as an alternative to avoid if possible. Conservation groups,
government officials, and ranchers have also been meeting to discuss
what to do about the grazing situation; attempting to force redesignation may undermine such efforts. Finally, although expectations are
clear concerning the protection of bears in MS 1 habitat, one underlying norm involved in the case of Bear 209 relates to the necessity of determining how to handle "nuisance" bears in national parks regardless
of MS zoning. Thus, redesignation could achieve the desired outcome
of protecting bears in a national park and could be an effective long-
term strategy, but it may not immediately address the prevailing expectations about management of bears within park borders regardless
of MS zoning designations.

In addition, one family with an expired grazing allotment "requested
an extension of their previously-held grazing privileges to avoid having
to sell the ranch. In return they committed to actively exploring options to preserve their ranchlands, including giving scenic easement to

the Jackson Hole Land Trust, seeking a conservation buyer and other options" (U.S. Senate, Committee on Energy and Natural Resources 1997:4). While conservationists oppose the link between the study and the extension of grazing privileges in bill S. 308 (Camenzind 1997b), "a coalition of conservation groups are interested in helping the ranching families in Jackson Hole find ways to continue their ranching operations, which in turn helps to protect open space and wildlife habitat" (Camenzind et al. 1997:1). The study and current dialogue around that study offer a potential strategy for ranchers, conservationists, and government officials to work together to clarify their own and other participants' perspectives, to clarify expectations and demands they have in common, and to find creative and integrative strategies to meet common interests.

One option that could be raised is the possibility of writing into existing leases that ranchers will receive compensation for confirmed depredations but cannot demand that bears be removed from or killed in the park. This would allow grazing to continue while meeting the demands of ranchers and conservationists to minimize contention while protecting both grazing and wildlife interests, and it would provide some time for ranchers with leases to make plans to move their cattle, phase out of ranching, or find other options with the help of rather than opposition from conservation groups. At the same time, continuing compensation programs, removing cattle carcasses so that bears are not attracted to them, and attempting other activities to deter bears from wandering onto allotments can minimize financial deprivations and concerns over respect and loyalty that ranchers might experience by complying with such an agreement. This strategy would also help avoid drawing attention from a less understanding national audience about the continuation of cattle grazing in a national park and about management decisions on the basis that grizzlies are a recovered species, not a recovering species.

Rewriting leases in the manner suggested would also make irrelevant the uncertainty that persists about the application of the management situation zoning system. The zone in which potential grizzly depredators live, depredate, or are captured would have no bearing on a decision to remove or kill as long as these activities occurred in the park. Finally, the WYGF would no longer have to make decisions on how to apply the zoning system in the park. The option of rewriting leases could also be prescribed and applied in a timely manner, in part because it is not likely to stimulate institutional resistance.

Other options such as conservation easements can be explored to help alleviate financial burdens, such as inheritance taxes on ranchers' children (Berry, interview 1997). The Jackson Hole Land Trust is exploring such options with the ranchers involved on the ERE (Berry, interview 1997). The Sonoran Institute, a nonprofit organization that promotes community-based strategies to meet ecological and economic needs, has produced a publication, "Preserving Working Ranches in the West," that discusses options such as conservation easements, estate planning, limited development, voluntary zoning districts, and collaborative planning more fully (Rosan 1997). This publication points out that a conservation easement is "a voluntary contract that permanently limits the type and intensity of future land use while allowing landowners to retain ownership and control of their property," and is "tailored to the needs of each landowner" (Rosan 1997:15). Such agreements lower the property value of, and thus the property and inheritance taxes on, the land while ensuring that development remains limited.[6]

Clarifying Expectations and Demands about Interagency Coordination Efforts and about Participation in Decision Making

When the outcome of coordinating agency efforts to reach the goal of grizzly bear recovery places obstacles to reaching that goal, the efforts—however well-meaning—need to be assessed. While parks do not exist in vacuums, and their political borders may seem arbitrary, certain expectations and demands that are not being met exist in the minds of conservationists and other concerned citizens regarding what happens within those borders. These expectations and demands —which include protecting endangered wildlife, especially in national parks and especially when the threat to the well-being of wildlife comes from special interests of individuals using park resources—do not change because agencies are attempting to coordinate efforts to increase the grizzly population. In addition, while cooperation across agencies is essential in thinking about grizzly bears as part of a larger ecosystem and not confined to political borders, the outcome of current efforts has been that borders designating property and jurisdiction have been redrawn with respect to the MS zoning system and expectations and demands of various participants are still not being considered adequately. More fully realized interagency cooperation requires consideration of prevailing expectations and demands about national parks and the relationship of those expectations and demands to meeting the goal of grizzly bear recovery.

It is important to create tolerance in rural communities for "good

bears" by handling "nuisance bears," but it is also necessary to be selective in management actions that disturb a grizzly's natural activity.
Conservationists have stated repeatedly that they would like to trust
government officials in their decisions to move or kill bears, and that
they are not opposed categorically to such management strategies.
However, when decisions such as the one to kill Bear 209 are made,
conservationists claim it is hard to trust that agency officials will consider their demands to protect wildlife, especially in national parks, for
offenses that do not endanger human life (Lichtman, interview 1997;
Taylor, interview 1997, and *Jackson Hole News* 1997). The strategy to
move bears when they come into conflict with humans—especially
when efforts to change human activity to deter such contact is minimal
—can and has lowered tolerance for management actions that alter
bear activity and has contributed to diminished trust in the decision-
making process.

This assessment that interagency efforts are failing to meet the
expectations and demands of all participants is made with the realization that there have been overall improvements in the grizzly bear
recovery policy process. Criticisms about data on human–grizzly bear
interactions have been answered, for example, by a dramatic improvement in the record-keeping of those interactions over the past decade.
In addition, while government officials in the past have refused to
participate in workshops aimed at generating dialogue among participants with conflicting perspectives and values, recently officials
from the forest service, the park service, and the Interagency Grizzly
Bear Committee were willing and enthusiastic participants in a problem-
solving workshop that discussed many contentious issues in the GYE.

In light of both the criticism and achievement of interagency efforts
discussed above, it is important to consider the many demands made
on government officials by various sectors of the public and the responses by officials to those sectors. In addition, the agency officials
with whom I have talked seem genuinely concerned with both grizzly
bears and the sector that wants to protect bears and to protect property and human safety. They are skilled and knowledgeable about matters of grizzly bear behavior and ecology and are aware of the contentious politics of grizzly bear conservation. However, there seems to
be a lack of training in policy sciences, social sciences, law, conflict
resolution, and other areas which focus systematically and empirically
on the human dimensions of grizzly bear conservation. Although agency
budgets may restrict the possibility for hiring people trained in understanding these human dimensions, there is a need to train officials in

areas such as understanding social and decision processes. Such training can help officials clarify their own and others' expectations and find ways to address everyone's demands.

Agency employees have a responsibility to and are constrained by the system in which they operate, a system that already places multiple demands on budgets and each individual's time and energy. "The system" is made up not only of written prescriptions and procedures, but also of individuals. These individuals include not only those who write prescriptions and make and enforce decisions based on those prescriptions, but also those who gather information that informs the prescriptions and decisions, those who influence decisions to be made in certain ways, and those who assess and challenge or support decisions— formally in court or informally through correspondence or in the press.

Cross-jurisdictional management and cooperative planning are necessary components in grizzly bear recovery and in ecosystem management in a more general sense (Mealey 1986; Primm and Clark 1996; Servheen 1993) because no one agency or organization "working alone, has the resources, such as expertise, funds, and authority, necessary or sufficient to get the job done" (Clark and Brunner 1996:1). However, interagency efforts as they now exist are in a sense both too much and not enough. They are too much because often in the genuine and taxing effort to coordinate across political boundaries—as in the case of Bear 209—the goals of grizzly bear recovery, achieving individual agency mandates, and meeting the interests of a substantial number of people to protect endangered species and to protect wildlife within national parks are not being met. The possibility exists, as in the case of Bear 209, for these latter goals to be subsumed by the goal of interagency coordination. This problem can be addressed by working less toward interagency coordination as a goal in and of itself and more toward the goal of grizzly bear recovery, with interagency coordination as one means to do so. Training agency personnel to conduct problem-oriented analyses, as discussed below, is one means to help place interagency efforts within the context of grizzly recovery goals (Clark 1997; Primm and Clark 1996).

Interagency efforts are not enough because the pool of participants involved in decision making extends beyond agency officials. Therefore, interagency efforts should be expanded to include nonagency citizens in the decision-making process. Researchers gather and analyze data presumably to inform officials so that they can make better decisions. As discussed above, many researchers have claimed that this process has been dominated by the IGBST, a group of agency personnel. An intelligence function that is dependable, open, and creative should allow for the inclusion of data from nonagency researchers. Government agencies can begin to work with outside researchers to determine why discrepancies in data exist, rather than arguing over whose data is correct.

In addition, although data on human–grizzly bear conflicts has improved significantly since the turn of the twentieth century when Yellowstone National Park officials began recording such data and since the late 1980s when data gathering became part of the IGBST activities, data gathered on grizzly bear policy is still lacking. Current research is dominated by natural science information. With few exceptions, little research has been done with respect to social or decision-making processes in the twenty-three-year story of grizzly bear recovery policy. Just as data about ecological issues, such as grizzly habitat, food distribution, population, and behavior, is critical in creating effective grizzly bear policy, so is information about such issues as the expectations and demands of the human population. Research coordinated by the IGBST should therefore actively seek not only nonagency data but also analyses of social and decision processes. Expanding interagency efforts to include nonagency expertise, such as people skilled in understanding social and decision processes, can improve the ability of agencies to meet grizzly bear recovery goals. The first place to start is intelligence.

For example, data concerning grizzly bear–human interactions and management actions has improved in the past decade. For each grizzly bear–human confrontation, this information includes any management action taken, and whether or not there was human-caused grizzly bear mortality, the date of the confrontation, the management situation zone in which the confrontation occurred, the type of land ownership (park service, forest service, private, etc.) and the location of the confrontation, the bear's identification number, the type of conflict resolution, and the sources of all the data. Data such as public or private responses —including, for example, the number of newspaper articles and editorials concerning a management action, petitions to government officials, and personal correspondence to various agencies regarding an action —to control actions or lack of control actions can be added to reports about grizzly bear–human confrontations. Such data might clarify people's expectations about when it is appropriate, for example, to move bears. Steven Primm and Tim Clark, two policy scientists who have conducted extensive research in the Greater Yellowstone Ecosystem, describe the importance of tracking trends: "Tracking trends . . . allows us to establish standards on key measures by evaluating what was considered acceptable in the past. Analyzing historical trends also gives us an idea of how much variation to expect on a regular basis" (Primm and Clark 1996:151). While Primm and Clark were referring here primarily to biological and ecological data, their observations hold true for political and sociological data. It is important to determine what was and is considered acceptable or not acceptable with respect to moving and killing bears in response to various actions such as livestock depredation. Such data should not be used to predetermine policy decisions, but they can help configure decisions that consider the expectations and demands of a variety of participants.

Research, however, is only part of decision making, and only one part in which those outside of government agencies should be included. Conservation groups, citizens, and ranchers have promoted various creative alternatives to help prevent cattle–grizzly bear interactions. These alternatives are sometimes ignored or blocked by those within agencies who focus on the difficult tasks of invoking and applying mandates put forth by the IGBC and creating tolerance among ranchers who "may take matters into their own hands" and kill all grizzlies on their property, not just problem bears (Bruscino, interview 1997; Moody, interview 1997; Puchlerz, interview 1997). For example, the refusal to participate in meetings, such as the one organized to resolve cattle–grizzly bear conflicts on Togwotee Pass, indicates a certain amount of resistance on the part of government officials to listen to alternatives such as collaborative problem-solving efforts (Primm, interview 1997). The removal of Bear 209 from GTNP in 1995 while the WYGF study was being conducted is an example of an agency ignoring its own alternative—studying the problem—to find ways to resolve the grizzly bear–cattle conflict. The rejection of the amendments proposed by conservationists to the open-space study and grazing bill is a more recent example of alternatives being precluded.

This contention is not universal, however. Agreements such as the one reached among agencies, ranchers, and conservationists to blow up cattle carcasses indicate that collaborative efforts are being made to prevent conflicts. While conservation groups, citizens, and ranchers point to such agreements as examples of how bears and cattle can coexist—and that bears and people can as well—some of these participants also hesitate to engage in dialogue because of the perception that previous attempts have failed, because certain personalities are more comfortable with one-on-one or small group interactions, or because of the erosion of trust in decision making caused by twenty-three years of contention. Those who promote dialogue must be willing to take risks to participate in it. All members must offer some trust in order to receive the trust of other members.

One way to try to create trust is to conduct workshops organized around a common interest, such as how to make better decisions through extensive problem orientation that examines natural science, social process, and decision process. Within such workshops, focusing in depth on specific issues is vital (Participants 1997), but including an emphasis on social and decision processes allows participants to step back from the details and their passions and look at the larger picture. The strategy helps refocus attention on the root causes of problems such as value deprivation, expected value deprivation, and built-up frustration. Connecting possible solutions with root or core problems in participatory workshops can help alleviate tensions, personalize the "opposition," and reduce the tendency to point fingers at one common enemy. It is also a way to combine practical problem solving with skill building, given limited agency funding and time. Assessment is a vital part of such workshops to determine what works and what areas can be improved.

In addition, this debate over bears and humans competing for the same resources lies within the context of a larger debate over increasing pressure on resources in the GYE and who should make and participate in making decisions about those resources. Lessons can be learned from efforts in related debates in the GYE. For example, a number of groups around the GYE have organized to find solutions to various policy problems involving bison in Jackson Hole and Yellowstone. The work of these groups provides valuable lessons for those who want to establish problem-solving groups. A group organized to work on the Jackson Hole bison herd problem—conventionally discussed as a problem of maintaining a self-sustaining population, of minimizing the potential for transmission of brucellosis, of reducing dependency of bison on feeding programs, and of minimizing potential for bison-human conflict—began with a year-long study group phase, which was composed of ten people, including wildlife biologists, policy scientists, environmental activists, local politicians, and artists. The group focused on clearly diagnosing the problem by examining the decision processes that had led to the current policy for the Jackson Hole bison herd. They diagnosed the problem not in techno-rational terms, but in terms of value conflict, malfunctions in the decision-making process, and insufficient public input. The group then ran a promotional activity, the Buffalo Jubilee, a one-day event that included art, Native American dancing, poetry reading, children's activities, and presentations on bison ecology and management, decision making, bison ranching, and human relations to wildlife (Curlee and Day 1997).

The group is now in the Buffalo Forum phase. This forum aims to "facilitate a community-based decision making process characterized by partnerships among stakeholders, constructive public involvement, credible social and physical science, cooperation and a genuine commitment to the goal of sustainable, wholistic [sic] natural resource management" (Curlee and Day 1997). The methods of achieving this goal include convening a Buffalo Forum decision seminar and working group to plan a series of public forums. The public forums, which are intended to include the general public, representatives of responsible agencies, and representatives of organized interests, will be designed to clarify goals and a long-term vision for the Jackson bison herd in the context of wildlife management as a larger issue. From these meetings, the group hopes to draft a proposal for future action based on well thought out intelligence and appraisal activities and on broad and balanced participation that can address common rather than special interests. The group will hold town meetings to gather input on a draft for future action and then present a formal report and recommendations to responsible agencies. The group plans to continue monitoring activities, carry out consistent promotional activities, document their efforts, and convene a second Buffalo Jubilee (Curlee and Day 1997).

Longer-term goals of similar working groups focusing on grizzly bear conservation could be to assess the management situation zoning system and its role in interagency efforts and the larger context of grizzly bear policy. Much contention has arisen over the use or misuse of this zoning system in applying the Endangered Species Act (ESA) to meet the goal of grizzly bear recovery. The death of Bear 209 indicates that expectations and demands related to this system remain inconsistent. While it is clear that grizzlies captured on MS 1 habitat cannot be removed, it is also clear that participants do not accept the MS zoning system as a solitary prescription that can be invoked in applying the ESA. Multiple demands, such as other agency mandates, and expectations, such as whether to allow cattle to graze in the park, must be considered. In addition, such a working group could try to determine how to apply and/or improve the MS zoning system. Groups could also address other normative issues, including the rewriting of leases, redesignation of land, treatment of bears as a species that is still listed, rationale for and effectiveness of interagency coordination, and decision process in general.

In summary, workshops that aim both to address specific issues, such as finding alternative methods to handle grizzly-cattle-human conflict, and to help participants to develop problem-solving skills can bring participants together to work toward three goals: clarifying their own expectations and demands; understanding the expectations and demands of other participants; and finding areas of common interest that might lead to solutions addressing the expectations and demands of all participants.

Conclusions

Bear 209 was one grizzly involved in a limited number of management actions, yet the events leading to the decision to kill him, his death, and subsequent reaction are indicative of problems in a larger policy arena—ecosystem management in the Greater Yellowstone Ecosystem. The Bear 209 incident involved three agencies directly—the National Park Service, the U.S. Fish and Wildlife Service, and the Wyoming Game and Fish Department—and one wildlife species in an area, Jackson Hole, with a human population of about 5,000. The GYE encompasses at least 28 local, state, and federal agencies and about 300,000 residents, and an additional 10 million visitors annually including groups as diverse as ranchers, recreationists, oil and gas companies, mining companies, developers, residents, tourists, and park and forest concessionaires competing for resources (Burroughs and Clark 1995). Thus, there exist many groups with differing myths, expectations, and demands concerning the use or conservation of resources such as grizzly bears and national parkland. As a result of these factors, both grizzly bear conservation and related policy discussion in the GYE have a long history of controversy, marred by a lack of trust among many participants (Burroughs and Clark 1995).

The Bear 209 case, and grizzly bear conservation more generally, provides lessons for improved decision making in the GYE. While change does not happen overnight, and it is difficult to alter the course of a twenty-two-year-long current of contention, there are encouraging trends indicating that efforts at coordination and dialogue may be working. The decision process workshop held in Jackson and interest in more such workshops, efforts made around the GYE to find common ground on bison policy, and continued meetings among ranchers, conservationists, and government officials involved in administering the grazing allotments in GTNP are a few examples of successes.

Creating more open and inclusive decision-making processes—in which participants can listen to and address one another's expectations and demands—is one necessary component to successful ecosystem management. Such efforts provide an opportunity for participants to work toward a shared goal of improving decision-making processes so that they can be more effective at meeting the stated goals of grizzly bear recovery. However, successful efforts at dialogue and inclusive decision making will require the building of trust among participants, the coordination of agencies to meet goals specified by those within and outside of agencies, and the cooperation, participation, and acceptance of responsibility by all interested participants.

ACKNOWLEDGMENTS

This study could not have been completed without the help of many people. First, I thank those who made this analysis possible by participating in interviews, including officials from the Wyoming Game and Fish Department, who were incredibly helpful in providing details regarding the case; government officials from Grand Teton National Park and the Bridger-Teton National Forest; members of conservation groups, including the Jackson Hole Alliance for Responsible Planning (now the Jackson Hole Conservation Alliance), the Greater Yellowstone Coalition, and the Sierra Club Legal Defense Fund (now Earth Justice); members of the ranches with grazing allotments in GTNP; private citizens who devoted their time and energy to thinking about this case; and researchers who have spent years working on grizzly bear conservation. In addition, this study would not have been possible without the financial support of the Denver Zoological Foundation, the Northern Rockies Conservation Cooperative (NRCC), and Yale University. I am also indebted to a number of people whose suggestions and counsel, coming from years of experience in the GYE and/or in policy sciences analysis, have enriched the study immeasurably: Tim Clark of the NRCC and Yale University School of Forestry and Environmental Studies initiated the study and provided guidance throughout; Andrew Willard of Yale Law School advised and guided the entire process of writing and analysis and spent countless hours discussing the case; and Steve Primm of the NRCC and the World Wildlife Fund and Peyton Curlee of the NRCC provided continual support and insight into the case throughout the research and writing process.

NOTES

1. For example, Adams 1996d, Barron 1996, Olson 1996, Stratton 1996, Thuermer 1996a, b.

2. Harold Lasswell (1966:9) has described political myths as "'fundamental assumptions' about political affairs." They include "basic expectations and demands concerning power relations and practices in the society," and an important component of myths is the elaboration of social norms, of what is right (Lasswell 1966:9–11). Myths also articulate concrete power patterns and symbols and slogans

that elaborate, repeat, and apply the political myth (Lasswell 1966). The debate over Bear 209, in the claims, expectations, and demands of participants, offers a window into the political myths and personal values involved in grizzly bear conservation.

3. Data gathered by the IGBST has led to improved management strategies in Yellowstone. For example, information about human–grizzly bear conflicts has helped in the development of strategies to minimize conflicts. In 1960, Yellowstone National Park instituted for the first time a bear management program to reduce the number of bear-caused human injuries and property damage. Most of this management involved the removal of bears rather than preventative measures such as reducing human food and garbage sources that attract bears. A program was implemented in 1970 that involved removing these sources from developed areas and roadsides. By 1973, human-caused injuries had dropped to about ten per year, down from an average of forty-eight between 1931 and 1959. A program implemented in 1983 emphasized habitat protection in backcountry areas and has led to an average of only one bear-caused human injury per year and an average of four grizzly bear translocations and one grizzly bear removed from the population per year (Gunther 1994). The accomplishments of the IGBST are also mentioned in Primm (1993) and Schullery (1986).

4. For a more complete picture of the history of the institutions involved in grizzly bear recovery, see Primm (1993), Mattson and Craighead (1994), and Schullery (1986).

5. This option, called aversive conditioning, did not work for Bear 209 because bear behavior varies from individual to individual. Bear 209 would leave for a few days and return to the cattle-grazing allotments, while Bear 203 routinely disappears for the remainder of the grazing season.

6. A number of organizations located in Wyoming—including the Wyoming Department of Game and Fish in Cheyenne; the Wyoming governor's office in Cheyenne; Wyoming State Forestry Division in Cheyenne; Wyoming Open Lands in Buffalo; the Jackson Hole Land Trust in Jackson; and the Wyoming Outdoor Council in Lander—are involved in helping ranchers create land trusts. A number of publications are available as well, including: *The Conservation Easement Handbook* by Janet Diehl and Thomas S. Barrett, published by the Trust for Public Land and the Land Trust Alliance in Washington, D.C., in 1988, and *Preserving Family Lands: Essential Tax Strategies for the Landowner* by Stephen J. Small, produced by Landowner Planning Center in Boston, Massachusetts, in 1992.

LITERATURE CITED

Adams, M. 1996a. "Forest Won't Move Togwotee Grizzlies, Cattle," *Jackson Hole Guide* July 10:A1.

———. 1996b. "Togwotee Pass Conflict Deeply Polarizes Two Sides," *Jackson Hole Guide* July 17:A7.

———. 1996c. "Bear 209 Out-smarts Government Biologists," *Jackson Hole Guide* July 31.

———. 1996d. "State Kills Grizzly in Grand Teton Park: Wildlife Officials Kill Bear due to Excessive Cattle Conflicts," *Jackson Hole Guide* August 7:A2.

———. 1997. "Grizzly Study Sparks Cattle vs. Bear Debate," *Jackson Hole Guide* April 2:A7.

Anderson, C. R., and D. S. Moody. 1997. Grizzly Bear-Cattle Interaction on Two Cattle Allotments in Northwest Wyoming. Lander, Wyoming, Trophy Game Section, Wyoming Game and Fish Department, Cheyenne.

Anonymous. 1995. "Grizzly Relocation Is Puzzling Affair," *Jackson Hole News* September 20:4A.

Barron, J. 1996. "Grand Teton Park Grizzly Euthanized: Wyoming G&F Handling Livestock Damage Claims," *Casper Star Tribune* August 6:B2.

Botkin, Daniel B. 1995. *Our Natural History: The Lessons of Lewis and Clark.* New York: Grosset/Putnam Publishers.

Burroughs, R. H., and T. W. Clark. 1995. "Ecosystem Management: A Comparison of Greater Yellowstone and Georges Ban," *Environmental Management* 195:649–663.

Camenzind, F. J. 1997a. Letter to Honorable Craig Thomas, Chairperson on the Subcommittee on National Parks, Historic Preservation, and Recreation Committee on Energy and Natural Resources. Jackson Hole Alliance for Responsible Planning, Jackson, Wyoming. July 9.

———. 1997b. Letter to Senator Thomas and Members of the Subcommittee on National Parks, Historic Preservation, and Recreation, Jackson Hole Alliance for Responsible Planning. Jackson Hole Alliance for Responsible Planning, Jackson, Wyoming. July 9.

Camenzind, F. J., and M. Taylor. 1996. Letter to Jack Neckels, Superintendent GTNP. Jackson Hole Alliance for Responsible Planning, Jackson. October 23.

Camenzind, F. J., et al. 1997. Continuation of Grazing Leases in GTNP. Greater Yellowstone Coalition, Jackson Hole Alliance for Responsible Planning, National Parks and Conservation Association. Press Release. Jackson, Wyoming: 2.

Carlton, J. 1995. "Grizzly Bear Wins in Court! Judge Orders Revision of Recovery Plan," *Predator Project Newsletter.* www.wildrockies.org:80/Wild . . . dProj/PPnews/Fall_95/PP_court.html.

Claar, J. J., et al. 1984. "Grizzly Bear Management on the Flathead Indian Reservation, Montana," paper presented at the International Conference on Bear Research and Management.

Clark, T. W. 1997. *Averting Extinction: Reconstructing Endangered Species Recovery.* New Haven: Yale University Press.

Clark, T. W., and R. D. Brunner. 1996. "Making Partnerships Work in Endangered Species Conservation: An Introduction to the Decision Process," *Endangered Species Update* 13(9):1–5.

Collins, K. 1995. "Court Decisions May Set Back Grizzly Delisting Effort," *Casper Star Tribune* October 26:E2.

Congressional Research Service. 1986. "Greater Yellowstone Ecosystem: An Analysis of Data Submitted by Federal and State Agencies," Committee on Interior and Insular Affairs, U.S. House of Representatives, Committee Report No. 6. Washington, DC: U.S. Government Printing Office.

Craighead, J. F. C. 1979. *Track of the Grizzly.* San Francisco: Sierra Club Books.

Craighead, J. J., et al. 1995. *The Grizzly Bears of Yellowstone: Their Ecology in the Yellowstone Ecosystem 1959–1992.* Washington, DC: Island Press.

Curlee, P., and C. Day. 1997. Reforming the Process/ Jackson Hole Bison Herd. Letter from the Jackson Hole Alliance for Responsible Planning Buffalo Forum, Jackson, Wyoming.

Eberhardt, L. L., and R. R. Knight. 1996. "How Many Grizzlies in Yellowstone," *Journal of Wildlife Management* 60(2):416–421.

Garland, S., and P. Lichtman. 1995 (Sept. 21). Letter (unpublished) from the Jackson Hole Alliance for Responsible Planning to Jack Neckels, superintendent of Grand Teton National Park.

Gunther, K. A. 1994. "Bear Management in Yellowstone National Park, 1960–93," International Conference on Bear Research and Management.

————. 1996. "Visitor Impact on Grizzly Bear Activity in Pelican Valley, Yellowstone National Park," International Conference on Bear Research and Management.

Gunther, K. A., and M. T. Bruscino. 1995. "Grizzly Bear-Human Conflicts, Confrontation, and Management Actions in the Yellowstone Ecosystem 1994," Interagency Grizzly Bear Committee, Yellowstone National Park, Wyoming.

————. 1997. "Grizzly Bear-Human Conflicts, Confrontations, and Management Actions in the Yellowstone Ecosystem 1996," Interagency Grizzly Bear Committee, Yellowstone National Park, Wyoming.

Hoak, J. H., T. W. Clark, and J. L. Weaver. 1981. "Grizzly Bear Distribution, Grand Teton National Park Area, Wyoming," *Northwest Science* 55(4):245–247.

Interagency Grizzly Bear Committee (IGBC). 1996 Briefing Document. IGBC briefing paper on grizzly bears. Missoula, Montana, Interagency Grizzly Bear Committee: 10.

Interagency Grizzly Bear Study Team (IGBST). 1991. Reports and publications list of the Interagency Grizzly Bear Study Team, Interagency Grizzly Bear Study Team, Missoula, Montana.

Keiter, R. B. 1991. "Observations on the Future Debate over 'Delisting' the Grizzly Bear in the Greater Yellowstone Ecosystem," *The Environmental Professional* 13:248–253.

Landreth, D. 1996. Epitaph for Griz# 209. Griztrax. http:wnc.com:80/griztrax/ requiem.htm: 1.

Lasswell, H. D. 1934. *Psychopathology and Politics*. Chicago: University of Chicago Press.

————. 1966. "The Language of Power," in H. D. Lasswell et al., eds., *Language of Politics*, pp. 3–19. Cambridge, MA: MIT Press.

Lichtman, P. 1997 (May 15). Letter regarding BRSC Environmental Assessment, Jackson Hole Alliance for Responsible Planning, Jackson, Wyoming.

MacIver, R. M. 1947. *The Web of Government*. New York: Macmillan.

Mattson, D. J., and J. J. Craighead. 1994. "The Yellowstone Grizzly Bear Recovery Program: Uncertain Information, Uncertain Policy," in T. W. Clark, R. P. Reading, and A. L. Clarke, eds., *Endangered Species Recovery: Finding the Lessons, Improving the Process*, pp. 101–129. Washington, DC: Island Press.

Mattson, D. J., et al. 1995. "Science and Management of Rocky Mountain Grizzly Bears," *Conservation Biology* 10:1013–1025.

McInnis, D. 1997. "The Selling of the West," *Range* Fall:44–45.

Mealey, S. P. 1986. "Interagency Grizzly Bear Guidelines," Interagency Grizzly Bear Committee.

————. 1994. *Principles of Conservation Biology*. Sunderland, MA: Sinauer Associates.

Miller, G. T. 1992. *Living in the Environment*. Belmont, CA: Wadsworth.

Murie, A. 1948. "Cattle on Grizzly Bear Range," *Journal of Wildlife Management* 12(1):57–72.

National Park Service. 1996. "Management Summary—Bear 209," Moose, Wyoming, Grand Teton National Park.

Neal, D. 1995. "Scientists Dispute Grizzly Population Trends," *Casper Star Tribune* September 28:E6.

Neckels, J. 1996. Letter from Jack Neckels, Grand Teton National Park Superintendent, to Diana Stratton, private citizen, National Park Service. December 23.

Olson, L. L. 1996. "Grizzly #209 Captured in Grand Teton National Park," news release, National Park Service. Jackson, Wyoming.

Participants in Workshop on Conservation Problem Solving. 1997. Workshop Evaluations. Workshop on Conservation Problem Solving, Jackson, Wyoming.

Primm, S. A. 1993. Grizzly Conservation in Greater Yellowstone. M.A. thesis, political science department. University of Colorado, Boulder.

———. 1994. "Grizzly-Livestock Conflicts on Togwotee Pass: Using Policy Research to Find Solutions," *NRCC News* 10, 17.

Primm, S. A., and T. W. Clark. 1996. "The Greater Yellowstone Policy Debate: What Is the Policy Problem?" *Policy Sciences* 29:137–166.

Raab, A. 1997. Photograph of Terry Schramm. *Range* Fall:35.

Reisman, W. M., and A. R. Willard, eds. 1988. *International Incidents: The Law that Counts in World Politics*. Princeton, NJ: Princeton University Press.

Rosan, L., ed. 1997. *Preserving Working Ranches in the West*. Tucson, Ariz.: The Sonoran Institute.

Schullery, P. 1986. *The Bears of Yellowstone*. Boulder, CO: Roberts Reinhart.

Servheen, D. C. 1993. Grizzly Bear Recovery Plan. Missoula, Mont.: U.S. Fish and Wildlife Service.

Shaffer, M. 1994. Letter to federal government officials from the science community.

Shanks, B. 1984. *This Land Is Your Land: The Struggle to Save America's Public Land*. San Francisco: Sierra Club Books.

Shelton, C. 1995a. "Judge Mauls Griz Recovery Plan," *Jackson Hole Guide* October 11:A1.

———. 1995b. "Griz Moved to Avoid More Cattle Conflicts," *Jackson Hole Guide* September 13:A1, A12.

Staff. 1995. "Groups File 60-Day Notice on Bridger-Teton Griz Removal," *Greater Yellowstone Report* Fall:20.

———. 1996. "More Legal Victories and Challenges," *Wild Forever* Fall:7.

Stevens, T. 1996. Bears and Cows—Home on the Range? Griztrax: http://wnc.com:80/griztrax.grizgraz.htm.

Stratton, D. 1996a. Letter to Tim Clark. November 4.

———. 1996b. "An unbearable death," *Jackson Hole Guide* August 14:A4.

Taylor, M. 1997. "Park Should Protect Its Own Property First," *Jackson Hole News* July 16:5A.

Thuermer, A. M. 1995a. "Griz Study May Cost 2 Bears' Lives in '96," *Jackson Hole News* October 11:12A.

———. 1995b. "Judge Rules against Federal Grizzly Plan," *Jackson Hole News* October 11:13A.

———. 1996a. "Traps Laid in Park for Cow-Eating Griz: Bear 209 Ate Seven Calves in Seven Nights," *Casper Star Tribune* July 29:1A.

———. 1996b. "Teton Grizzly Is Killed to End Its Cow-Eating: Bear Was Considered a Problem According to Legal Guidelines," *Jackson Hole News* August 7:14A.

———. 1997a. "Senate Eyes Open Space," *Jackson Hole Guide* July 2:A1.

———. 1997b. "Greens Criticize Park Open Space Study," *Jackson Hole News* July 9:3A.

Turner, J. 1997. "Forum: Grizzly Bear Number 209," *Wild Duck Review* 3(1):11–14.

U.S. Forest Service. 1997. Blackrock/Spread Creek Allotment Management Plan Environmental Assessment. Moran, Wyoming, United States Department of Agriculture Bridger-Teton National Forest Buffalo Ranger District, Jackson, Wyoming.

U.S. Senate. Committee on Energy and Natural Resources. 1997. "Grand Teton National Park Grazing Study," Report No. 105–64. Washington, DC.

Weaver, J. L. 1996. "John and Frank Craighead," *Wildlife Society Bulletin* 24(4):767–769.

Wilkinson, T. 1995. "Critic: Grizzly Needs Continued Protection," *Jackson Hole News* October 18:4A.

9 Appraising Ecotourism in Conserving Biodiversity

Eva J. Garen

Introduction

Over the past few decades, ecotourism, a form of specialty travel, has emerged as a popular strategy for protecting biodiversity in many regions throughout the world (Brandon 1996; Mendelsohn 1994). While there are several competing definitions of the industry, ecotourism is defined commonly as "purposeful travel to natural areas to understand the culture and natural history of the environment; taking care not to alter the integrity of the ecosystem; producing economic opportunities that make the conservation of natural resources beneficial to local people" (Ecotourism Society 1994). In essence, an ecotourism program has the potential to protect biodiversity by providing resource-dependent communities with the economic means to participate in natural resource conservation (Jabcobson et al. 1992; Mendelsohn 1994; Munn 1992; Taylor 1994).

In practice, however, it appears that many ecotourism programs are failing to protect the biodiversity upon which they depend. Although local participation is a major premise of an ecotourism program, resource-dependent communities expected to switch to tourism often are excluded from the more profitable positions within the industry. With-

Research for this project was conducted intermittently between 1995 and 1997.

out economic compensation or opportunities for involvement, such communities often do not have the means or desire to participate in conservation efforts. Ecotourism programs must also have guidelines to limit negative tourist impacts on local cultures and biodiversity, as well as the means with which to effectively apply protective prescriptions. Unregulated tourist behavior frequently causes negative impacts to local communities and irreparable damage to natural resources.

Despite these shortcomings, ecotourism appears to be heralded as a relative success in natural resource conservation. With pressures increasing to incorporate resource-dependent communities into conservation efforts, ecotourism is embraced widely by the international conservation community as an effective integrative strategy. For those with financial objectives, ecotourism is a powerful marketing strategy for attracting tourists who wish to believe they are participating in a socially and environmentally responsible form of travel. Regardless of the motive behind ecotourism promotion, it appears as if many programs are doing more harm than good. Although recently there have been numerous claims questioning the effectiveness of ecotourism as a conservation strategy (Brandon 1996; Padgett and Begley 1996; West and Brechin 1991; Whelan 1991), little effort has been made to identify successes and failures systematically.

By examining how decisions are made in ecotourism development, this paper explores why many ecotourism programs may not be protecting biodiversity. Using the policy sciences (Lasswell 1971) as an analytical framework, an appraisal of how program development should proceed, in light of trends and conditions, is presented. Examples from my field research in Roátan, Honduras, Costa Rica, Belize, and Jackson Hole, Wyoming, as well as my survey of ecotourism literature, show how exclusive and acontextual decision making may impede conservation efforts. Additionally, an appraisal of the Mountain Gorilla Project (MGP) in Rwanda, considered one of the most successful ecotourism programs, demonstrates how decision making can produce outcomes that closely reflect the common interest. Recommendations to improve ecotourism are given based on my analysis of the industry and the case material.

Standpoint

My interest in ecotourism and integrative problem solving evolved during my undergraduate tenure as a political science major and biology minor. While studying the evolution of language in primates, I was impressed by the strategy used in Rwanda to protect mountain gorillas and their habitat. Conservationists incorporated the region's social, political, and economic realities into a tightly monitored tourism program based on viewing mountain gorillas in their natural environment. Although political instability has currently halted tourism within the region, this strategy is credited as one of the major factors supporting the continued existence of mountain gorillas in the wild. Although I have concerns about tourism impacts on resident people and natural areas, ecotourism has become a popular conservation strategy and must be examined to determine potential benefits. I am focusing my graduate and professional research on exploring alternative approaches to natural resource management, with emphasis on examination of ecotourism as an integrative conservation strategy.

Data for this paper were obtained from published literature, primarily journal, newspaper, and magazine articles, as well as books and book chapters. Information was also gathered from discussions with people involved with the ecotourism industry, such as Megan Epler Wood, executive director of the Ecotourism Society, during the Ecotourism Conference hosted at the Yale School of Forestry and Environmental Studies in spring 1996. An interview with Dr. William Weber, one of the primary architects of the Mountain Gorilla Project, provided information for the case appraisal. Unpublished information is included throughout the paper from my field research in Roátan, Honduras, Costa Rica, Belize, and Jackson Hole, Wyoming.

The Ecotourism Industry: An Overview

Ecotourism emerged as a conservation strategy in response to the frequent failure of traditional natural resource conservation approaches (Brandon and Wells 1992). The traditional approach, commonly referred to as "protectionist" or "preservationist" (Machlis and Tichnell 1985; West and Brechin 1991), did not incorporate local or proximate resident peoples into conservation efforts; rather, parks and protected areas were established by the creation of "impenetrable" boundaries around threatened biodiversity and natural resources. Residents within these areas often were expelled from their land, and traditional uses on or within these areas, such as hunting, logging, agriculture, or gathering plants and herbs for medicinal or spiritual uses, were no longer permitted within park or reserve boundaries. Affected residents rarely were compensated for such losses of land and traditional ways of life (Boo 1990; Craighead 1991; Poff 1996; Wells and Brandon 1992; West and Brechin 1991).

Without alternative means of survival or compensation for traditional use, communities surrounding parks or protected areas often were forced to exploit these areas through illegal and destructive resource-extractive activities, such as poaching and land clearing (Olindo 1991; Whelan 1991). As conflicts between people and protected species and habitats increase in many areas throughout the world, natural resource managers have come to recognize the urgency of incorporating the needs of human populations into conservation efforts (West and Brechin 1991). Such projects that attempt to link conservation and development goals, commonly referred to as integrated conservation development projects (ICDPs), emphasize local participation with natural resource protection (Wells and Brandon 1992). The underlying premise of an ICDP is that resident peoples will support conservation efforts if they are involved with their development and can benefit from them (Wells and Brandon 1992).

A number of promising trends in the international tourism industry provided natural resource managers with an opportunity to integrate residents with conservation efforts. With improvement in international transportation and public health standards (Eadington and Smith 1992), as well as an increased global interest in environmental issues (Brandon 1996), tourists now spend substantial amounts of money to travel to remote and unique natural areas in exchange for intimate experiences with exotic aspects of the natural world and "indigenous cultures." Rather than permitting these tourists to roam freely through natural areas, a tourism infrastructure could be established so that economic revenue is generated through entrance fees, tour guide costs, the hospitality industry, and the selling of souvenirs. Residents living in or near an ecotourism site can tap into this revenue stream through tourism-related job opportunities, thereby creating the means and incentives for local participation in conservation efforts.

Tourism as a means of incorporating residents into conservation efforts is now one of the most popular strategies used for integrating conservation and development goals. Travel and tourism have become the world's largest industry (Brandon 1996), and ecotourism has gained international recognition as the industry's fastest growing segment, constituting 10% and 30%, respectively, of the international tourism market (Boo 1990; Whelan 1991). Ecotourism is not, however, a solution without problems. Although ecotourism projects frequently are funded by the international conservation community and are marketed widely by the international tourism industry, rigorous program evaluations must take place to determine whether or not programs are accomplishing the goal of biodiversity protection. Such evaluations must also examine the appropriateness of ecotourism as a conservation strategy in many regions.

Goal Clarification and Problem Definition

As currently defined by the international conservation community, ecotourism is an integrative conservation strategy that works to protect
biodiversity by balancing the needs of resident people with conservation efforts. Although several questions must be considered in program development, my review of the existing ecotourism literature
and my field research suggest that accomplishing the following two
objectives is essential for protecting biodiversity. Such programs must
gain, maintain, and intensify local support for natural resource conservation and must limit negative impacts on natural resources and local
cultures. Comprehensive decision making is required to develop management guidelines that, when applied, accomplish these two objectives.
In essence, ecotourism programs must be developed and applied in a
manner that reflects the common interest in biodiversity protection.

My preliminary findings suggest that because of inadequate decision
making, ecotourism programs may not be protecting biodiversity. With
biodiversity protection as the driving force, a goal that may not be shared
by resident groups living in proximity to the natural area in question,
program development and application should account for the values
and perspectives of all program participants, such as the rural poor,
local elite, local and national government, foreign tour operators, tourists, and local and international conservationists. Decision making often
appears to be monopolized, however, by participant groups interested
in personal gains in terms of wealth and power. Although many programs may be marketed under the ecotourism label, biodiversity protection and local participation often are not shared goals of program
development. While the success of an ecotourism program is dependent upon gaining the support of resource-dependent communities,
more powerful participants, such as local elite and foreign tour operators, often receive the most benefits while remaining resident groups
may be excluded from the industry (Wells and Brandon 1992).

Although it may seem obvious that ecotourism success is dependent upon a healthy and intact natural area, inadequate decision making may contribute significantly to biodiversity destruction within an
ecotourism site. Programs that are controlled by participant groups
with economic interests often do not include guidelines to protect biodiversity. When protection guidelines are established, structures with
which to effectively apply such protective measures, such as guards
and administrators, are often ineffective or nonexistent (Barker 1996;
Herliczek 1996). Without protective measures, direct tourist contact
with natural resources, as well as tourism-related development in or

around the natural area in question, often causes extensive damage to the biodiversity upon which a program depends.

Appraising the Ecotourism Industry

To explore the hypothesis that inadequate decision making may be responsible in large part for the apparent failure of many ecotourism programs to protect biodiversity, it is essential to appraise decision making throughout program development. In my view, ecotourism programs should establish and maintain a decision process through which participants formulate, apply, and appraise program guidelines that reflect the common interest in protecting biodiversity. To assess how well ecotourism programs are meeting this goal, I used policy sciences as an analytic framework (Lasswell 1971). According to the policy sciences, seven interrelated functions make up any decision process, and comprehensive and rational decision making should carry out each function. Appraising decision making in ecotourism development will identify successes and failures that contribute to or impede biodiversity protection.

The following discussion provides a description of each of the seven decision functions, as well as a description of how ecotourism decision making should proceed in reference to each function. Table 9.1 lists each function and corresponding activity, as well as criteria for the adequate performance of each.

Trends in decision making drawn primarily from two ecotourism programs are presented to demonstrate how well and in what ways the criteria for each function were addressed. In particular, examples from my field research from the island of Roátan, off the northern coast of Honduras, are provided to illustrate trends and conditions of exclusive and acontextual decision making. For comparative purposes, the subsequent discussion emphasizes decision making successes, with illustrative material drawn from the Mountain Gorilla Project in Rwanda.

Appraisal 1: Lessons from Exclusive Decision Making

Intelligence Function

The intelligence function involves gathering site-specific information, clarifying program goals, and planning and predicting outcomes. If an ecotourism program is developed to protect biodiversity, decision makers must gather, process, and disseminate information for the use of all who participate in program development and applica-

Table 9.1. Decision-Making Functions, Activities, and Criteria

Decision Functions	Activities	Criteria
Intelligence	Gather reliable site-specific information Clarify program goals Plan and predict outcomes and alternatives	Dependable, comprehensive, selective, creative, available
Promotion	Discuss management alternatives	Rational, integrative, comprehensive, effective
Prescription	Set rules, policy, and guidelines	Effective, rational, inclusive
Invocation	Provisionally characterize particular instances	Timely, dependable, rational, nonprovocative
Application	Finally characterize interactions	Open, inclusive, rational, uniform
Termination	Discontinue existing prescriptions	Timely, dependable, comprehensive, ameliorative
Appraisal	Evaluate successes and failures with regard to desired goals	Comprehensive, selective, independent, continuous

Source: Adapted from H. D. Lasswell, *A Pre-View of Policy Sciences*, New York, American Elsevier, 1971; and H. D. Lasswell and M. McDougal, *Jurisprudence for a Free Society*, New Haven, CT, New Haven Press, 1992.

tion. Reliable site-specific social, political, economic, and ecological information should be obtained in a manner that is dependable, comprehensive, selective, creative, and available for all participants involved with decision making. The intelligence function should help program developers to clarify objectives, describe trends, analyze conditions, project developments, and formulate policy alternatives (McDougal et al. 1981).

Several important questions should be addressed with reference to ecotourism development. For example, what are present threats to the natural area in question? Why is a natural area being degraded? How do we know if a natural area is being degraded? What are the perspectives and values of all ecotourism participants representing local, national, and international levels? What is the larger social, political, and economic context of the region in question? What is the current land tenure system and how does it apply to the natural area in question? How can the goal of biodiversity protection be achieved in an integrative and equitable fashion? Finally, is ecotourism an appropriate strategy in this regard? Information should be collected from all relevant sources and can be gathered through background research, field work, social and biological surveys, discussions, interviews, as well as a variety of Participatory Rapid Appraisal methods, such as community mapping and group interviews (Freudenberger 1997).

Information gathered in this way will have tremendous impact on all other decision functions. For example, a social analysis can provide insight into the relations between resident people and their natural resources. If surrounding residents are economically dependent on natural resources for survival, the development of a well-managed, lucrative ecotourism industry may provide the impetus for local participation in conservation efforts. Social analyses may also reveal power relationships among or between resident groups that may impede community-based conservation efforts or may even suggest that ecotourism is an appropriate solution. Comprehensive ecological surveys are also essential for providing baseline information to determine visitor guidelines and monitor tourist impacts (Norris 1992). It is also possible that a properly operating intelligence function may indicate that ecotourism is a poor strategy for protecting biodiversity in particular contexts.

In practice, most ecotourism programs appear to be established with inadequate intelligence. On Roátan, for example, tourism focused on the surrounding coral reefs appears to be expanding rapidly without prior planning efforts. Program development appears to be monopolized by elite groups primarily interested in personal gains in terms of wealth and power and not in biodiversity protection. Although the area is widely marketed as an "ecotourism" destination by tour operators, business owners, and the Honduran government, it appears as though coordinated efforts to obtain social or ecological information before the influx of tourists did not take place. Careful examination of the relations between resident people and the surrounding reefs, as well as power struggles among different residents groups representing local, national, and international interests, did not occur. Consequently, the reef has suffered extensive damage since the inception of the island's ecotourism industry (Forest 1994).

Promotion Function

The promotion function involves, among many strategies, discussion of alternatives, and the clarification of participants' expectations and demands. With site-specific, contextual information, management alternatives must be discussed in a manner that is both socially and contextually inclusive. At this stage, program developers should have a comprehensive understanding of all participant perspectives. Resources, data, and opinions already gathered can be employed to discuss management alternatives through open forums. Issues regarding perspectives promoted by various participants as well as which participants will benefit and which will not should be discussed. Recommended policy alternatives should reflect the common interest and should, therefore, be rational, integrative, comprehensive, and effective (Lasswell and McDougal 1992).

As a further refinement of information obtained during intelligence, several important issues should be addressed. For example, can differing participant values and perspectives be incorporated into an ecotourism program that accomplishes the goal of biodiversity protection? What measures should be taken to ensure that local communities benefit from an ecotourism program? Do existing power structures permit ecotourism development? Do existing institutions have the capacity to support and manage an ecotourism program? How can an ecotourism program help to finance conservation efforts? What measures should be taken to limit tourists' impact on natural resources and surrounding communities?

In actuality, ecotourism is a potentially lucrative industry that often is monopolized by a small group of participants with interests in short-term financial gain. Consequently, discussions of management alternatives appear to be biased and acontextual because they are controlled by a select group of wealthy and powerful participants. On Roátan, special interests seem to control decision making so that a narrow set of participants receive the majority of benefits. Despite efforts by the general public and conservation groups to develop protection guidelines, their perspectives often are ignored during discussions of management alternatives. Although those in control of decision making may recognize the importance of protecting the island's surrounding reefs for ecotourism success, they often do not follow through with such realizations during program application.

Prescription

Prescription involves the setting of rules, policies, or guidelines for action. Program guidelines should be established that crystallize the expectations and demands of all participants to accomplish biodiversity protection. Prescriptions should be characterized by norms that reflect the common interest, and behaviors that contravene such norms should be sanctioned (McDougal et al. 1981). If developed in a comprehensive manner, prescriptions will be effective, rational, and inclusive. The establishment of equitable and comprehensive ecotourism guidelines, a prospective activity, will have a tremendous impact on overall program success.

Several issues require careful consideration. Guidelines to ensure that the resident human population benefits from tourism must be established if communities located in or around an ecotourism site are expected to shift their livelihood to tourism. For example, program developers can require that a portion of tourism-generated revenue remain within the host community. Residents can be trained as program guides and managers so that they have the opportunity to hold profitable and respected positions in the industry. Guidelines to protect the biodiversity on which an ecotourism program depends should also be determined. Without local support and measures to limit direct tourist impact and unsustainable development, an ecotourism program likely will fail to protect the biodiversity on which it depends.

In practice, ecotourism guidelines that protect biodiversity and natural areas are either nonexistent or ineffectively applied (Brandon 1996; Herliczek 1996). Ecotourism guidelines, both formal and informal, are either carried out or ignored in a manner that appears to reflect the needs and interests of a select group of participants, rather than the common interest. On Roátan, although local communities are trying to establish protective guidelines that limit the negative impacts of tourism on surrounding coral reefs, local elite and foreign tour operators appear to undermine their efforts and apply tourism practices that yield the highest short-term, personal gains. For example, although prescriptions were established by members of a community-based marine reserve to monitor tourist behavior, wealthy and powerful resort owners often ignored these guidelines to accommodate what they felt were tourist desires.

Invocation

Invocation is the provisional characterization of particular instances of interaction in terms of prescription(s). This function may stimulate or initiate application. With the development of program prescriptions that reflect the common interest in biodiversity protection, structures or "enforcement bodies" must be established through which participants can effectively invoke, or make reference to, these guidelines when challenged in concrete situations. The provisional characterization of prescription deviations may have significant value consequences. Invokers should, therefore, be authorized, motivated, and well equipped, as well as timely, dependable, rational, and nonprovocative (McDougal et al. 1981).

Several issues must be considered when invoking ecotourism prescriptions. For example, if ecotourism guidelines were established to prohibit direct tourist contact with certain portions of a natural area, program officials should be hired to enforce such rules. Program guards who witness inappropriate behavior should make a provisional characterization of the incident, as well as a provisional assertion of control to prevent the action. Likewise, if a prescription is not being carried out, appropriate persons should be able to acknowledge that prescription conformity is not taking place and should report such an incident to the proper authorities.

In practice, the invocation of prescriptions often appears to be nonexistent or ineffective. For example, a community group in Roátan tried to establish a marine reserve focused on a small portion of the reef to minimize damaging tourist impacts. Prescriptions were established through a series of open and inclusive town meetings to limit direct tourist impacts, to determine where tourists could snorkel and deep-sea dive, and to minimize shoreline development. The community, however, did not have adequate funding, so their efforts were dependent entirely on funding from the local elite. Although these elite groups appeared to support conservation efforts, they did not give reserve staff the authority to invoke protective prescriptions. For example, although local residents were hired as patrol officers, they rarely were taken seriously in their attempts to make provisional citations or assertions of control.

Application

Application is the final characterization of particular instances in terms of prescription(s). Application puts invoked prescriptions into effect in specific instances. When ecotourism prescriptions are challenged and/or invoked, application is the subsequent function that transforms authoritatively prescribed policy into a controlling event. Application should be a uniform function that is guided by the common interest formulated in prescription. Methods for enacting program guidelines should be decided in an open and inclusive fashion. In essence, application represents the distribution of values and the conformity of behavior to ecotourism policies that reflect the common interest in biodiversity protection. Application should be characterized by authoritative and controlling enforcement, although it should not be monopolized by special interests.

Application is essential in ecotourism development. If a prescription is established that requires ecotourism developers to return a portion of tourism revenue to local communities, the application of this policy will be manifest in the actual distribution of revenue. Without the means of effectively distributing such revenue, prescriptions are meaningless. If a park guard or guide witnesses tourists trampling an area in which they are forbidden to enter, predetermined penalties should be applied if the guard or guide invokes this behavior and reports it to park or protected area administrators. Penalties determined for inappropriate tourist behavior should be applied by appropriate administrative agencies. Once again, without effective application of such invoked prescriptions, an ecotourism program may not effectively protect biodiversity.

Several problems arise during the application of prescriptions that are developed by a select group of participants. For example, although ecotourism programs must provide local communities with tourism-related economic opportunities, when actually applied, few ecotourism programs require that tourism-generated revenue remain in either the host country or local community (Brandon and Wells 1992). While resident groups often refer to formal or informal prescriptions entitling them to tourism generated revenue, the World Bank estimates that approximately 55% of tourism revenue "leaks" back to the developed world, while additional studies report that leakage up to 80 and 90% may be more common for countries capitalized by foreign-owned airlines, hotels, and transportation services (Brandon 1996), and by tour operators (Munn 1992).

Biodiversity and natural resources are also adversely affected by the inability of claimants to apply invoked prescriptions. When decision making is dominated by select participant groups, program guidelines to protect biodiversity often are ignored if they impede the generation of wealth and power. In Roátan, guidelines to mitigate tourist behavior and tourism-related development were established with the development of the community-based marine reserve. However, when a local elite family that had been active with the development of the reserve wanted to build a home within the protected region, efforts to prohibit their actions were ignored. Although it is clear that prescriptions did exist that forbade such development within marine reserve boundaries, it was not clear if the reserve staff, who had tried to apply the invoked protective prescriptions, had been taken seriously when construction of the family's private home took place during the summer of 1996.

Termination

Termination covers the repeal of prescriptions, the adjustment of claims grounded on expectations crystallized when the prescriptions were in effect, and the adjusting of claims that pertain to the ending of permissible continuing relations. The termination function refers to the discontinuance of existing prescriptions and/or organizational arrangements established in conformity with such prescriptions; termination must make provisions for claims arising from disruption of an expected and demanded arrangement. Prescriptions that do not contribute to the common interest in accomplishing biodiversity protection should be ended in a manner that is prompt, respectful of all participants, dependable, comprehensive, balanced, and ameliorative.

At least three distinct forms of termination may take place in ecotourism programs. First, ecotourism often relies on the termination of local practices and traditional ways of life, such as reef fishing, hunting within the rainforest, or the use of biodiversity or natural areas for symbolic purposes. Such relationships must be recognized by decision makers, and their termination, if necessary, should be properly compensated. More important, program developers should carefully consider the ethical implications of such termination. Second, traditional power relationships among resident groups are also likely to end. For example, in a region where colonization has afforded tremendous power to elite groups, ecotourism success is likely to be dependent on the willingness of such groups to relinquish power and control. Last, various program prescriptions will most likely end during some point of program application and should take place in an equitable and ameliorative fashion.

The termination of local practices and traditional ways of life, power relationships, and program prescriptions are seldom accomplished, however, in a comprehensive, balanced, and ameliorative manner. First, communities surrounding the biodiversity or natural area that is the focus of an ecotourism program rarely are compensated for terminating traditional practices within the natural area in question. While an ecotourism program is supposed to provide economic compensation to affected communities, elite participant groups often impede the application of this necessary prescription. In regions where surrounding communities attach values other than wealth to their natural resources, such as symbolic and aesthetic values, little or no effort is made to compensate for these losses, let alone identify such relationships during the formative phases of program development. Destructive tourism impacts that may have a negative effect on community composition, structure, and stability also are rarely recognized.

Second, the termination of traditional power relationships may be essential if an ecotourism program is going to accomplish biodiversity protection. Elite groups appear to monopolize ecotourism development and often refuse to share wealth and power benefits that may accrue from program application. In Roátan, local elite and foreign tour operators monopolize the tourism industry and will not allow the general public to become genuine and active participants in program development. Government officials in Rwanda, the primary decision makers for mountain gorilla tourism, have repeatedly rejected claims to return tourism revenue to residents who have been forced to terminate traditional hunting and agricultural practices for conservation purposes (Weber 1981). Little information exists on ways to overcome such power struggles in ecotourism, let alone on the structure and causes of power-sharing conflicts.

Third, elements in an ongoing ecotourism program should end in an equitable manner if they are not contributing to the goal of biodiversity protection. Yet rarely does this form of termination proceed in such a manner. For example, the community-based marine reserve in Roátan was terminated abruptly by local elite who withdrew reserve funding despite intense objections from community members. According to a local elite family, the reserve was terminated because it was being managed inappropriately. Community members contend, however, that the reserve was terminated because the local elite were denied permission to build within marine reserve boundaries. In essence, termination appears to have been dominated by a participant group primarily interested in personal gains.

Appraisal

The appraisal function evaluates successes and failures with regard to achieving desired goals. Particular emphasis is given to the performance of each of the seven decision functions. During appraisal, a review of decision process successes and failures with regard to achieving the goal of biodiversity protection should be conducted. An independent examination of past decision function activity will reveal performance quality. Program appraisals should be both formal and informal, as well as internal and external. Participants affected by an ecotourism program must be identified and those responsible for outcomes held accountable. Appraisals should be dependable, realistic, ongoing, unbiased, and fully contextual (Clark 1997a; Lasswell 1971).

The majority of ecotourism literature and research is focused on reporting apparent tourism impacts on natural resources and local communities. While rich in anecdotal information, few case analyses use a conceptual framework with which to examine ecotourism development contextually and to measure program success (Brandon and Margoluis 1996). Similarly, program developers and managers often do not appraise field performance systematically. While some program developers have used social and ecological surveys to monitor local communities and ecological conditions (Jabcobson 1992; Weber 1993), field efforts have not been appraised by involved participants using an analytic framework to examine decision making.

When program appraisals are said to have taken place, however, such appraisals may have been initiated and carried out by elite participant groups—hardly unbiased participants. In Roátan, local elite and foreign tour operators have been accused of conducting a closed appraisal of the previously discussed community-based marine reserve. Although their method of inquiry was not inclusive, they decided, as previously indicated, that the reserve was managed improperly and that they would no longer provide economic support for its functions. In the absence of an unbiased and independent appraisal, it appears as if the appraisal function was used for the purpose of personal gain.

Appraisal 2: The Mountain Gorilla Project, Rwanda

In contrast to the previous analysis, the following discussion explores decision making during the development and application of one of the most successful ecotourism programs, the Mountain Gorilla Project (MGP) in Rwanda. Although the original purpose of this analysis was to use the MGP as a constructive exemplar of ecotourism decision making for comparative purposes, it became evident that the published literature does not provide sufficient detail for this purpose.

While the MGP literature highlights the consequences of a seemingly inclusive decision-making process, such as the development of an ecotourism program that has incorporated some of the needs of residents, the process through which these outcomes were achieved often is not entirely clear.

Without comprehensive documentation of how decisions were made during program development and application—a common and unfortunate occurrence with many ecotourism programs—the MGP cannot be used as a prototype for ecotourism decision making. Rather, this appraisal attempts to identify positive attributes of the MGP by making inferences about decision making from the published literature. By using the previously described criteria for decision making as a metric (Table 9.1), positive lessons from the MGP can be highlighted from the literature as well as the relevant questions for which information is lacking. Despite the fact that several criteria may not be addressed adequately, the purpose of this analysis is to learn as much as possible from the information available and to illustrate the importance of conducting systematic program appraisals.

Background

The slopes of the Virunga volcanic mountain range, a montane rainforest ecosystem consisting of eight volcanoes in east central Africa, are home to the only population of mountain gorillas in the world, a highly endangered species with approximately 350 surviving individuals (Weber 1981). This region also contains some of the most fertile mining and agricultural land in Africa and has attracted large human settlements for centuries. In 1925, the Belgian colonial government designated the eight volcanic mountains as the Parc National Albert, the first national park in Africa, in an effort to protect the gorillas from increasing anthropogenic pressures, such as poaching and land clearing for agricultural purposes. Park boundaries were created, however, with little regard for the resident human population, who depended on the area's natural resources. Without an alternative means of survival, the surrounding communities and government continued to threaten the gorilla population as they hunted and cleared land.

In the early 1960s, Rwanda gained independence from Belgium, which severely affected the future of the park and the mountain gorillas. The Parc National Albert was split into three different entities: the Parc National des Volcans in Rwanda; the Parc des Virungas in Zaire; and the Kigezi Gorilla Sanctuary, at the eastern corner of the range in

Uganda (Schaller 1989). As a result of this division, mountain gorilla protection was no longer controlled by Belgian authorities, and the gorilla population became increasingly vulnerable to invasion and destruction. The once united range was now under the control of separate governments that did not give a high priority to the protection of the gorillas and their habitat. Consequently, cattle-grazing and land-clearing practices escalated within park boundaries (Norton 1990). For example, in 1970 the Rwandan government converted almost half the Parc National des Volcans into farmland to harvest pyrethrum, a natural pesticide made from marigolds.

During the 1960s and 1970s, mountain gorilla populations reached a low of approximately 250 individuals. Surrounding human populations were expanding rapidly, and the consequent increase of hunting and poaching practices within park boundaries had devastating effects on the remaining gorilla population (Schaller 1995). Activities initiated by government officials, such as land clearing for agricultural purposes, also threatened mountain gorillas and their habitat. Although western researchers, such as George Schaller (1963) and Dian Fossey (1983), worked diligently to learn more about the ecology and condition of the gorillas, they did not address the most pressing issue threatening gorilla survival: the lack of support for gorilla conservation among resident groups and government officials.

In 1978, the New York Zoological Society sent zoologists Amy Vedder and Bill Weber to Rwanda to work with the Office of Tourism and National Parks to determine why the gorilla population was declining and to recommend "corrective action" (Weber 1993). Their research had four primary goals: to study the land-use practices of gorillas and humans; to determine attitudes and values of local peoples toward the park and gorillas; to develop a tourism program focused on the gorillas; and to initiate a conservation education program within the villages surrounding the park and gorillas (Schaller 1989). The following year their work was incorporated into a research initiative entitled the Mountain Gorilla Project, an integrative conservation project financed by an international consortium of conservation organizations (Schaller 1995; Weber 1981).

Although the extinction of this rare and impressive species seemed imminent, the careful development and application of the MGP was the turning point. Recognizing that the continued existence of mountain gorillas was contingent on the support of surrounding human communities, researchers attempted to integrate the needs of local people into conservation efforts through the development of an ecotourism program based on viewing mountain gorillas in their natural habitat. If managed properly, researchers recognized, mountain gorilla tourism could provide viable economic alternatives to local destruction of the gorillas and their habitat. Although the MGP is no longer in existence owing to a civil war and political instability within the region, a great deal can be learned about effective ecotourism development and application by examining MGP decision making.

The following discussion explores MGP decision making using the seven decision functions as a basis for analysis (see Table 9.1). For comparative purposes, this appraisal will follow a format similar to that of the previous appraisal section. This section is based upon my interpretation of the information provided in the published literature and may not represent as completely as possible what occurred during program development. For example, my discussion with Dr. Bill Weber, one of the primary program creators of the MGP, took place before I became interested in examining how decisions are made during ecotourism development. Consequently, Dr. Weber and others involved with the MGP may have answers to several of the questions I pose in this analysis, but they have not been incorporated into this discussion. The purpose of this analysis is to illustrate the kinds of questions that should be addressed by program appraisals so that the strengths and weaknesses of decision making can be used by scholars and practitioners.

Intelligence

The Mountain Gorilla Project was developed primarily by two western researchers, Dr. Bill Weber and Dr. Amy Vedder. Their goal, similar to that of other researchers who worked in the region, was to protect the mountain gorillas from human encroachment. Weber and Vedder's approach differed, however, in that they wanted to devise a conservation strategy that would incorporate the needs of resident groups living in proximity to the gorillas and their habitat. With poaching and land-clearing pressures from surrounding communities a primary threat to gorilla survival, they recognized the importance of incorporating the needs of these communities with conservation efforts.

To develop a clear and comprehensive understanding of the obstacles facing mountain gorilla conservation, Weber and Vedder first gathered information about the condition of the gorillas and their habitat. They conducted a new census of the remaining gorilla population, which revealed that it was relatively stable at approximately 268 individuals. They also found, however, a continued decline in the percentage of young gorillas because of poaching and other disturbances (Crouse 1988; Weber 1993). Vedder examined gorilla habitat and feeding habits to determine whether or not the remaining forest could sustain the existing population. Although a significant amount of habitat had been lost to agricultural land clearing, she found that enough habitat remained to support gorilla population growth if destructive practices, such as deforestation and poaching, could be controlled (Crouse 1988; Weber 1993).

With relatively promising results from the ecological analysis, Weber (1981) examined public knowledge and perceptions of the park and gorillas. Surveys measuring local attitudes and awareness found that there was little support for the protection of mountain gorillas and their habitat. The rich volcanic soils of the region had attracted a large human settlement with an alarming growth rate of 3.7% per year, resulting in a population of 6 million residents and a reduction of the average farm size to barely one acre per family with poor agricultural potential. Land acquisition was particularly intense because of a lack of nonagricultural employment opportunities in a rural economy where over 95% of the population lived by farming (Weber 1993). Weber found that with such a high human population and scarcity of land, Rwandan attitudes toward protecting the park and the gorillas largely were negative (Crouse 1988; Weber 1993).

Additional background research revealed the importance of developing conservation efforts in conjunction with the Rwandan government. In recent decades, government officials generally had not been supportive of mountain gorilla conservation. With pressures from the exploding human population for jobs and economic expansion, government officials were threatening to clear a large portion of the remaining forested area, which included the richest gorilla habitat, for cattle-raising purposes (Weber 1993). To stimulate and solidify government resistance to park conversion, program developers recognized, government officials should be involved directly with program development and application.

The information obtained by Weber and Vedder during the ecological and social analyses appeared to be dependable, comprehensive, selective, and creative, and helped them to clarify objectives, describe trends, analyze conditions, project developments, and formulate policy alternatives. For example, it appears that Weber recognized that there could be tremendous economic gains if tourists were brought by guides to see the mountain gorillas during their visits (Weber 1981). It also appears that he recognized that a successful ecotourism industry would have to be designed and applied carefully so that: the mountain gorillas were not harmed by tourism; significant economic revenue would be generated; local people would benefit from the industry; and the Rwandan government would place a high priority on mountain gorilla protection.

While it is clear from the published literature that Weber and Vedder gathered extensive data on the gorillas, the values and perspectives of surrounding resident groups, and the regions' social, political, and economic context, there is little or no explanation of the process by which decisions were made. At a very basic level, who was involved with the development of the MGP? Why was the MGP established in Rwanda instead of the neighboring countries of Zaire and Uganda? How are natural resource decisions made in Rwanda? Who participated in the decision to conduct social surveys? Who was involved with analyzing the information obtained from the social and ecological surveys? Who was involved with interpreting the results? How was the Rwandan government approached? How were local residents involved with program development? How and with whom did Weber and Vedder clarify objectives, describe trends, analyze conditions, project developments, and formulate policy alternatives?

Promotion

Without a basic description of the process by which natural resource decisions are made in Rwanda, it is difficult to deduce the way in which management options were discussed and formulated during the formative stages of the MGP. The extent to which all participant perspectives were included in the discussion and formulation of management alternatives is also unclear. It can be inferred, however, that the resources, data, and opinions gathered during the ecological and social analyses were considered by Weber and Vedder as they developed and discussed management alternatives. For example, the surveys revealed that resident groups living in proximity to the park and gorillas placed a high wealth value on the gorillas and their habitat, and decision makers recognized the importance of developing a conservation strategy that would provide local residents with tangible economic revenue. The negative attitudes toward the park and gorillas that were identified during the social surveys also prompted Weber to consider implementing conservation education programs within surrounding communities.

It can also be inferred that the information obtained during the ecological surveys was extremely important during program development. With mountain gorilla protection as their primary goal, ecological surveys revealed the importance of limiting direct tourist impacts on gorilla groups that would be habituated to human presence. Social and ecological surveys also illustrated the need for establishing protective measures to eliminate immediate threats from poaching and illegal land clearing. Once again, however, it is not clear how management options were discussed, or even the different kinds of options that may have been considered. For example, who was involved with the development of management alternatives and what were the different options considered during discussions? Nevertheless, it appears as though the promotion function was carried out in a manner that was integrative, comprehensive, and effective.

Prescription

In 1979, after spending approximately one year obtaining site-specific information and developing management options, Weber and Vedder launched the action phase of the MGP (Weber 1993). With increased poaching and forest clearing identified as the primary and immediate threat to the gorillas and their habitat, the first prescriptive action taken by program developers was the decision to heighten border patrols. Once poaching and land clearing pressures were alleviated, Weber and Vedder appeared to focus their attention on the development of the ecotourism program. Tourism would be based on the viewing of mountain gorillas in the wild, but strict program guidelines were established to minimize tourist impacts. Based on information obtained in the ecological analyses, disturbance was to be controlled by: limiting the number of visitors to six per gorilla group each day; making no more than one visit to each group per day; timing the visits to coincide roughly with regular late-morning gorilla rest periods; and excluding direct physical contact with the gorillas. In addition, children, the most frequent germ carriers, and visitors with colds and other infections were not to be permitted to visit the gorillas (Weber 1993).

Although this was not a formal prescription, both local residents surrounding the park and the Rwandan government were to be actively involved in the tourism program. Instead of employing foreign tour operators, a practice that is typical of many ecotourism programs, the program trained and employed local people as workers and tour guides (Weber 1993). At the same time, revenues generated from substantial entrance fees, $170.00 per visit, were to be funneled back into the surrounding communities by increasing park guard and guide salaries. Program developers also hoped that the Rwandan government would earmark a certain portion of additional tourism revenue to be returned to local communities.

Weber decided to develop a conservation education program within local communities. Based on the results of the social analysis, the educational program had three main objectives: to promote a better understanding of the park's benefits to the local people; to create support for the park in local villages; and to sensitize the local people to regional environmental problems and the means to reduce these problems (Schaller 1989; Weber 1995).

While careful examination of the published literature does describe many MGP prescriptions, the way in which these prescriptions were developed is not discussed. Who was involved with the development of MGP prescriptions? How were prescriptions decided upon? Although program guidelines appear to have been inclusive, the extent of inclusiveness is unclear without a comprehensive understanding of all participants involved with or affected by tourism development. MGP prescriptions appear to have been effective, but did they reflect the common interest in the protection of the mountain gorillas?

Invocation and Application

Although issues relating to the invocation and application of program prescriptions are rarely discussed in the literature, it appears that those hired to make reference to previously established guidelines encountered little resistance. For example, the border patrols established during the formative stages of program development appeared to be successful because there were no reported poaching incidents once they were in place. Weber habituated four gorilla groups (out of thirty-one total groups in the reserve) to human presence for tourism purposes, and tourist guidelines were strictly enforced by local guides and guards who had received extensive training for dealing with international tourists. Although this was not a formal program prescription, approximately fifty guards and several tour guides were hired from surrounding villages to ensure that a portion of economic revenue remained within local communities. Conservation education efforts also appeared to encounter little resistance from surrounding communities.

One of the few documented conflicts during invocation and application occurred when program developers invoked and tried to apply the informal prescription that government officials should provide local communities with an additional portion of tourism revenue. While program developers argued for the need to establish a revenue-sharing system so that local communities would receive more financial benefits from the tourism industry, government officials systematically rejected such proposals (Weber 1995). It is still unclear, however, whether the invocation function was carried out in a timely, dependable, rational, and nonprovocative manner. Similarly, although application does appear to have been authoritative and controlling, it may have been controlled by special interests.

Termination

The published literature does not address explicitly the three aspects of termination as described in the previous analysis: the termination of local practices and traditional ways of life; the termination of power structures; and the termination of program prescriptions. For example, how were families and individuals compensated for ending local practices, such as poaching and land clearing? While some individuals clearly benefited from the tourism industry through job opportunities, was this an adequate form of compensation? How were traditional power struggles addressed during program development and application? While the consequence of such struggles was evident when government officials refused to earmark tourism revenue for local communities, did these struggles affect the entire decision-making process? If so, in what ways? Furthermore, if such struggles or conflicts existed, the ways in which they were resolved are not clear.

The literature does, however, indicate that the termination of program prescriptions was not always done in an ameliorative or equitable manner. For example, a popular component of the education initiative, complimentary student visits to see the gorillas, ended abruptly because the Rwandan government believed that "these visits were limiting sites for foreign visitors to visit and thereby causing the government to lose precious foreign revenue" (Weber 1995:40). The manner in which additional program prescriptions were terminated is less clear. In addition, with reference to the termination criteria listed in Table 9.1, it is difficult to ascertain the extent to which the termination function was carried out in a timely, dependable, comprehensive, or ameliorative manner.

Appraisal

Within a few years after program application, program developers started to appraise the effectiveness of the MGP. The project proved to be an economic success by boosting park revenues some 2,000% and by making tourism Rwanda's third largest producer of foreign capital behind coffee and tea (Crouse 1988). At the program's peak, approximately 7,000 visitors came to Rwanda to see the gorillas each year (Crouse 1988). Park entrance fees were increased to try to curb demand, but tours continued to sell out and annual revenues in 1989 surpassed $1 million for direct entry fees alone (Weber 1993). Structured economic studies were not done to measure economic impact on local communities, but more than 100 full-time jobs were created by the MGP, and guard and guide salaries were estimated to be four times the national average (Weber 1995). It is also estimated that an additional $3 to $5 million dollars remained in local communities through tourist expenditures on lodging, food, and souvenirs (Weber 1993).

To monitor program effectiveness, follow-up knowledge and attitude surveys were conducted among the original survey participants approximately five years after program commencement. The surveys revealed a much greater recognition of nonconsumptive values toward the park and gorillas and that the forest was better recognized for its ecological importance and tourism value (Weber 1995). The Rwandan government also became an avid supporter of mountain gorilla conservation. Program developers no longer had to explain the importance of the gorillas and their habitat to government officials, because the tourism industry brought tangible justification for conservation. This strong political support was displayed during Rwanda's recent political conflicts. Although the tourism program ended abruptly in 1989 with the outbreak of civil war, both warring parties pledged to the international conservation community that they would not harm the gorillas while fighting (Schaller 1995).

Most important, a gorilla population census revealed that the MGP was protecting the gorillas and their habitat. Park guards brought a rapid decline in gorilla hunting, with no known poaching incidents after 1984. Gorilla numbers increased dramatically from a low of approximately 254 individuals at the beginning of the program to 320 in a 1989 census. The percentage of young also increased from 31% to more than 51% of the total population. In fact, the four gorilla groups visited by tourists had the highest percentage of young in the 1989 census (Weber 1995).

While it is clear that Weber and Vedder appraised certain aspects of the MGP, the process by which these appraisals took place is not described in the literature. Were appraisals independent? Were they formal or informal, as well as internal and external? Were appraisals dependable, realistic, ongoing, unbiased, and fully contextual? Who was involved with conducting these appraisals? Without a systematic and explicit approach, it is difficult to determine exactly how MGP appraisals were designed and implemented.

Lessons Learned

The primary purpose of this paper is to illustrate the importance of using a contextual, systematic approach when analyzing ecotourism successes and failures. The decision process analyses in this paper are an exercise in using a systematic and contextual framework for analysis and provide a preliminary illustration of the processes by which ecotourism decisions are made and applied in two different contexts: ecotourism development in Roátan, Honduras, and mountain gorilla tourism in Rwanda. These two case studies were chosen to compare decision processes between a relatively successful ecotourism program that appeared to provide adequate protection for mountain gorillas and a seemingly less successful program that is causing extensive damage to the region's natural resources. Table 9.2 summarizes how each program measures up to the previously described criteria for the adequate performance of each decision function (see Table 9.1).

Several important lessons can be learned from these two case study analyses. First, it is important to recognize the value of examining the processes through which decisions are made within each context, using the same comprehensive and systematic framework of inquiry. As illustrated in Table 9.2, such analyses can generate information for making very basic comparisons between each program. For example, ecotourism development in Roátan appears to be monopolized by special interests and does not appear to be accomplishing the goal of biodiversity protection, while mountain gorilla tourism in Rwanda appeared to have been developed with an inclusive decision process and provided the means and incentives to protect mountain gorillas and their habitat. Although this paper is a first attempt at examining ecotourism decision making in such a manner, any subsequent appraisals will be more comprehensive and effective if program developers are able to analyze their field efforts using this framework.

Second, the decision process analyses have revealed the need for better understanding of the social process dimensions of the ecotourism industry (Clark and Wallace 1998). Program developers must have a comprehensive understanding of the values and perspectives of all participants involved with or affected by ecotourism development. More specifically, if ecotourism is to provide residents with the economic means to participate in natural resource protection, efforts must be made to understand how human populations interact with and perceive their natural resources. For example, developers of the MGP clearly grasped the importance of understanding how resident people interact with and perceive their natural resources during the for-

Table 9.2. Decision Process Comparisons (Inferences from Field Research and the Published Literature)

	Roátan	Mountain Gorilla Project
Intelligence	Sporadic and acontextual; virtually nonexistent prior to program application	Background research; social and ecological surveys during formative stages of program development; dependable, comprehensive, selective, creative
Promotion	Exclusive and acontextual; appears to be controlled by wealthy and powerful participants	Appeared to be socially and contextually inclusive, as well as integrative, comprehensive, effective
Prescription	Appears to be effective and inclusive in some cases but was not effectively invoked or applied	Appeared to be effective, rational, and inclusive
Invocation	Elite groups appear to control the industry; exclusive and ineffective	Not stated explicitly in the literature, but appeared to be timely, dependable, and rational; problems, however, with earmarking funds for local communities
Application	Appears to be controlled by wealthy elite; nonexistent or ineffective	Not stated explicitly in the published literature, but appeared to be open, inclusive, rational, and uniform
Termination	Nonameliorative; seemingly controlled by a select group of participants	Unclear from the published literature; some compensation for local practices via tourism revenue; conservation education program ended abruptly
Appraisal	Biased, sporadic, and exclusive	Appeared to be comprehensive and selective, but unclear about independence and continuity

mative stages of program development. Program developers on Roátan, however, did not take the time to understand the relations between residents and the surrounding reefs as well as the effects of increased, unrestricted tourism and development.

An additional aspect of social process mapping is identifying and understanding how power struggles among or between disparate participants groups—particularly local and international elite, foreign tour operators, and the rural poor—often affect the application of many ecotourism programs. While MGP developers appeared to have had a better understanding of the socioeconomic and political context of the region, there is little information regarding the kinds of power struggles that may have existed during program development and application. Although the literature mentions that the Rwandan government would not earmark an additional portion of tourism revenue to be returned to local communities, program developers do not provide a contextual understanding of this conflict. In Roátan, power struggles between the rural poor and local elite appear to be the primary obstacles to a successful ecotourism industry, and significant attention should be focused on better understanding such conflicts.

Third, both decision process analyses illustrate the need for better documentation of the actual process by which natural resource decisions are made in a particular region. Is there an existing decision process through which ecotourism decisions are made? Who are the participants involved with or affected by ecotourism development? What are the values and perceptions of each participant group? Who is in charge of decision making? How do disparate participant groups interact during decision making? How are management alternative discussed? Who establishes rules and policies? How are prescriptions invoked and applied? Who is in charge of termination and how is it carried out? How is decision making appraised to ensure that program goals are being met in an equitable manner?

Last, this project has identified the need to take a step back and examine the appropriateness of ecotourism as an integrative strategy in natural resource conservation. Is biodiversity protection an appropriate goal for an ecotourism program? Can tourism development and biodiversity protection be merged effectively in an ecotourism program? Is ecotourism another strategy to impose western ideals of natural resource conservation on developing countries, or do programs really strive to maximize human dignity? Is biodiversity protection a suitable goal for maximizing human dignity? For whom does ecotourism provide successes and failures? The international conservation community should examine the ecotourism industry within this larger context. Drawing upon the related disciplines of anthropology and political ecology, I believe that the ultimate success of natural resource conservation will depend upon our ability to examine how human needs and concerns, including those of the international conservation community, may define and ultimately impede conservation efforts.

Recommendations

With an increasing number of ecotourism programs in a variety of different regions throughout the world, there is ample opportunity to systematically appraise decision making in many different contexts. Similar to the case studies presented in this paper, existing ecotourism programs should be appraised systematically using a practiced-based approach (Brunner and Clark 1997; Clark 1996) to gain a better understanding of the strengths and weaknesses of program development and application. More specifically, programs should be evaluated using policy sciences as an analytic framework in order to determine whether, and how, they are accomplishing their goals in a manner that reflects the common interest. While any attempt to examine decision processes in the ecotourism industry using the policy sciences would be beneficial, it would appear to be most advantageous if program developers were able to appraise their field efforts using a systematic framework for analysis.

Practical recommendations can be extracted from appraisals and used for developing ecotourism prototypes (Clark 1997a, b). Based on information obtained from decision process appraisals, prototypes can provide recommendations for how decision making should proceed and practical considerations for how to accomplish each decision function. Appraisals will undoubtedly be controversial because they assign responsibility for program successes and failures (Lasswell and McDougal 1992), but such accountability is essential for making overall improvements. For example, lessons learned from program appraisals can be examined in a creative fashion by developing decision process prototypes for the ecotourism industry and testing them in a field setting using an adaptive management approach (Brunner and Clark 1997). Such an approach would be most applicable in a region in which ecotourism is emerging so that researchers can propose to work with practitioners to test particular decision process prototypes. Ongoing learning is encouraged because prototypes are flexible and can be altered and adapted after an adequate trial period (Clark 1997a, b).

Conclusion

In this paper, I have attempted to understand why ecotourism programs often do not appear to be protecting the biodiversity upon which programs depend. Ecotourism is a complex issue, and the policy sciences have been instrumental in my attempt to analyze why such problems occur and how to work toward developing effective solutions.

While my analysis does not provide solutions to the problems I have identified in Roátan or answer the questions posed about the MGP in Rwanda, I hope that it can provide insight into the utility of conducting program appraisals using a comprehensive and systematic framework for analysis. Such analyses may not yield immediate answers to conflicts associated with ecotourism programs, but perhaps a collection of program appraisals will begin to reveal some of the underlying problems that need to be addressed, such as the complex relations between people and natural resources, power struggles between disparate participant groups, and a very basic understanding of the process by which decisions are made. Most important, program appraisals may force the international conservation community to examine the appropriateness of ecotourism as an integrative conservation strategy.

ACKNOWLEDGMENTS

I thank Dr. Tim Clark and Andrew Willard for introducing me to the policy sciences and encouraging this project. Their tireless devotion to teaching the policy sciences and promotion of creative thinking is inspiring. This project would not have been possible without the help and support of Andrew Willard, and I greatly appreciate his thoughtful and inexhaustible input and encouragement. Dr. Clark and I also have spent hours discussing ecotourism and the policy sciences, and I am truly grateful to have him as a mentor. I also thank Denise Casey of the Northern Rockies Conservation Cooperative (NRCC) for her support and thoughtful editing on several drafts of this paper. Many thanks also to Peyton Curlee (of the NRCC) for allowing me to work in her office and for providing constant encouragement.

LITERATURE CITED

Arnold, D. 1995. "Surviving Human Chaos," *World View* 8:5–8.

Barker, K. . 1996. "To Ecotour or not to Ecotour: Unpacking the Impacts and Business Realities of Tourism Development in Sana and Tariquia Reserves in Tarja, Bolivia," *School of Forestry and Environmental Studies Bulletin Series* 99:251–262.

Boo, Elizabeth. 1990. *Ecotourism: Potentials and Pitfalls.* Washington, DC: World Wildlife Fund—US.

Brandon, K. 1996. *Ecotourism and Conservation: A Review of Key Issues.* Washington, DC: The World Bank.

Brandon, K., and Richard Margoluis. 1996. "The Bottom Line: Getting Biodiversity Conservation Back into Ecotourism," *School of Forestry and Environmental Studies Bulletin Series* 99:28–38.

Brandon, K., and Michael Wells. 1992. *Parks and People: Linking Protected Areas Management with Local Communities.* Washington, DC: The World Bank.

Brunner, R. D., and T. W. Clark. 1997. "A Practice-Based Approach to Ecosystem Management," *Conservation Biology* 11(1):48–58.

Clark, T. W. 1996. "Appraising Threatened Species Recovery Efforts: Practical Recommendations," in S. Stephens and S. Maxwell, eds., *Back from the Brink: Refining the Threatened Species Recovery Process*. Australia: Surrey Beatty & Sons.

———. 1997a. "Conservation Biologists in the Policy Process," in G. K. Meffe and C. R. Carroll, eds., *Principles of Conservation Biology*. Sunderland, MA: Sinaur Associates.

———. 1997b. *Averting Extinction: Reconstructing Endangered Species Recovery*. New Haven: Yale University Press.

Clark, T. W., G. N. Backhouse, and R. P. Reading. 1995. "Prototyping in Endangered Species Recovery Programmes: The Eastern Barred Bandicoot Experience," in A. Bennett, G. Backhouse, and Tim Clark, eds., *People and Nature Conservation. Transactions of the Royal Zoological Society of New South Wales*. Australia: Surrey Beatty & Sons.

Clark, T. W., and Richard Wallace. 1998. "Understanding the Human Factor in Endangered Species Recovery: An Introduction to Human Social Process," *Endangered Species Update* 15(1):2–9.

Cope, G. 1996. "Nature Travel and Rainforests," *School of Forestry and Environmental Studies, Bulletin Series* 99:43–48.

Craighead, J. 1991. "Yellowstone in Transition," in R. B. Ketier and M. S. Boyce, eds., *The Greater Yellowstone Ecosystem: Redefining America's Wilderness Heritage*, pp. 27–39. New Haven: Yale University Press.

Crouse, D. 1988. "Up Close with the Gorillas," *International Wildlife*. November/December:5–11.

The Ecotourism Society. 1994. Fall Newsletter 4(4):6.

Forest, N. B. 1995. Coastal Regulation and Implementation: Case Study of Roatan, Islas de la Bahia. Master's thesis, University of California at Berkeley.

Freudenberger, K. S. 1997. Rapid Rural Appraisal, Participatory Rural Appraisal. Notes to Accompany an Introductory Course. Yale School of Forestry and Environmental Studies, New Haven, Connecticut.

Herliczek, J. 1996. "Where Is Ecotourism Going?" *Amicus Journal* Spring:31–35.

Jacobson, S. K. 1992. "Ecotourism, Sustainable Development, and Conservation Education: Development of a Tour Guide Training Program in Tortuguero, Costa Rica," *Environmental Management* 16(6):701–713.

Lasswell, H. D. 1971. *A Pre-View of Policy Sciences*. New York: American Elsevier.

Lasswell, H. D., and M. McDougal. 1992. *Jurisprudence for a Free Society: Studies in Law, Science and Policy*. New Haven, CT: New Haven Press.

Machlis, G., and David Tichnell. 1985. *The State of the World's Parks: An International Assessment for Resource Management, Policy, and Research*. Boulder, CO: Westview Press.

Mendelsohn, R. 1994. "The Role of Ecotourism in Sustainable Development," in G. K. Meffe and C. R. Carroll, eds., *Principles of Conservation Biology*, pp. 511–515. Sunderland, MA: Sinauer Associates.

Munn, C. 1992. "Macaw Biology and Ecotourism, or 'When a Bird in the Bush is Worth Two in the Hand,'" in S. R. Beissinger and N. Snyder, eds., *New World Parrots in Crisis*, pp. 47–72. Washington, DC: Smithsonian Institution Press.

Norris, Ruth. 1992. "It's Green, It's Trendy, Can Ecotourism Save Natural Areas?" *National Parks* 66(1–2):30–34.

Norton, B. 1990. *The Mountain Gorilla*. Stillwater, MN: Voyageur Press.

Olindo, P. 1991. "The Old Man of Nature Tourism: Kenya," in T. Whelan, ed., *Nature Tourism*, pp. 23–38. Washington, DC: Island Press.

Padgett, T., and Begley, S. 1996. "Beware of the Humans," *Newsweek* February: 52–54.

Poff, C. 1996. Protected Areas Management Options for the Next Century. Unpublished Report.

Schaller, G. B. 1963. *The Mountain Gorilla: Ecology and Behavior.* Chicago: University of Chicago Press.

———. 1989. *Gorilla: Struggle for Survival in the Virungas.* New York: Aperature Foundation.

———. 1995. "Gentle Gorillas, Turbulent Times," *National Geographic* 188:65–68.

Taylor, M. R. 1994. "Go, but Go Softly," *Wildlife Conservation* March/April:12–18.

Weber, B. 1981. Conservation of the Gorillas: A Socio-Economic Perspective on Habitat and Wildlife Preservation in Rwanda. Ph.D dissertation, University of Wisconsin.

———. 1993. "Primate Conservation and Ecotourism in Africa," in C. S. Potter, J. I. Cohen, and D. Janczewski, eds., *Perspectives on Biodiversity: Case Studies of Genetic Resource Conservation and Development,* pp. 129–147. Washington, DC: AAAS Press.

———. 1995. "Monitoring Awareness and Attitude in Conservation Education: The Mountain Gorilla Project in Rwanda," in S. Jacobson, ed., *Conserving Wildlife: International Education and Communication Approaches,* pp. 28–48. New York: Columbia University Press.

West, P. C., and S. R. Brechin, eds. 1991. *Resident Peoples and National Parks: Social Dilemmas and Strategies in International Conservation.* Tucson: University of Arizona Press.

Whelan, Tensie. 1991. *Nature Tourism.* Washington, DC: Island Press.

10 An Integrated Approach to Conservation and Human Development in the Management of Kyabobo Range National Park, Ghana

David Lyon

The new Kyabobo Range National Park (KRNP), in the Volta Region of Ghana, was set up by the governmental Wildlife Department (WD) between 1991 and 1996 to conserve a rare mountainous forest/savanna ecosystem. The WD acknowledges that "individual protected areas are rarely self-sustaining ecological isolates but integral components of a larger ecological continua and that if protected areas are to be developed sustainably they should aspire to meet the basic needs of local people in an equitable way" (Government of Ghana 1994:1). The policy problem is that local, national, and international institutions have had a limited capacity (or understanding) to incorporate many of the underlying ideologies of other key actors affected by or involved in KRNP. As a result many local perspectives are not expressed sufficiently in the decision-making process of the park management. Therefore, the WD principle that "meaningful protected area management should be integrated into the dynamic process of rural development" (Government of Ghana 1994:2) has not been fully realized.

This paper examines and appraises the decision functions involved in setting up and managing conservation activities in and around KRNP. The study investigates the roles of different participants in establishing arenas of power and access to these arenas. The goal of this

This paper was written in 1993.

study is to make recommendations for how local perspectives can be represented in the decision process to enable both conservation and development. Much of the information used is influenced by personal perceptions of the problem and does not necessarily reflect the views of either the WD or the Kyabobo Conservation Project (KCP), for whom I worked while in Ghana. Therefore, it is important to clarify my standpoint to explain the basis of this study.

In 1994, I was a biologist involved in the first surveys conducted around the proposed Kyabobo Range National Park in the Volta Region of Ghana (KCP 1995). As a result of this study, I became more aware of resource depletion and degradation in and around the National Park. My awareness was conceptualized through a myth that "the overriding purpose of science is prediction with precision, scope and accuracy" (Brunner and Ascher 1992:296). Therefore my perspective was narrow and inevitably resulted in a highly selective characterization of the environmental problems I was studying. Research and methods to collect the data for this study had been influenced by an upbringing in the positivist, rational field of biological science. From 1994 to 1997, I began viewing the role of science in conservation as "freedom through insight" (Brunner and Ascher 1992:311). Through fieldwork in this period, especially my involvement in the boundary assessment for KRNP, my perspective has shifted from appreciating the environment mainly for its nonuse values to appreciating the importance of the environment for improving people's quality of life. This has led to a strong belief in the need for local participation in the conservation of ecosystems.

In addition to the perspectives just described, my role in this policy problem is shaped by other factors. I am a co-director of the KCP. This is a small nonprofit set up by three graduates, myself included, from the United Kingdom with strong support from Ghanaian partners. The close collaboration between the KCP and the Wildlife Department is realized by most Ghanaians affected by the National Park. This can bias the information provided by local people.

Although I have engaged in two periods of research working in villages surrounding KRNP, I am still a foreigner. Therefore, factors of local culture may be overlooked. In addition, I am aware of using certain strategies and involving myself in decisions that could put me at risk of being a neocolonialist. I naturally have strong views and often promote them.

Therefore my standpoint in 1994 was that of a student. I was collecting biophysical information, and conducting a few background interviews on natural resource use. In 1995, I had added to my role of student that of director of the KCP. This involved being a participant observer, an interviewer, and a project manager.

The information obtained for this report is a combination of all the research conducted in the area. This includes biological surveys, structured interviews, participatory research approaches, semistructured interviews, group discussions, and personal observation. Information has also been collected from the correspondence between the KCP, WD, local people, and the local government. Finally, the work involves literature research of similar projects in West Africa. This does not describe the current situation at KRNP because only information up to the end of 1996 is included and KRNP is an ongoing project with continually changing conditioning factors.

The Policy Problem

In addition to the great benefits that Kyabobo Range National Park will bring to Nkwanta District, the park's establishment might have led to value deprivation of certain villagers surrounding the park as a result of lost traditional land and limited access to resources. In addition, if the perceived benefits from the park such as rural development, tourism, and employment are not realized, then local support for the park will most likely decrease.

An underlying problem is the difference between the government and some local participants in conceptualizing the merits of alternative land-use regimes. For example, whether protection or shifting cultivation is perceived as beneficial or harmful varies among participants. The problem is not only an ideological one, but also one of incorporating different participants' demands into the decision process.

Therefore, the goal for this study is to investigate how (or whether) conservation and rural development can both be integrated in KRNP. Accomplishing this goal would involve developing a decision process through which participants can attempt to clarify and secure their common interest. Security of the common interest will require contextuality in place of fragmentation (Lasswell 1971a:439) through the use of "problem-oriented perspectives" (Brunner 1987:10).

Clarification of the Problem

The Social Context Behind the Establishment and
Management of Kyabobo Range National Park

There is wide variation in the perspectives of participants involved in this policy process. Table 10.1 summarizes the perspectives, situations, base values, strategies, desired outcomes, and effects of some of the key participants involved. These differences between local people, the KCP, and the WD are related to the way in which environmental change is perceived: Local people do not necessarily view biodiversity conservation (in the western sense) as a priority (Ledward and Bowes-Lyon 1996:57). With a western positivist, rationalist myth, the effects of present events are often evaluated in terms of their future impact. Conservation is often perceived as conserving resources for the next generation. This is illustrated in many current publications such as Bruntland's definition of sustainability (WCED 1987). Around Kyabobo, the forest "represents the ancestors and protects people from fire, hunger, and death" (Ledward and Bowes-Lyon 1996:57). Therefore, restricting general forest access is regarded as not respecting the ancestors who struggled to farm in the first place. However, within the category "local people," there are many different perspectives. In general, this paper has simplified the term *local people* to refer to the Akyode tribe, the largest group in the area and the group most affected by the park.

Differences in perspectives also influence the way intelligence is gathered and processed and an alternative promoted. For example, the KCP and WD collected information for the 1996 boundary survey, such as where people farmed, and displayed this information on maps (O'Keefe et al. 1997). Maps have the potential to "dismember an exceptionally complex set of relationships and processes which are poorly understood in order to isolate a single element of instrumental value" (Scott 1998). On the other hand, depicting information on maps could be the only method available to the WD and KCP for organizing the information, given the constraints. The differences in perspectives can account for some of the "indicators of malfunction" (Lasswell 1971b:96) such as comprehensiveness and openness (see Table 10.3).

Situations vary. Table 10.1 shows that different participants have power bases at different locations, such as Nkwanta (district capital), Accra (capital of Ghana), the villages surrounding the park, and Great Britain. The dispersion of participants means that the criteria of "openness in intelligence" is difficult to maintain, as some farms can be more than a seven-hour walk from the nearest road. Situations also limit the continuity of appraisal. In addition, different locations create different benefits among the range of participants. For example, a thick ravine forest can mean timber revenue for rural elites, a sacred area for members of the religious cult, a biodiversity hot spot for the KCP and WD, or an important area for local women to harvest oil palm nuts for soup.

The influence of each participant in terms of "value position and potential" is extremely varied and complex (Lasswell and Kaplan 1950:60).

Table 10.1. Summary of the Social Context of the Establishment and Management of Kyabobo Range National Park

	Wildlife Department (part of national government)	Local Communities	Nkwanta District Assembly	Kyabobo Conservation Project
Perspectives	Conservation will help rural development; conservation is important for national heritage. *Demand:* Areas are protected for future generations. *Expectation:* People are prepared to support; the Kyabobo ecosystem will be conserved.	Emphasize respect for ancestors and traditional culture. *Demand:* Park should consider human dimensions such as respect for ancestors. *Expectation:* Park will improve rural development but may affect traditional forms of livelihood.	"Modernization" is important for Nkwanta to develop. *Demand:* Nkwanta District must develop and increase in significance nationally. *Expectations:* KRNP can provide the necessary development.	The importance of conservation and development. KCP also believes in strong communication links between participants. *Demand:* Local participation can lead to conservation. *Expectation:* KCP is a short-term NGO which will facilitate other organizations to work in the area to help all participants.
Situation	Organized bureaucracy but not monolithic group. Substantial amount of decision making conducted in Accra. Remaining power in hands of reserve officers.	Not a monolithic group. They are scattered in farmlands, based in villages, directly affected, or tenant farmers. But most decisions made in village.	Based in the district capital. Officials are appointed by central government, which is voted in. This is not a monolithic group.	Substantial amount of time spent in UK promoting work conducted in KRNP. Present in KRNP during 1994 and 1996. Some work also conducted in main Ghanaian towns, Accra and Kumasi.
Base values (anticipated assets used)	Power, respect, enlightenment, skill, rectitude, affection, well-being, wealth.	Rectitude, enlightenment (of local resource management), skill, affection.	Respect, power, wealth, enlightenment, skill, affection, and rectitude.	Wealth, enlightenment, respect (from WD and some local people), rectitude, skill, affection (power informally).

Table 10.1. continued

	Wildlife Department (part of national government)	Local Communities	Nkwanta District Assembly	Kyabobo Conservation Project
Strategies	Noncoercive with all participants (e.g. local government), sometimes coercive with others (local communities). Ideologic, diplomatic, militaristic.	Noncoercive: ideological, emotional pleas and diplomatic. Coercive: passive forms of resistance.	Coercive: can arrest citizens breaking laws. Noncoercive: works to ensure diplomatic strategies. Also economic, ideological: e.g., explains benefits of tourism.	Address problems through research, propose small projects to address problems of participation. Ideological and diplomatic strategies are important.
Outcome (in terms of demands and expectation)	Park boundary (prescription) set up, local staff employed, conservation initiatives (application). Potential gain of power, respect, wealth, affection, enlightenment, skill, rectitude. Potential loss of affection, rectitude.	Lost resource access, mixed outcomes depending on institution. Potential loss of wealth, respect, well-being, affection, power, and rectitude. Potential gain of enlightenment, skill, wealth, and respect.	A well-managed park providing benefits to the local government. Potential gain of respect, power, wealth, skill, and affection (nationally). Potential loss of respect (locally).	A long-term sustainable management plan with equitable value indulgences and deprivations. Potential gain of enlightenment, skill, rectitude. Potential loss of well-being, wealth.
Effects	Ghana has value indulgences in international respect, wealth, rectitude, etc. WD has increased power, rectitude, wealth. It also has increased enlightenment with respect to people's participation.	Distribution of base values with local communities varies with both indulgences and deprivations. Social structure in some villages affected.	Increase in enlightenment re: conservation issues. The significance of Nkwanta District increases as a result of the park.	Increased dialogue between local people and WD; increased local, national, and international awareness of KRNP. Improved experience benefits directors and results in increased enlightenment, skill, and rectitude.

Note: KCP = Kyabobo Conservation Project; KRNP = Kyabobo Range National Park; NGO = nongovernmental organization; WD = Wildlife Department.

Dramatic shifts in values as a result of the park could include increased power to conserve biodiversity and enlightenment for the Wildlife Department. The value shift for local people could include respect (if farmers must move from being landowners to tenants), wealth (if areas of valuable timber become part of the National Park), and rectitude (if members of the traditional religious organization, or cult, are forced to find farmland away from their shrine). However, there are also potential indulgences for local people. For example, enlightenment could increase if people are more likely to go to schools; well-being could also increase if people are closer to clinics and transport is improved.

At the larger scale, the local government and local members of parliament could increase the profile of Nkwanta within Ghana, which would improve the respect, power, and wealth values of the district. The co-directors of the KCP could gain enlightenment and rectitude from the work they have done. It should be realized that different individuals within an institution have varying degrees of base values depending on their position within an organization, their social background, and their relationships with other participants. The summary provided in Table 10.1 should not obscure the high variability within each participant category.

Despite these differences, each of the participants has, to a certain degree, a "pattern of solidarity and cooperation" (Lasswell and Kaplan 1950:31). The difference between each participant in terms of influence affects the ability of the decision functions to operate efficiently and equitably. For example, the overpowering influence of a particular value used by certain participants can limit promotion by decreasing rationality; it can limit appraisal by removing independence.

The strategies employed to conduct different decision functions vary with respect to participants and circumstance. The coercive strategies stem from the WD's ability to control the area through legal backing for activities such as the relocation of farmers currently inside the park. Local people can use coercive strategies by threatening passive resistance. However, all the participants generally use strategies that are noncoercive and rely heavily on diplomatic and ideological strategies (especially during the promotion, intelligence, and appraisal functions). The wildlife officer in charge of KRNP (K. L. Bahian) spends a considerable amount of time in informal discussions with local people in order to use a collaborative approach. The roles of different strategies in different functions of the decision process vary, which can often create problems. For example, ideological or militaristic strategies to promote local people's perspectives may not result in the desired action.

The common desired long-term outcome of all participants is the conservation of resources for use by present and future generations including local people who wish to express respect for their ancestors. However, the emphasis on satisfying present rural development strategies varies among participants. Although many local people do not want to lose access to particular resources, no participant wants the trends of decreasing animal populations and other resource degradation to continue.

Decision Making Behind the Establishment and Management of Kyabobo Range National Park

A management plan has yet to be drawn up for Kyabobo Range National Park. At the time of this writing, the work being conducted in the area includes gathering information for management purposes, promoting the planned management of the park, and continuing the appraisal of this work. In addition to an eventual management plan, various prescriptions are already in place. These include an executive instrument that allows the Wildlife Department to enforce various National Park rules prior to KRNP being gazetted (e.g., logging laws). WD has also begun to apply natural resource laws in the area (e.g., hunting laws). This study focuses primarily on the intelligence and promotion functions. The information used and promoted for the management plan will strongly influence how successful the park is in securing the common interest. The boundary prescription and application of wildlife legislation are fixed variables. Establishment of the park will also result in a substantial termination of the village arena of power that controls the areas inside the park. Tables 10.2 and 10.3 summarize how the decision process has been conducted thus far and some "indicators of malfunction" that have contributed to the policy problem (Lasswell 1971b:96).

Final Determination of Problems and Goals

Local people, the national government, and local government tend to perceive conservation and development in different ways, and attribute varying importance to different desired or scope values. Each participant (Wildlife Department, local people, local government, the KCP) has different degrees of control and authority over the process of shaping the park's management plan. Each participant also employs a range of strategies. These differences could result in a problem that might limit effective management for conservation and development

Table 10.2. Participant Roles in Decision Functions

Decision Function	Participant's Role in Decision Function
Intelligence	Official intelligence gathering conducted by WD. All participants have access to certain local information. WD can access governmental information to which local people have less access. KCP conducts socioeconomic and biological research in the area. All participants have access except less educated local people who are unable to understand the research. Local people's intelligence gathering is more informal. Access to local intelligence arena is limited, owing to its diffuse nature as a result of the situation and different perspectives. International intelligence has contributed to the knowledge of conservation and development. Knowledge of landmarks such as the UNCED conference in 1992 has encouraged KCP and WD to promote a policy involving local people.
Promotion	Promotion by WD is often supported by local government and has used coercive as well as noncoercive strategies. There is a strong reliance on diplomatic and ideological strategies. WD promotion is from both Accra and Nkwanta. Access to this arena is possible for local government, KCP, and some local groups. Local people also promote different viewpoints and primarily use ideological and diplomatic strategies. Local promotion is diffuse. Access to the local arena is limited (though this varies among villages).
Prescription	Prescription is primarily in the hands of WD. Broad goals are decided in Accra; other objectives are decided in Nkwanta (though this can be influenced by other participants). The main prescription at present is the park boundary (promoted from the 1996 KCP/WD boundary survey). Informally, different participants have been involved in prescription, e.g., discussions between MPs and WD.
Invocation and Application	There is no invocation and application at present, except for the policing of the park boundary and the application of wildlife laws. The park is *anticipated* by local people to be managed by WD with access to local people restricted, though some local involvement is expected.

Table 10.2. continued

Decision Function	Participant's Role in Decision Function
Termination	WD and other government departments will terminate various local management practices inside the park and those outside related to certain resource uses. The established arena is the government, with few strategies available for local people to access this arena.
Appraisal	WD conducts regular group meetings, village discussions. KCP and local people's research monitors changes inside KRNP. There is no formal appraisal system. Appraisals are primarily the participants' promotions and the potential prescriptions. Appraisals will eventually be linked to national park activities for management purposes. The arena for formal appraisal is currently not established. Appraisal has been conducted by WD, villagers, and KCP.

Note: MPs = members of parliament.

in the area. In addition, the social process could also result in many malfunctions in the decision process, preventing the common interest from being clarified and secured.

From the analysis of the social and decision process, the final goal is the suitable involvement of local people in the carrying out of each decision function for the purposes of conservation and human development in and around KRNP. This involves acknowledgment and incorporation of different perspectives and their expression.

Analysis of the Problem

The Establishment and Management of Kyabobo Range National Park, 1991–1996: History Behind the Problem

The Wildlife Department (then called the Game and Wildlife Department) was originally drawn to the area because of several factors. First, many bushmeat licenses were issued by the Nkwanta District Assembly, which indicated the existence of substantial large-mammal populations in the area. Second, the area is a unique ecosystem currently unprotected in Ghana. Third, the Fazao-Malfacassa National Park (FMNP) lies on the Togo-Ghana border in Nkwanta District, presenting opportunities for international cooperation. Fourth, support for the National Park came from the highest level of government.

Table 10.3. Summary of Relevant Decision Functions in Establishing and Managing KRNP

Decision Function	Criteria for Function	Selected Impressions of How Each Function Is Performing	
		Satisfactory	Unsatisfactory
Intelligence	Dependable	The majority of research conducted by KCP and WD is dependable.	Guides used to indicate important biological areas in 3d boundary assessment are not dependent on previous research. Limited knowledge.
	Comprehensive	Surveys of all areas where locals farm conducted in boundary assessment.	
	Open	Results and conclusions of all research are discussed with villages.	The research is not always understood by all participants.
	Selective	Boundary survey focused on area of farmland as best "objective" criteria.	Desire to avoid internal politics meant some vital components of the issue were overlooked.
	Creative	Boundary redemarcation had joint local/WD/KCP team to understand perspectives.	
Promotion	Rational	Environmental education working for the common interest.	Some local participants try to promote using coercive strategies such as refusing to talk to outsiders.
	Integrative	WD/KCP boundary proposal was not viewed as a patchwork of compromises by most participants.	Different perspectives resulted in participants overlooking issues, e.g., importance of conservation outside the park to realize its purposes.
	Comprehensive	All villages have been consulted in the gazetting of the park.	

Table 10.3. continued

Decision Function	Criteria for Function	Selected Impressions of How Each Function Is Performing	
		Satisfactory	Unsatisfactory
Prescription	Stability of expectation Rational and comprehensive	Close dialogue among participants through group meetings. Management plan has not yet been formulated.	Local people expect to promote another boundary demarcation. Management plan has not yet been formulated.
Invocation and Application	Timely, dependable, rational, nonprovocative, realistic, uniform	The operations currently occurring around KRNP will attempt to ensure that when a plan is applied, there will be sufficient understanding of all participants to ensure criteria are met. This is accomplished through discussions and environmental awareness.	
Termination	Timely Dependable, comprehensive Balanced, ameliorative	Mammal protection is occurring rapidly in the area. Local people are aware of effects of termination. Boundary determined to minimize negative and destructive change.	A major time lag exists between reports of environmental degradation and action. Some aspects of termination effects may not be considered. Uncertainty exists as to how to provide adequate compensation.
Appraisal	Dependable, rational Comprehensive and selective, independent Continuous	Appraisals conducted in every community surrounding the park.	Policies and criteria are not agreed upon. Explanatory analyses are not explicit. Information collected depends on collector's perspective. A few influential people often dominate local, national appraisal. Because of lack of funds and infrastructure, appraisals are not continuous.

Once the WD decided to study the area, the first field analyses investigated biological phenomena. Until 1991, local participants and various other "outsiders" collected information regarding animal populations in an informal manner. For example, a Dutch Catholic priest working in the area during 1975 stated that "26 years ago, lorry loads of bushmeat from this area went down to Ho and Hohoe almost daily and even as far as Accra and Kumasi. Now however, it is even hard to find a grasscutter (a large rodent) on the local market in Nkwanta." According to the priest, "farming and the ever increasing pressure of population had resulted in the killing of most if not all, the game in the area" (Merz 1975). Interviews with Nana Oberko Agyei II (the paramount chief) also revealed that many animals are much more difficult to catch now than in the past.

However, this anecdotal information involved no formal gathering, processing, or disseminating. In addition, it could be argued in these intelligence exercises that "investigators did not manifest enough self-awareness to distinguish the trends they purport to describe from their own preferences about events" (see McDougal and Reisman 1981:271). For example, some local people believe that the lower levels of bushmeat were primarily the result of "animals migrating to Togo" (Ledward and Bowes-Lyon 1996:61).

In 1991, the WD preliminary reconnaissance was conducted. The WD research was based on the species of animals present and the potential of the area to be a National Park (Ankudey 1991). The information collected was not completely comprehensive, as its main focus was on large mammal species. Other factors highlighted in the report were also biophysical: "In addition to the potential high populations of animals it can contain, the unique topography and vegetation should act as an attraction for the park" (Ankudey 1991). The results of this study suggest that it would be highly desirable to protect KRNP. Given the circumstances and situation outlined in this preliminary research, it was suggested that the area should be set up as a National Park by using the following management objectives:

a. To perpetuate and enhance the natural and scenic values of these areas of recreational, educational, scientific and rational use and to provide suitable facilities for such purposes through appropriate zoning of the park's assets.

b. To enforce the respect of those features leading to the establishment of the National Park by preventing/regulating the unwarranted extraction or destruction of any wildlife species, community or natural features and controlling all development which will detract from these values.

c. To increase public awareness of the value and nondestructive use of the park's resources.

d. To develop opportunities for local communities to develop from the park through integrating its management into the process of rural economic development (for example, by permitting harvest of certain wildlife products).

e. To ensure that any management interventions such as culling for overpopulations, species introductions are strictly in accord with an approved management plan and are scientifically justified.

f. To establish buffer zones in which community wildlife-related projects can be developed. (Policy for Wildlife and Protected Areas, Government of Ghana 1994:13)

The biological/geographical research was continued by a group of undergraduate students from Oxford University, United Kingdom, and Kanton Luri Bahian from the WD in 1994. (The students, myself included, later founded the nonprofit Kyabobo Conservation Project. K. L. Bahian later became the wildlife officer in charge of KRNP.) The research covered the mammal and vegetation components extensively (KCP 1995). This research did not focus on the deep-rooted functional and symbolic significance of wild animals in the local people's lives.

The trends described above relating to animal populations also relate to perceptions of changes in plant populations. Many people were aware that the abundance of important timber trees had declined, that there were more fires, and that a large area of the district had been invaded by an exotic plant; the KCP and WD mentioned the impacts of logging odum (mahogany) (KCP 1995:41). Local people were aware of the changes in plant distributions, pointing out that they have to walk farther to collect resources (Ledward and Bowes-Lyon 1996:61).

Thus a wide and comprehensive amount of information on the biological situation was collected. However, in many cases the local gathering of information was not disseminated to other participants. It was primarily the initial intelligence gathered by the WD that influenced the first boundary demarcation on a broad scale. The work by the KCP and WD was significant for the third boundary demarcation. The reliance on information collected by the KCP and WD, who emphasize the importance of biodiversity conservation, had substantial implications for the establishment of KRNP.

Following the initial WD biological surveys in 1991, KRNP was demarcated (Nuhujinbiina 1991). It was 360 square kilometers and occurred between latitudes 8°20' and 8°30' North, and longitudes 0°30' and 0°40' West. This was before the new WD policy had been prescribed, hence the demarcation was carried out in the conventional fashion.[1] No formal assessment of local needs was made and communication with communities was minimal. The documentation on the boundary demarcation (Nuhujinbiina 1991) does not explain sufficiently the intelligence behind the determination of the boundary. However, it is clear the demarcation was researched and implemented by a skilled cartographer. In 1993 an executive instrument was passed which legally recognized the area as a proposed National Park.

Subsequently, local people made a number of appeals concerning the number of farms that lay inside the National Park. This was despite the fact that the WD had been informed (from various local sources) that there were virtually no farms in the area. There were also complaints from local people regarding the lack of information on how the government was to set up the park and compensate farmers and landowners. At this point there was no wildlife officer based in the area and there was limited communication between the local people and the WD.

However, in 1994 a WD team visited Nkwanta District to help resolve these issues. The team aimed to promote the park using an ideological strategy and planned to "educate the communities around the proposed Kyabobo Range National Park." The district secretary (political head of local government) requested the team due to "rumors circulating among the youth in communities bordering the reserve . . . that the paramount chief with his sub-chiefs had sold the stool land to the government without either consulting the people or asking for compensation to those whose lands have been taken."[2] The secretary showed his "personal interest, commitment and willingness and that of the District Assembly to get the reserve established." He was concerned that the "pleas and complaints be carefully examined as they might not be the true situation." The district administrative officer (administrative head of local government) then explained that "most farmlands have not been taken as they [the people] claim. The area demarcated for the reserve is generally unsuitable for agriculture as the area is mountainous" (Nyarko and Fordjour 1994). However, in further discussions with local chiefs and government officials, the need for a review of the boundary was highlighted. These views were reinforced by the joint Oxford University and WD biological survey of KRNP conducted in 1994 (KCP 1995).

From these discussions, it is evident that the base-values and desired outcomes of the local government influenced the promotion phase of setting up the park. The record of this period, from 1994 to 1995, shows the variety of strategies employed by different participants. The form of promotion was essentially communicational, being both diplomatic (e.g., the paramount chief's discussions with the WD) and ideological (e.g., the Nyarko and Fordjour education project in 1994). As a result of the WD's new, more flexible policy, the visits by the WD also acted as informal appraisal exercises.

Following the education project in 1994, a meeting between WD staff and local community representatives addressed the problems of farmland inside the park and compensation. In contrast to other parks in Ghana, the WD agreed to redemarcate the park to exclude many of the farms currently inside the park. Prior to this second demarcation, which took place in 1995, a brief survey of farmland inside the park was made, but because it was not recorded it is impossible to compare it with subsequent farmland assessments. The re-demarcation was possible because of shifts in the WD policy which were in line with international concerns to integrate local people into biodiversity conservation (United Nations Environment Programme 1992). It is also likely that additional participants with sufficient power also helped promote the redemarcation (such as the national politicians).

In 1995 the Kyabobo Conservation Project was formed as a result of the 1994 Oxford University team winning a conservation award. This is a nongovernmental organization (NGO) that aims to facilitate community-oriented conservation activities in and around KRNP. Preliminary KCP meetings with village chiefs, elders, and other opinion leaders revealed a high level of dissatisfaction with the current boundary and the inability to draw up a boundary that the WD could accept. Although this area was not included in its original objectives, its codirectors, in conjunction with the WD (especially, the first wildlife officer to be stationed in Nkwanta), decided that the KCP should assist with a review of the current park boundary. The level of discontent was such that many of the proposed KCP activities such as environmental awareness and social research would not have effectively continued had the boundary issue not been addressed. With the agreement and collaboration of the WD, the KCP offered to carry out research with the department into the location of farmland around KRNP. It was also agreed that the KCP would help propose a new boundary for KRNP if this was deemed necessary (O'Keefe et al. 1997).

Some local people were still not satisfied with the second demarcation and again lodged protests with the WD. The department did not have funds for a further survey and demarcation, however. Therefore, they advised those unhappy with the second demarcation to find funds for an independent demarcation for review by the WD. Also working against those who were advocating a third demarcation was the fact that the WD wanted KRNP to be fully recognized (by having a legislative instrument passed through parliament) as soon as possible. This would allow the area to be fully protected. From the WD's (and local government's) perspective, this was urgently required, as hunting had continued on a large scale, and, knowing that the area would soon be protected, many local and commercial loggers were rapidly removing timber (largely illegally). Also the WD had been trying to establish the park for over five years and there was pressure to speed up the process.

The Wildlife Department and the KCP proposed a new boundary for KRNP using "rational objective criteria" drawn up by the KCP and WD team (see Table 10.3) (O'Keefe et al. 1997). In drawing up these criteria, local people had no direct input. This was principally because the KCP's and WD's resources were limited in terms of time, money, and experience. The information used to draft the third demarcation was gathered through group discussions in each village, interviews with various key people, and walks through and surveys of the villagers' farmlands. Of the nine villages surrounding the park, seven approved of the proposed boundary and actually helped demarcate the park. One village, Shiare, did not agree, as it would lose valuable farmlands. Another village, Kyillinga, also objected, probably to show solidarity with Shiare. The disputed area of Shiare's farmlands had been designated as part of the park for several reasons. The following list is extracted from the report of the senior wildlife officer in charge of KRNP to the chief wildlife officer upon completion of the boundary survey (Bahian 1996):

> (1) These areas are in watershed areas. Farming can be seen to be on very steep slopes with removal of vegetation.
> (2) If these farms were outside the park, they would be surrounded on all sides by the national park and would suffer from crop-raiding.
> (3) Of the land removed from the park between the 2nd and 3rd re-demarcation, over 60% is for Shiare villagers.
> (4) The few scattered farmlands that are not considered are more or less hunting camps. As indicated from the survey, there were many snares and traps set up.
> (5) Walking for five or more miles to cultivate an area of about a hectare or two at distant intervals on rocky terrain is both economically and agriculturally unsustainable.
> And (6) An additional factor that came up in these discussions was that it would be difficult for the WD to patrol and maintain a boundary going through the mountains (rather than along the foot of them).

There is no doubt that with such a complicated policy problem, some value deprivations were inevitable. Following the third demarcation, the KCP assessed these deprivations by conducting further socioeconomic research (Ledward and Bowes-Lyon 1996). Unfortunately this could not be conducted in Shiare village, as several villagers were unwilling to communicate with any outsiders. The project also conducted environmental awareness activities in the schools surrounding the park. In June 1996, the park was demarcated around the villages of Kue, Pawa, Gekorong, Odome, and Kromase by a joint team of local people and WD staff. The 1996 demarcation set a precedent for this type of collaboration. However, the villages of Shiare and Kyillinga were unwilling to cooperate. Therefore the boundary remained at the same location as in the second demarcation around these two villages.

During the process of drawing up this proposed third demarcation, many pleas issued from Shiare. Several local people suggested that in writing the proposal, the knowledge of Shiare's social process and the potential outcomes for the people had been limited. However, Shiare had often been unwilling to enter into a dialogue despite the attempts of WD. An excerpt of a letter from some elders in Shiare to the KCP directors illustrates the local response:

> WD staff is preaching to us that if we let go of our farmland and allow the project to continue, it will benefit the local people, the Nkwanta District, and our country as a whole. That may be true for the most part. But what shall we, the local inhabitants, do to get food? Shall we be resettled at the government's expense? Where shall we make our farms? We believe and recognize that our father is the Ghana government. If we don't eat, can we serve our father? No! Without our farmland to give us food, we shall all perish and our father will be left alone with all the development areas nicely preserved.

Since the boundary was redemarcated, the Wildlife Officer in charge of KRNP has been receiving further training in wildlife management and community development in Tanzania. The KCP participants have also been gaining more training and have been evaluating and publishing their work. In the Nkwanta area, a temporary WD officer has been employed who has started applying WD laws. At the time of writing, the members of KCP and WD had just returned to Nkwanta. The more recent history is included in the postscript at the end of this paper (preceding the Acknowledgments section).

Summary of Pertinent Factors Currently Influencing Decision Making in and Around Kyabobo Range National Park

Perspectives

The perspectives of those who performed the intelligence function were a substantial influence on the criteria used to establish the park (Table 10.3). The initial intelligence was insufficiently contextual (leading to local complaints about the boundary). Conceptualization of local values of the land using only spatial criteria still did not completely resolve the problem. The research in 1996 does show that the KCP and WD were becoming more contextual. Over time, more people have become involved in the intelligence, promotion, and appraisal functions. The WD and KCP have continued to view the importance of local participation in the park as paramount to the success of the park. The continued environmental awareness program and ranger activities in villages have helped local people understand the perspectives of the WD and vice versa. These conditions are paralleled by international concerns that local people participate in conservation activities (UNEP 1992).

However, there are still problems with the wide-ranging perceptions of the primary determinants of environmental change. Many people often fail to integrate natural and social phenomena in their inquiries. Some attempts to "consider local inhabitants' representation of habitats in terms of modern western categories obscure local perspectives on social and ecological relations" (Fairhead and Leach 1996). Although degradation is often perceived to be caused by local practices, the factors responsible may be beyond the control of the farmers who are affected by the park (e.g., the migration of Konkomba farmers onto Akyode traditional lands to the north).

All participants emphasize the importance of rural development. However, the conceptualization of what is meant by "development" differs. This is manifested in different spatial, economic, and temporal scales. For example, the WD manages resources for Ghana as a whole for the twenty-first century, whereas local people manage resources for local needs to enhance the survival prospects of a family, lineage, or clan. Local people typically equate "development" with "security" (in terms of all values).

Situation

The park is now established in the mountains, which includes both forest and savanna vegetation types. Continued hunting and resource use in the area affect the resource base. There is also a small number of farms belonging to Shiare residents, from a few clans, in one part of the park.

Particular conditioning factors relate to situations that affect local lifestyles. For example, local people are concerned primarily with having fertile land to farm. However, the definition of "good land" differs depending on the situation of the farmer. Of the Shiare farmers who farm in the mountains, the land is located on forest soils and the crops and lifestyle there are quite different from the other villages surrounding the park in the savanna areas. However, a recent shift in agricultural practices has encouraged savanna farming and the production of yams and cassava.

Many local values are related to situations. For example, among all people living around the park, respect values are conferred on those who own their own land and are not tenants. This will be affected if any of the farmers are relocated. Many local people are also concerned with the religious values (rectitude) and affection values of the area, such as ancestors, gods, and sacred groves. The sacred grove (or fetish) in Shiare confers substantial power on the initiated. The local people hold a wide range of different beliefs relating to conservation. The leopard, for example, is sacred in some villages but not in others. In the indigenous Akyode tribe, the python is the basis of the women's cult but not the men's cult. Local situations have a substantial influence on local involvement in decision functions. Farms are scattered. In the mountains, the only mode of transportation is by foot. In the plains, despite the existence of roads, many people are unable or unwilling to travel to Nkwanta or other villages.

The WD manages the park daily from its headquarters, which is currently located in the main district town of Nkwanta. However, a permanent headquarters several hectares in area will be established near Kromase village. The local WD has a vehicle and the junior staff members have bicycles, so it is possible to have access to the villages. In addition, the vast majority of the junior staff are from the local villages. However, with the exception of local level planning, most major decisions are approved (or are directed from) the WD office in Accra. The KCP is managed in both the United Kingdom and in Ghana. When working in KRNP, KCP members stay in Nkwanta and in the local villages.

Base Values

The current values that influence the participants' ability to promote and secure their demands are highly complex. It must be stressed that mapping these values for presentation in the following summary has been a complicated process and cannot be covered comprehensively with the data available.

• *Power.* The WD has the power to conduct activities mandated by it inside the park. However, the local government, based in the district capital, Nkwanta, does influence activity. Local people manifest their power primarily through the use of passive resistance and diplomatic strategies. The KCP has indirect power through its informal influence over other participants and its involvement in activities such as the boundary re-demarcation.

It is also important to note that there is variation in the degree of power within the different institutions outlined in this report. Among local people, for example, men often have considerable power over women, and members of the cult have power over other villagers in Shiare. However, the distribution of traditional authority has shifted as a result of westernization and the spread of Christianity. The power structure of the WD is hierarchical, where laborers are answerable to the senior wildlife officer in KRNP, who in turn is answerable to the chief wildlife officer. The KCP is organized such that the three trustees have equal power over KCP activities.

• *Wealth.* Wealth deprivation is common among all participants. The WD does not have enough money to fulfill all its objectives. Local people typically have only enough money to subsist. The KCP is an NGO that relies on small donations, and its directors are volunteers. Shifts in the distribution of wealth do and will occur, however, as a result of KRNP—in the form of compensation to local people and employment opportunities.

• *Enlightenment and skill.* Indulgences and deprivations of enlightenment and skill vary depending on the perspective used. Local people are indulged with respect to understanding the local ecology (in indigenous terms) and understanding how it can be manipulated for farming and other local needs. They are also indulged in understanding their spiritual values of the area and have the skill to acknowledge these values. The KCP and the higher-ranked employees of the WD are enlightened in modern western science and conservation biology as practiced internationally. They also have the skill to conduct surveys and create written prescriptions.

• *Well-being.* Many of the participants have contracted illnesses in the area. The ability to cure them varies considerably. This is related to wealth (payment for health care) and situation (proximity to medical facilities). However, well-being is also strongly related to rectitude, as many local participants believe that well-being is affected by the action of benign and malignant spirits.

• *Affection.* Local people gain affection from close family and village ties. The WD and KCP gain affection from recognition within Ghana and internationally for their attempts to pursue conservation and human development.

• *Respect.* As a result of the park, the wildlife officer in charge of KRNP has earned considerable respect from most of the villagers surrounding the park. The indulgences and deprivations of respect among local people and other participants depend to a considerable extent on the local individuals involved. It is widely accepted that a participant who is not respected will be unable to apply or influence policy decisions.

• *Rectitude.* Both WD officers and the KCP are receiving low (or no) salaries. Much of their work is motivated by rectitude. Local people are indulged with rectitude from the perspective of helping out other members of the same tribe. The traditional religions link the people to the land.

Strategies

Strategies that stimulated the Wildlife Department's initial interest in establishing KRNP were primarily ideological and diplomatic. Since initiation of the idea for creating KRNP, the WD has used diplomatic and economic strategies to try to resolve problems. However, ideological strategies are still important. They form the basis for aspects of park establishment such as training and environmental awareness. With the third demarcation, coercive strategies were initially used to get people to accept the boundary plan. Within the WD, strategies vary, often depending on the base values and scope values of specific individuals (for example, enlightenment and desire for respect).

The KCP has used ideological strategies for promoting its intelligence and appraisals. It has also used economic strategies by employing local people and members of the WD, and diplomatic strategies as part of the KRNP boundary re-demarcation team.

Local people used primarily ideological strategies initially, in the form of pleas for park reestablishment. As the dispute over the park boundaries evolved, they employed diplomatic strategies. Occasionally, they have threatened to use coercive strategies.

The present strategies used to manage KRNP include the development of consultative strategies, with the goal of progressing to more participatory approaches. In this case, research is done with people in their villages and farms as well as the other arenas in Nkwanta and Accra.

The Impact of the Continued Establishment and Management of Kyabobo Range National Park: Projections

From a local standpoint, the combination of the trends outlined and the present conditioning factors had, potentially, a substantial influence on the local people. The following list is adapted from a report by various members of the Akyode tribe (April 1996):

1. Once the park boundary is applied, the farmlands of two complete clans from Shiare village will be removed. The rest of the community, as members of different clans, is not obliged to help at all. The dispossession will aggravate tensions between the different sides of the Shiare village.

2. A large exodus of people from Shiare will completely change the community structure and religious practice. Many of the elders of the soon-to-be-displaced clans are Shrine priests of Brukung and must regularly visit their mountain shrine.

3. The farmers moving from Shiare will have to convert from a mountain style of farming to a lowland/plains form of farming. This involves shifts in methods of farming, the crops grown, and the climatic conditions. Even though the farming areas are close in terms of location, the differences will affect not only the farming but also the lifestyle of the people. In addition, there is at present no system for appraisal of the impact of the park.

4. Difficulty in finding land involves not only its location but also its true availability. Many areas seemingly empty are actually owned by people with specific plans for its use. Under the land tenure system, it is possible to beg land, but this will result in a substantial shift in status.

In contrast, from the WD standpoint it can be argued that there are only a few farmers involved in this issue, and that it will be possible for relocation to occur without substantial value deprivation:

1. The government will be able to provide monetary compensation to offset the value of the land. (However, there have been a few cases elsewhere in Ghana where monetary compensation has not been adequate to help farmers get re-established.)

2. The overriding support from seven of the nine villages will contribute to successful management of the park.

3. Issues will be addressed adequately, as has already been shown by "the willingness and the preparedness of WD to compromise with local communities on various issues" and "the willingness and preparedness of the District Chief Executive to serve problems emanating from the park" (report of the senior wildlife officer in charge of KRNP to chief wildlife officer). For example, the WD is now saying it may allow villagers access into the park to harvest crops and to access fetish groves. However, such activities may have to be carried out in the presence of a ranger. Farmers will also be given a generous period of time in which to move.

From the author's standpoint, if current conditioning factors remain in place, then the outcomes for each participant could include the following: once the park is established the WD will have increased respect nationally and internationally as a result of conserving a unique ecosystem. It will also increase the power and rectitude of the WD. There will be substantial shifts in local people's value indulgences. This will depend on situations and perspectives (see conditioning factors). The park may also act as a catalyst for development activities in Nkwanta District. The members of the KCP will stop working in the area by 1998 (for financial and personal reasons) and will have gained enlightenment and rectitude.

Using the policy sciences framework, a projection of the decision-making process in KRNP can be summarized (Lasswell 1971b). The following could occur:

1. *Intelligence* for biological information could continue to be collected using positivist rational criteria. Information on the socioeconomic situation of local people will be collected and processed by the KCP and WD (though local input in information content could continue). The numerous discussions between the WD and local people will lead to greater understanding of one another.

2. *Promotion* of different practices will vary by participant. It is expected that the differences in perspectives will restrict the promotion of comprehensive and integrative strategies.

3. The arena to *prescribe* will be primarily with the WD, in both Accra and Nkwanta. Local people will continue to have unofficial access to this arena.

4. *Invocation* at the park will begin by focusing development activities in the Laboum area and by invoking wildlife laws. It is uncertain how the broader *invocation and application* prescriptions will progress.

5. *Termination* of local resource use inside the park will take place over a period of time, using coercive strategies. Local reactions could include passive resistance.

6. *Appraisal* will be conducted by the KCP as research for a short period. This will be taken over by the WD. WD staff will document illegal activities and park boundary patrol reports.

It is important to consider whether these projections will result in the achievement of the aims that the KCP and WD team have identified and which are presented in Table 10.3. Given the current conditions and resources available for each participant, it should be possible to achieve the aims of conserving the ecological resources of KRNP for a reasonable period of time in an effective and efficient manner. The WD has thirty-eight rangers who can enforce prescriptions. The backing of the government will, in the short term, ensure the protection of the area. However, given the conditioning factors as of 1996, achieving some of the other aims will be problematic. In particular, if there were to be resentment against the park from within Shiare village, villagers will be tempted to encroach and poach in the park whenever possible.

Some Shiare villagers would argue that biological conservation was prioritized over human development in assessing the boundary around Shiare, and that many of the aims presented in Table 10.4 will not be met. However, around other villages bordering the park, it can be argued that all the criteria for the boundary outlined in Table 10.4 have been achieved. This is despite the fact that some of the human development and conservation aims listed in the table appear to be in conflict.

In the following section I discuss potential alternative strategies that could contribute to an integration or compromise of the dual goals of conservation and human development.

Recommendations and Alternatives

An integration of conservation and development concerns should strive for "openness of participation to all interested actors in the constitutive process" (McDougal and Reisman 1981:270). This involves participation in all stages of the decision process. Therefore, alternatives are discussed in terms of the decision process.

The objectives of conservation and human development are often perceived to be opposing. Whereas many alternatives may support conservation at the expense of human development and vice versa, it is also possible to argue that a middle ground can be secured whereby both objectives may be satisfactorily addressed. Table 10.5 presents a scenario of such a middle ground.

Table 10.4. Excerpt of Criteria Used for the Third Boundary Proposal

Aims of Setting up Kyabobo Range National Park	Criteria for Boundary Location
1. To conserve the fauna and flora of the Kyabobo area: • as representative of the block of forest, • as consisting of endemic species, • as representative of the unique mixture of forest and savanna vegetation	Include areas of undisturbed forest. Include all representative habitats in the area.
2. To conserve the mountainous environment of the Kyabobo area	Include representative and significant mountains of the area.
3. To conserve the area for research, tourism and education (local, national and international)	Include prime sites suitable for research, tourism and education. Ensure that these sites are easily accessible.
4. To ensure management of the park is effective and efficient	Boundary line should be simple; straight and/or follow geographic features. Minimum need for access to resources. Minimum temptation to encroach and poach inside the park. Most areas of the park (in particular the boundary) are easily accessible.
5. For the management of the park to act as an example to other WD reserves: • to promote international co-operation in conservation efforts • to implement the new WD policy • to promote community collaboration • to use the scientific principles of conservation biology for the effective management of the park (for example, edge effects and minimum viable populations) • to test new management practices (for example, the range system rather than the camp system for antipoaching patrols)	Boundary line is adjacent to Fazao-Malfacassa National Park. Ensure that the park is conducive to working harmoniously with local people. No islands of park surrounded by farmland or vice versa. Park is large enough to include viable populations of wildlife. Ensure that the park is conducive to possible new management practices (for example, suitable sites for rest camps, all areas within one day's journey of park headquarters).

Table 10.4.   continued

Aims of Setting up Kyabobo Range National Park	Criteria for Boundary Location
6. To minimize the negative and maximize the positive impacts of the park on local people (preferably to the household level) • to act as a reservoir for large mammal species for bushmeat • to conserve endangered natural resources and promote their sustainable use • potential sources of income and other financial benefits • for increased profile of Nkwanta District • to protect important watershed areas	Park is large enough to contain sufficient populations of large mammals to provide a "surplus" to surrounding areas. Key areas of currently unused natural resources are included. Important watershed areas are included. Minimize eviction of people from their land. Potential positive and negative impact of park is as equitable as possible (in order of priority; between tribes, villages, clans, families, and households). Maximize opportunities for cooperation with local communities.
7. To promote conservation outside the park through: • protection of sacred groves • changing agricultural practices (for example, agroforestry) • tree planting • protection of existing areas of conservation value (for example, riverine forest)	Minimize negative effects on areas outside the park.

Source: E. O'Keefe, D. Bowes-Lyon, and J. Tordoff, "An Assessment of the Boundary of Kyabobo Range National Park," KCP Report Series No. 1. Cambridge, UK: KCP Publications, 1997.

Alternatives for Intelligence Function

Option 1

For the management of KRNP, further knowledge of the local base and scope values could be collected. With local people actually participating in the intelligence gathering, the information would be collected using a range of methods and would come from different local institutions and perspectives. The intelligence gathered could integrate biological and social information to treat local people as part of the KRNP ecosystem.

Table 10.5. Potential Selection of Alternatives that Could Address the Problem of Integrating the Goals of Conservation and Human Development

Decision Function	Preferred Alternative
Intelligence	Local people could be involved in collecting some data related to their relations with the park. To assess the validity of the data, KCP or WD could concurrently collect information to corroborate. This would entail continuing with the KCP/WD research program until a system of joint local/WD information gathering is supported by all participants. In some cases certain information will need to be collected from specialists. It is important that a method for appraising and disseminating this information (and clarifying the standpoint of the information collector) is developed.
Promotion	The different participants must continue to have access to arenas for promotion. However, participants should base their promotions on sufficient knowledge of the issues. Therefore, an environmental and social awareness program and continued training is important for all participants.
Prescription	Theoretically, prescriptions drawn up by all participants together are more likely to secure the common interest. Therefore, in certain circumstances, this is an option. However, in other cases, joint management may not address the overall conservation goals behind the establishment of KRNP. In these cases, WD should have the authority and power to dominate the arena.
Invocation	Some activities could benefit from being invoked in small areas (such as resource harvesting from KRNP); other activities should be invoked on a larger scale. The determination will depend on WD experience and the outcomes of intelligence and appraisal activities.
Application	Applications should not be conducted before adequate intelligence, promotion, and prescription have been conducted and a management plan has been invoked sensibly. It is not appropriate to summarize alternatives for application as a result of insufficient information. However, discussions between WD and local people on anticipated application should continue.
Termination	Termination of local activities within the park should not be conducted immediately or too slowly. It is important to assess the potential impact of not terminating certain activities and the potentials for zoning different management areas of varying local importance.
Appraisal	Activities that have a direct impact on local people's livelihoods could be assessed by local people (and they should be backed up by KCP/WD appraisal). Other forms of appraisal (e.g., wildlife populations inside the park) should be conducted by KCP/WD.

Evaluation. This form of intelligence gathering would ensure that the data were comprehensive and many of the problems discussed in the projections could be addressed. It would also be creative and open because many participants would be included. However, it might not be sufficiently selective and could be dependent on more assumptions. With many people involved in the process, more perspectives are involved and the intelligence gathering may not be able to define the underlying problems clearly. In addition, different participants may not recognize the validity of some data generated by other participants. There would also be restrictions. For example, the WD would not want local people to collect information inside KRNP without WD staff; local people might not have time or the motivation to be involved in such activity.

Option 2

The KCP and WD could continue collecting social and biological data. In drawing up research plans, they could attempt to be sufficiently comprehensive by using their collective knowledge of the area and by having group discussions with local people to confirm that the problems being addressed are contextual. They could also ensure that they have no dependent assumptions and would be selective. The research could be creative by investigating other projects, and taking advice from local people. The work would be open, as the results would be discussed and shown to all participants.

Evaluation. This approach would be the easiest for the KCP or WD to carry out and would possibly secure much of the information required to manage the park (except that it relies on a continuing source of funding). However, there is a danger that if the local people are not defining and addressing research questions themselves (e.g., by using participatory rural appraisal methods as discussed in Chambers 1994), the common interest may not be secured. If one participant controls the intelligence process, there is the risk of not being comprehensive as a particular myth could dominate the research. Indeed, it can be seen from the trend analysis that variants of this alternative have often resulted in problems stemming from a lack of contextuality.

Option 3

Responsibility for intelligence gathering could be undertaken separately by different participants. Some participants could focus on information that concerns them the most. For example, the WD could control mammal sampling and local people could map sacred groves.

Evaluation. Whereas this method could be comprehensive (many forms of information may be obtained only through this alternative) and creative to a certain extent, ensuring that the information gathered is sufficiently selective is problematic. In addition, the extent to which the information collected is dependable may not be known. As a result, some information collected could be discounted by some participants. There are also problems of openness, as some participants may have difficulty disseminating their information.

Alternatives for Promotion Function

Option 1

An education program for local people could help explain the benefits of the park. The materials used would be developed with the local community to ensure that a suitable perspective is used. The KCP and WD are currently pursuing this alternative.

Evaluation. This program could be rational, integrated and comprehensive. However, such a program carries many risks. It is likely that not all participants will have access to this program (many farmers live in the hills); it may not be applicable, as many of the problems of local people may not be addressed by this proposal. Finally, it may not be rational or comprehensive if it fails to understand and accommodate different perspectives and the expression of these perspectives. These limitations will be accentuated if the environmental awareness teams are not from the area.

Option 2

A series of training workshops with the WD rangers covering community development could be carried out to help promote the importance of considering the needs of local people. This would ensure that the WD projections are more likely to be realized as an adequate community liaison is established. This alternative has already been conducted by the KCP (O'Keefe 1997).

Evaluation. This alternative undoubtedly ameliorates many of the projected problems and would ensure that the promotional activities of the lower WD staff do not increase problems. However, on its own, this alternative is unlikely to be comprehensive, and there is a risk that it would not be integrated and rational.

Option 3

Existing arenas for discussion could be strengthened and could encourage promotion activities by different participants. This option
would involve encouraging ideological and diplomatic strategies and
ensuring that the arena for discussion is open to all local individuals
and institutions and would use existing local institutions. This could
involve continuing the group meetings between the WD and local people. Such activities could be experimental and use different styles and
noncoercive strategies in meetings to strive for a rational promotion. A
potential option of having a policy sciences workshop (as discussed in
Willard and Norchi 1993) is logistically difficult but could be instrumental in clarifying common goals to promote.

Evaluation.   A similar formal program was proposed by the KCP in
1996. In that evaluation, it was decided by the KCP co-directors that
such a program would be too difficult to conduct. In particular, these
meetings may lack comprehensiveness: typically, they do not reach all
people (especially women and other nonelites) who spend most of their
time working on the farms. In order to participate, local people have to
perceive a real benefit from group discussions. In addition, integration
problems may also emerge: if the different participants have directly
conflicting viewpoints, no such promotional exercise will work.

Option 4

The WD could be the main participant promoting activities related to the management of KRNP, but the information collected in intelligence phases could be used to create a rational foundation for
promotion.

Evaluation.   To an extent, this alternative has been conducted during the establishment of KRNP. There is a substantial risk that the intelligence will not be appropriate or comprehensive if it is not gathered
and interpreted adequately. This is a problem, given the differing perspectives (see Table 10.1). However, this option could ensure that promotion is integrated and the participants do not view solutions as a
patchwork of compromises (Lasswell 1971b:88). Also, this method reduces local access to arenas, and coercive strategies can have the potential to dominate. However, in some circumstances, this alternative
is necessary; local participants may not understand some information
(for example, the effect of international politics affecting conservation
policies), and it could be argued that some participants should not be
involved in promotions of subjects with which they are not acquainted.

The use of this argument can make promotion not work to the common interest.

Alternatives for Prescription Function

Various prescriptive alternatives could be proposed to address the problems outlined in the projections. Although the common interest cannot be completely secured when participants have different perspectives, any prescription should ensure some consensus on what to expect (the degree of doubt about the lawfulness and enforceability of policy in different contexts), rationality (prescriptions are formulated to give effect to common interest), and comprehensiveness (sanctions are developed that deter and resist nonconformity) (Lasswell 1971b:90). The two possibilities below illustrate two contrasting strategies to address the goals through different institutions.

Option 1

Continue with the conventional method of resolving problems where the government acts on its promotions, which are based on its perceptions of the common interest. For example, in the case of the Shiare farmers who still farm inside KRNP, the WD could give them four years to move out of the area and establish new farms. They could be allowed periodic access to KRNP with rangers to visit sacred groves and collect resources such as fruit. This would be a coercive alternative.

Evaluation. This alternative would create consensus on what to expect, and local people would understand that WD activities are lawful and that the department has the sanctions to prescribe. However, this alternative may not work toward the common interest if the WD does not fully understand the important components of the social process. On the other hand, this alternative could be used to avoid having the WD get involved in internal Shiare politics. Such involvement could result in the WD being accused of bias toward some clans.

Option 2

A joint management system could be developed whereby prescriptions are drawn up equally by both the WD and other participants. This collaboration would ensure that local needs and perspectives are represented in the management plan and the different participants could work together to secure the common interest. If deemed necessary, the boundary of the park could be reconsidered.

Evaluation. This alternative addresses the common interest, yet potential problems exist. With many potential participants involved, the community heterogeneity could result in expectations that are not stable and prescriptions that are not comprehensive. This option was chosen by the WD in 1998 and has resolved the policy problem.

Alternatives for Invocation Function

Option 1

A small area or community around the park could be used for invocation and assessing the impact of the prescription. This is currently being conducted in the Laboum valley.

Evaluation. This alternative is an essential component of the WD's policy of using the Laboum area as a test case for community collaboration in park management. It is also timely, as it addresses some local concerns about KRNP such as loss of farmland and use of resources. However, there is a risk that this alternative may not work toward the common interest depending on how local people are involved in drawing up the prescription. Also, although this approach is nonprovocative because the people have received farmland in return for their involvement, it could be provocative to Shiare people not living in this valley.

Option 2

The wildlife policy (Government of Ghana 1994) has been invoked by applying wildlife prescriptions (and other natural resource laws) in the whole area in and around the park. For example, certain hunters have had their bushmeat confiscated and chainsaw operators have had wood and machines confiscated.

Evaluation. This alternative is very timely for certain aspects of the management plan. For example, as a result of the awareness that the park is being established, the levels of commercial resource extraction have increased. Immediate invocation has been essential. In other aspects of park management, however, activities might be provocative in the future (e.g., the dispossession of land). The success of this method of invocation does depend on which prescription is invoked.

Alternatives for Application Function

The alternatives available for the application function are highly dependent on what prescriptions are developed. Once a prescription has been developed, alternative applications should aim to be rational, realistic, and uniform. The arguments and lessons from invocations should be used. Essential to the rational application of prescription is the continued appraisal of activities.

Alternatives for Termination Function

The development of KRNP will result in a termination of certain local village activities on what is now parkland. Any alternative should ensure that adequate compensation and alternative farmland will be provided to dispossessed farmers. If this is not possible, then it is important to reconsider the park prescription and determine an alternative plan that supports local people's dignity (such as a redemarcation). The two options provided below are on two ends of the spectrum of different alternatives for this termination. From the evaluation it can be seen that both are undesirable, but an intermediate alternative could be suitable. With such an alternative, it could be possible to investigate the possibility of having different management guidelines for different areas of the park, such as those areas containing farms, those containing only resources, and those containing sacred groves.

Option 1

All local use of the park could be terminated as soon as possible and sever local links to resources and abandoned farms. This could ensure that the ecological "integrity" of the park is secured as soon as possible.

Evaluation. From the biodiversity conservation viewpoint, this alternative will be timely in that it does begin conservation immediately. However, it is not timely with respect to the rural development needs of local people who need time to secure new farms. It is also a very dramatic change that may not be balanced or likely to minimize the destructive impacts of change. In addition, it is unlikely to contribute to securing the common interest.

Option 2

The termination of local management inside the park could be prevented, and continued farming and resource use could be allowed.

Evaluation. This alternative will not be acceptable to the WD and would be unlikely to contribute toward the park's aim of conserving biodiversity. This alternative might not be dependable, comprehensive, or balanced, as local activities could degrade KRNP.

Alternatives for Appraisal Function

The alternatives for the appraisal function are similar to those of the intelligence function. However, additional factors should inform the formulation of alternatives. As well as being dependable, rational (working toward the common interest), comprehensive, and selective, an appraisal should be independent and continuous. Therefore this section evaluates the potential of different assessment possibilities to be independent and continuous. All the options are dependent on there being sufficient resources available for appraisal.

Option 1

Local people and the WD could perform a joint appraisal.

Evaluation.   A joint team might increase independence because an individual participant's perspective will not dominate the activity. However, there may be problems with continuity if two participants have to coordinate the appraisal. Many of the problems discussed in the alternative for the intelligence function also apply here.

Option 2

The KCP or WD could carry out regular appraisals.

Evaluation.   Although the KCP is officially independent, it is too ephemeral to be involved in the appraisal function. However, the WD could appraise continuously. This may not be completely independent, as one perspective could dominate the procedure.

Option 3

Different participants could be responsible for conducting different components of the appraisal function.

Evaluation.   This option has the same problems as the other alternatives. It is not necessarily independent, and if responsibility for different appraisal components is not clarified, the function might not be continuous. It would also be difficult to ensure that the appraisal is comprehensive and selective and the data is reliable.

Conclusions

The goal of having an integrated human development and conservation project encounters many conflicts, which means that securing the common interest is challenging. The analysis in this paper has focused on the decision process, as this is an efficient way to evaluate the processes involved in establishing KRNP. The analysis has shown that many participants with different perspectives have been involved in the decision process. Fundamental to this problem is the different conceptualization of the meaning of development and conservation to different participants. Analysis of the problem has also shown that although the WD has attempted to maximize human dignity while still contributing to its mandate of preserving representative samples of ecosystems, there are many challenges in the future. The set of strategies outlined for each decision function help determine a middle ground for seemingly conflicting goals. However, integration of the differing goals might not be achieved even if a middle ground is identified and established. It is important to understand that the difficult history in the area and the current conditions make small, effective, adaptive advances the most appropriate ones. As the Ghanaian proverb says: "Kakra kakra acoco ben non nsua" ("Small small like a chicken drinks water").

POSTSCRIPT: 2000

Since the writing of this paper, the development of KRNP has progressed very positively. The WD has had numerous meetings with local people and recently redemarcated the park for a final time. All the villages around the park are now supportive of the WD. The WD staff are continuing this pioneering work of collaboration with local people to develop a new National Park.

ACKNOWLEDGMENTS

I thank all the people from Nkwanta District, the Wildlife Department staff, and my colleagues who helped me gather the information discussed here. Also, thanks to Andy Willard and Tim Clark for their advice on clarifying and analyzing the problem.

NOTES

1. The new policy represents a shift away from the traditionally preservationist approach to wildlife to one of conservation through sustainable use. This paradigm shift is due to factors such as Agenda 21, the drive to create an environmental action plan and have an "enlightened and pragmatic approach to the sustainable use of wildlife resources" (Government of Ghana 1994:1).

2. In Ghana, the title to land is often vested in stools. In this case all the land is owned by the ethnic group. The stool is a representation of the power of the chief.

LITERATURE CITED

Ankudey, N. K. 1991. "Preliminary Reconnaissance Survey of Nkwanta Area in Northern Volta Region for a Game Reserve." Wildlife Department, Accra, Ghana.

Bahian, K. L. 1996. Redemarcation of Kyabobo Range National Park. WD, Accra. Unpublished report. Held on file by author.

Brunner, R. D. 1987. *Conceptual Tools for Policy Analysis.* Center for Public Policy Research. Boulder: University of Colorado.

Brunner, R. D., and W. Ascher. 1992. "Science and Social Responsibility," *Policy Sciences* 25:295–331.

Chambers, R. 1994. "Participatory Rural Appraisal (PRA): Analysis of Experience," *World Development* 22(9):1253–1268.

Fairhead, J., and M. Leach. 1996. *Misreading the African Landscape. Society and Ecology in a Forest Savanna Mosaic.* Cambridge, UK: Cambridge University Press.

Government of Ghana. 1994. Policy for Wildlife and Protected Areas. Revised Policy. Unpublished report. Accra, Ghana.

Kyabobo Conservation Project. 1995. *Kyabobo '94: Final Report.* Cambridge, UK: KCP Publications.

Lasswell, H. D., and A. Kaplan. 1950. *Power and Society. A Framework for Political Inquiry.* New Haven: Yale University Press.

———. 1971a. "From Fragmentation to Configuration," *Policy Sciences* 2:439–446.

———. 1971b. *A Pre-View of Policy Sciences.* New York: American Elsevier.

Ledward, A., and D. Bowes-Lyon. 1996. "The Relationship Between the Akyode People and Their Local Environment: A Study in Northern Volta Region, Ghana," Kyabobo Report Series No. 2. Cambridge, UK: KCP Publications.

McDougal, M. S., and W. M. Reisman. 1981. "Constitutive Process," in *International Law Essays: A Supplement to International Law in Contemporary Perspective,* pp. 269–286. Mineola, NY: Foundation Press.

Mertz, A. 1975. Unpublished report of trip to north Volta Region. Held on file by author.

Nuhujinbiina, N. A. 1991. Demarcation Survey of the Proposed Koue Hills National Park in the Nkwanta District of Volta Region. Unpublished report. Ghana Wildlife Department, Accra, Ghana.

Nyarko, S. J., and S. J. K. Fordjour. 1994. "Report on the Education of Communities in the Proposed Kyabobo National Park Area." Wildlife Department, Accra, Ghana.

O'Keefe, E. 1997. "A Training Course for Wildlife Department Staff in Kyabobo Range National Park," KCP Report Series No. 3. Cambridge, UK: KCP Publications.

O'Keefe, E., D. Bowes-Lyon, K. L. Bahian, and J. Tordoff. 1997. "An Assessment of the Boundary of Kyabobo Range National Park," KCP Report Series No. 1. Cambridge, UK: KCP Publications.

Scott, J. C. 1998. *State Simplifications: Nature Space and People.* New Haven: Yale University Press.

United Nations Environment Programme. 1992. "Convention on Biological Diversity, 1992." Environmental Law and Institutions Programme Activity Centre, London, UK.

Willard, A., and C. Norchi. 1993. "The Decision Seminar as an Instrument of Power and Enlightenment," *Political Psychology* 14(4):576–606.

11 Local Participation in Conservation and Development Projects

Ends, Means, and Power Dynamics

Peter R. Wilshusen

Since the mid-1980s, conservationists and development practitioners have attempted to protect important natural areas while promoting the welfare of local agrarian communities situated within or in close proximity to these areas. This strategy of linking conservation and development has attained wide currency among the most prominent environmental nongovernmental organizations (NGOs) including the World Wildlife Fund (WWF), the World Conservation Union (IUCN), Conservation International (CI), and The Nature Conservancy (TNC). Currently, conservation and development projects are under way in all regions of the developing world. As of mid-1995, WWF's United States–based offices had allocated 41% of their resources to some thirty field projects that integrate conservation and development goals (WWF—US 1995).

One of the principal unexamined tenets of conservation and development projects is the expectation that local participation in decision making is required to achieve the objectives of environmental protection and human development. Historically, however, conservation projects have disregarded the needs of local people and in several cases, such as the creation of Kidepo National Park in Uganda,

Research for this project was conducted intermittently between October 1994 and June 1996.

have forcefully relocated communities, causing irreversible social impacts (see West and Brechin 1991). Existing appraisals suggest that most conservation and development projects have achieved only superficial participation by local communities through consultation, monetary compensation, or low-level employment. In general, local populations have had little, if any, decision-making power during the formative phases of project development including problem definition, identification of goals, establishment of monitoring procedures and design of activities. At the same time, in recent years lead conservation and development organizations have made more concerted attempts to incorporate local people in the operational phases of project decision making (cf. Western et al. 1994). I will argue that lead organizations need to tailor community involvement to local conditions and structure interventions such that local people share decision-making power with outside groups in the formative phases of project development.

In this paper, I explore why the majority of conservation and development projects have not fostered a greater degree of participation among local people. Moreover, I contrast projects that have failed to generate participation with examples of interventions that appear to have succeeded in this respect in order to uncover constraints and lessons regarding local involvement. First, I present a case study from Thailand to illustrate the evolution of a project that is representative of the kinds of actors, decisions, issues, and outcomes that emerge when trying to link conservation and development. Second, I examine the roles and practices of the diverse participants involved in the Thai project. Third, I analyze the successes and failures of several projects that have attempted to involve local people. In addition to the Thailand case, I draw on examples from Nepal, Colombia, Madagascar, and Costa Rica. I discuss major trends in local participation, offer explanations as to why these trends emerged, and project how the trends will continue into the near future if no changes are made. Finally, I offer recommendations for increasing the degree and improving the quality of local participation in conservation and development projects. It is my hope that the recommendations will contribute to the ongoing debate regarding local participation and that conservation and development practitioners will find them useful for designing future interventions or reorienting current projects.

I approach this study searching for a more nuanced understanding of the complexities underlying local participation in conservation and development interventions. The examples and observations I use come from existing appraisals as well as one case study on Colombia's *Bio-*

pacífico project that I prepared. My aim in studying local participation in conservation and development projects is to elevate consideration of human dignity in the process of elaborating goals for intervention such that this ideal attains the same importance as environmental protection. I propose that maximizing human dignity will open a pathway to environmental protection that satisfies all participants' demands, while emphasis on environmental protection alone will tend only to serve specific environmental interests and, in the end, will compromise both conservation and development goals.

The Problem of Local Participation in Conservation and Development Projects

The guiding principle of conservation and development projects is sustainable development. Sustainable development is broadly understood as a process that guarantees the long-term prosperity of human communities and the natural environment that supports them (World Commission on Environment and Development 1987, hereafter WCED). I submit that although conservation and development projects have advanced toward that goal, typically they do not establish clear goals for sustainable development, critically analyze their strategies with respect to participation, or consider the political implications of power sharing in relation to specific projects. Local communities often do not buy into project goals, accept strategies, or feel that they possess any power to influence project outcomes. This occurs, in part, because local people are not involved in the initiation of activities when decision-making structures are established. As a result they often choose not to participate in project application and, in some instances, even oppose conservation and development projects altogether.

In the context of conservation and development projects, sustainable development refers to improved social and economic conditions for natural resource–dependent rural communities while protecting ecologically valuable habitats. Proponents of this type of project reason that local communities will degrade forests and other areas less if they are organized to take action, have control or access to the natural resource base, possess adequate information and knowledge, believe that their welfare is grounded in sound resource management, and perceive that their economic and social situations will improve (WWF 1995). In many cases, conservation and development projects combine local social and economic development with management of protected areas and their buffer zones. These types of projects are often referred to as integrated conservation and development projects or ICDPs. Those projects that are not associated with protected areas carry the more generic label of community-based conservation.

The strategic focus on protected areas and their buffer zones reflects priorities established by the international environmental community to preserve ecologically important habitats worldwide. Local social and economic development involves a range of activities including income generation opportunities such as low-interest, revolving loan programs or nature-based tourism, health and sanitation improvement, small-scale agriculture, and agro-forestry. Protected areas such as national parks and wildlife reserves are largely undisturbed tracts of habitat that house diverse, and often rare and endangered, plants and animals. Protected area management emphasizes monitoring, protection, and maintenance of those habitats. Buffer zones are areas surrounding national parks and reserves where land use is partially restricted. Buffer zones add a layer of protection to the protected area and can serve as a staging ground for environmentally beneficial activities such as agro-forestry, organic agriculture, and environmental education.

The following discussion presents the case of Thailand's Khao Yai National Park in order to illustrate the social and decision processes relating to local participation in conservation and development projects. The Sup Tai Rural Development Conservation Project, located on the park's northwestern border, provides a representative illustration of the dynamics of participation including types of actors, decisions, issues, and outcomes.

Conservation and Development at Thailand's Khao Yai National Park

Khao Yai National Park is located 200 kilometers northeast of Bangkok and embraces a land area of 2,168 square kilometers. Khao Yai, created in 1962 to protect valuable wildlife habitat, includes some the largest remaining areas of tropical moist forest in mainland Asia and houses diverse plant and animal species including an estimated 2,500 plants and over 60 mammals (Griffin 1994). Because of its accessibility from Bangkok, Khao Yai has become a hot spot for tourist activity. According to 1990 figures, the park attracts annually between 250,000 and 400,000 Thai and foreign visitors who spend some US$5 million on admission, lodging fees, transportation, food, and other services (Dixon and Sherman 1990). An array of high-cost facilities within Khao Yai, including luxury hotels and two golf courses, gives parts of the park a country club image.

In addition to tourism, other human activities such as hunting, burning, logging, cultivation, and extraction of forest products are persistent threats to biological diversity in the Khao Yai region. These more visible threats are accompanied by trends in population growth, land-lessness, regional migration, and distribution of wealth, which indi-cate that large numbers of poor people have moved to the area sur-rounding Khao Yai National Park in search of productive land and other natural resources. As of 1992, an estimated 53,000 people lived in 150 villages around the park (Wells and Brandon 1992). Of the local residents, many have no guaranteed land-use rights and those living in protected areas are considered illegal squatters. One study states that by 1984, 5% of the park's total area had been transformed into agri-cultural land by the landless poor (Dobias 1988).

During the late 1970s and early 1980s, attempts to enforce laws protecting Khao Yai National Park produced animosity between local villagers and personnel of the Royal Forestry Department's National Parks Division. Intensified policing led repeatedly to armed conflicts with casualties inflicted on both sides (Wells and Brandon 1992). The number of enforcement officers was inadequate to police all the vil-lages around Khao Yai on a regular basis such that it was impossible to curb illegal activities significantly.

As early as 1984, the National Parks Division began experimenting with a new incentive-based approach designed to address the socioeco-nomic factors that fuel hunting, agricultural expansion, and illegal log-ging. Park staff initiated a pilot ecotourism project in the village of Sup Tai to offer alternative sources of income to villagers. Although suc-cessfully launched, the project was applied irregularly and did not serve as a genuine alternative to exploiting park resources. As a result of this first experience, the National Parks Division turned to a Thai NGO called the Population and Community Development Association (PDA) for as-sistance. Presumably, the National Parks Division felt that it lacked the experience and know-how to carry out a time-consuming community development project. The PDA's participation brought about the Sup Tai Rural Development for Conservation Project. Sup Tai was selected for the project because it is located along one of Khao Yai National Park's borders that has experienced high amounts of illegal hunting and con-version of forest to farmland (von Loebenstein et al. 1993).

Sup Tai Rural Development for Conservation Project

Two Thai NGOs—the PDA and Wildlife Fund Thailand (WFT)—began the Sup Tai project in 1985. The PDA is Thailand's largest NGO, and since initiating rural development projects in 1974 it has carried out health, family planning, and income generation projects in some 16,000 villages. WFT was founded in 1983 and is a WWF affiliate. It is a small organization that focuses on Thailand's pressing environmental issues. Prior to the Sup Tai project, WFT had no experience with managing conservation and development projects (Wells and Brandon 1992).

As an initial step, the NGOs completed a preliminary survey of socioeconomic conditions in Sup Tai village. Results of the inquiry showed that Sup Tai residents had lower incomes than villagers in comparable regions and that over 80% were heavily in debt to local moneylenders. In addition, health and sanitation levels were low, literacy was rare, and one-third of the villagers (mainly recent immigrants) possessed no legal land titles. Many locals acknowledged illegal hunting and logging inside Khao Yai National Park as a means to supplement income (Wells and Brandon 1992). The moneylenders controlled the local economy under a sharecropper system whereby cash-poor villagers took out high-interest loans (5 to 10% *per month*) to pay for household materials, agricultural implements, and other goods. In return, households paid out 33% of their annual income to service the debts. When villagers were unable to repay the loans, they forfeited their lands to the loan sharks. The cycle of dependence was firmly entrenched because the intermediaries purchased villagers' products at a reduced price and sold them goods such as fertilizers and pesticides at inflated prices under the credit system (von Loebenstein et al. 1993).

The NGOs designed the Sup Tai project to reduce the amount of exploitation of park resources by local villagers through improved income-generating opportunities. The PDA in particular concluded that villagers could pursue community development only if they were free of the heavy debt burden imposed by the moneylenders. Representatives from the NGO reasoned that with easier access to low-interest loans, villagers would be better able to invest in diverse income-generating activities. Furthermore, the loans could serve as incentives for villagers to curtail environmentally destructive activities within Khao Yai National Park.

The core of the project thus involved the establishment and operation of a village-level institution called the environmental protection society (EPS), which functioned as a credit cooperative. An elected village committee administered the society with the participation of a full-time project manager (representing the PDA). In exchange for commitments to abide by park regulations, members of the environmental protection society were permitted to take out low-interest loans from a revolving fund. All loans were to be repaid within one year, and most were based on annual interest rates that varied between 9 and 14% depending on the activity being financed. Normal bank interest rates in Thailand at that time ranged from 16 to 20% per annum. The

PDA incorporated scaled interest rates within the revolving loan system to encourage environmentally beneficial activities. Rules for borrowing included higher interest rates for cash crop production, lower rates for activities such as agro-forestry, and no interest for activities associated with community improvement. From 1985 to 1989, 436 loans totaling some 2.1 million baht ($75,000) were made to Sup Tai residents and all of them were repaid in full and on time (Wells and Brandon 1992).

In addition to the loan program, the two NGOs promoted other conservation and development activities as part of the Sup Tai project including livestock and fish-raising, fruit tree cultivation, soil conservation, establishment of cooperative stores and daycare centers, improved sanitation and health practices, and small park trekking programs for tourists. Environmental education programs focusing on the importance of Khao Yai National Park were developed for environmental protection society members, youth groups, and schoolchildren.

By 1987 the two lead groups, the PDA and WFT, decided to work separately and to expand the project by initiating activities in other communities along the park's borders. WFT withdrew from the Sup Tai project and initiated the Environmental Awareness and Development Mobilization (TEAM) project in ten villages on the eastern side of Khao Yai, with conservation education activities in another forty villages. PDA continued work at Sup Tai and extended the project to two nearby villages. By 1991, the PDA had established separate environmental protection societies in six villages. Although the National Parks Division initiated the activities when it contacted the PDA, the government agency had no significant involvement in any stage of project development.

Initial outside funding for the project (1985 to 1990) came from Agro Action, a German foundation, which provided US$180,000. The PDA contributed US$36,000 of its own funds. The United States Agency for International Development (USAID) approved a three-year grant of $190,000 for Wildlife Fund Thailand's TEAM project but then discontinued its funding early in 1990 after WFT declined to expand the project to include more villages (Wells and Brandon 1992:23).[1]

Early Impact of the Sup Tai Project

Two assessments of the Sup Tai project carried out in 1992 and 1993 found that, despite important advances in generating investment capital and income producing opportunities for local villagers, the principal objective of improving biodiversity conservation at Khao Yai National Park was not being achieved (von Loebenstein et al. 1993; Wells and Brandon, 1992).[2] Results of village surveys performed in 1990 by the PDA showed that awareness of the park had increased but that illegal activities such as logging and hunting continued. The PDA report concluded that illegal activities within the park had dropped since the project had begun but that the reduction could not necessarily be attributed to the project's educational or other components. Among those villagers interviewed, most cited a fear of punishment by park rangers as their principal motivation for curbing illegal activities (PDA 1990). Most villagers did not make a connection between the credit program and reduced poaching, timber extraction and other illegal activities in Khao Yai National Park (von Loebenstein et al. 1993). Therefore, PDA's promotion of economic incentives in exchange for promises from villagers that they would refrain from resource exploitation within Khao Yai did not generate the desired results.

Regarding social and economic development at Sup Tai village, a German development agency (GTZ) report concluded that the EPS loans provided important investment capital to local villagers but, as of 1993, they had not freed residents from moneylenders. Moreover, the report called into question the EPS's ability to function without the PDA (von Loebenstein et al. 1993).

Perhaps most important, the rate of development at Sup Tai led to a "landslide" of unforeseen consequences that ran against the project's long-term goals. By 1987, the project's high visibility inspired a visit by Thailand's prime minister, which in turn led the provincial government to carry out road improvements and electrification. With the infrastructure improvements, land speculation increased such that land prices grew sixfold from 1985 to 1989 (Wells and Brandon 1992). Many farmers sold their properties and became either renters or moved elsewhere; the number of absentee landlords increased considerably. As a result, individual landowners became wealthier but the community as a whole became less cohesive as outside investors arrived in search of profit-making opportunities.

Participants in Conservation and Development Projects

Conservation and development projects bring together diverse participants including government agencies, national NGOs, international NGOs (such as the WWF), international development agencies (like the United States Agency for International Development, or USAID), and local communities. Each organization or community exhibits structural characteristics and behavior that set it apart as a participant in the project's social process. Table 11.1 presents a general participant analysis for Thailand's Sup Tai project.

Participation Dynamics at Sup Tai

At Sup Tai, the Population and Community Development Association controlled project decision making. The National Parks Division invited the PDA to undertake the project presumably because of the latter's extensive experience with population planning and community development. The association performed the initial assessment and thus controlled the process of problem definition. The project's goals reflect the PDA's organizational emphasis on community development. Wildlife Fund Thailand brought an environmental focus to the project, but when WFT and the PDA decided to part ways in 1987, each organization separately pursued activities that fell within its domain of expertise. While the PDA was able to draw on its wealth of experience in rural development and family planning, it was less successful in promoting environmental activities. WFT, on the other hand, chose to focus its energies on environmental education and deemphasized community development. Exactly why the PDA and WFT discontinued their working relationship at Sup Tai is unknown, and reports do not indicate any overt animosity between the two groups. Yet the vast differential in experience and distinct organizational goals may have led the PDA and WFT to a convenient separation. WFT may have decided that in order to exert more control over decisions and to emphasize environmental conservation goals more heavily, it would do better to start its own project on the other side of the park.

Unlike many conservation and development projects, the Thai National Parks Division did not participate in the Sup Tai project. Lack of government participation perhaps diminished villager distrust of the project stemming from a history of negative relations and even violent exchanges with the parks division officials. Parks division administrators did not have the resources, staff, or jurisdiction to operate outside of Khao Yai's borders. On the other hand, the absence of park representatives in decision making may have meant that the Sup Tai project was largely disconnected from Khao Yai National Park. In part as a result of the absence of park representatives, the PDA was unable to integrate conservation and development activities as originally planned.

It is difficult to assess exactly how villagers responded to the Sup Tai project. The PDA created the Environmental Protection Society (EPS) to establish a community-administered revolving fund. Sup Tai villagers reacted positively to the financial incentive. However, the administra-

Table 11.1. Participant Analysis for Thailand's Sup Tai Project (1985–1992)

Participants	Arenas	Expectations	Demands	Strategies	Outcomes
PDA	20 years' experience in 16,000 villages throughout Thailand.	Family planning, health and income generation projects.	Increase villagers' incomes. Increase own visibility and power.	Organize and implement rural development projects. Conduct appraisals.	Environmental protection societies.
WFT	Founded in 1983. Environmental advocacy in Bangkok. Sup Tai first action project.	Environmental issues; political action.	Increase experience and credibility. Promote environmentalism. Increase power.	Conservation projects. Environmental education. Affiliate with WWF. Political lobbying.	Environmental protection societies. Environmental education programs.
National Parks Division	Government agency responsible for protected areas management nationwide.	Protected areas management.	Alleviate conflict with villagers. Maintain power base.	Enforce national protected area laws. Scientific land management.	Historically, conflict with local people. Completed research. Management plans.
International donors: WWF, USAID, Agro Action (Germany)	Based in Europe and U.S. Support projects globally.	Support conservation and development activities.	Increase own visibility and power. Protect biodiversity.	Provide financial and technical support to partner organizations. Conduct appraisals.	Funded projects. Published appraisals.
Sup Tai Villagers	Mix of established residents and landless migrants in Sup Tai region.	Rural livelihoods, income generation.	Increase material well-being and autonomy. Secure land tenure.	Participate in projects. "Illegally" extract resources.	Increased family incomes; sold land; extracted resources.

Note: PDA = Population and Community Development Association; USAID = United States Agency for International Development; WFT = Wildlife Fund Thailand; WWF = World Wide Fund for Nature.

tion and control of the EPS, as of 1992, lay mostly with the PDA project manager, and it appeared that for the near term the EPS would not persist on its own without a PDA representative to shepherd the process (Wells and Brandon 1992). It is possible that local people saw the revolving fund as belonging to the PDA and therefore felt no control over the institution. Based on the fact that exploitative activities continued within the park after the project was in place for five years, it appears that villagers did not initially fulfill the verbal agreement of accepting low-interest loans in return for nonexploitative behavior (Wells and Brandon 1992).

The Sup Tai case indicates that one-third of local inhabitants were landless migrants and that inflated real estate prices (indirectly caused by the project) led people who did own land to sell their properties to make a profit. The first point suggests that at least one-third of the people in Sup Tai village had very little power and control over the means of agricultural production and thus few incentives to invest in land management. These landless migrants probably had little or no emotional attachment to Sup Tai given that they came from other regions searching for resources. The second point shows that Sup Tai villagers sought to increase their monetary wealth, probably as a means of pursuing new economic opportunities.

International donor agencies or NGOs participate in projects by providing funds and technical support. At Sup Tai, the two national NGOs took the lead, and international organizations offered primarily financial support. The German foundation Agro Action and USAID did not have a formal role in decisions regarding project application. However, because they funded activities at Sup Tai, the two organizations influenced methods, time frames, and evaluation and reporting requirements. None of the above effects is necessarily negative, although real or perceived pressure to produce visible results often leads recipients to produce overoptimistic analyses in proposals, reports, and evaluations (Wells and Brandon 1992:54; see also Ascher and Healy 1990:162).

Participants and Power

In general, lead organizations that manage conservation and development projects have included local inhabitants only superficially in any part of the decision process. Wells and Brandon (1992:47) conclude that of the twenty-three conservation and development projects they examined, "most treated local people as passive beneficiaries of project activities and have failed to involve people in the process of change and their own development. As a result, the targets of the proj-

ects often have no stake in or commitment to the activities being promoted. None of the projects based on this beneficiary approach has
demonstrated significant progress toward its goals."

Similarly, Little (1994) concludes that most of the twelve conservation and development projects reviewed for a 1993 workshop approached local participation as a means of protecting national parks
and other areas rather than as an objective for local empowerment. As
a result, lead organizations tended to choose strategies such as economic compensation that give people material benefits in exchange for
not exploiting resources within protected areas. Little argues that even
though local people derive benefits from the project under this form of
"participation," they do not have any decision-making power.

Pimbert and Pretty (1995) distinguish different levels of participation ranging from passive reception of information to self-mobilization,
where people act on their own behalf, independent of outside agents
(Table 11.2). It is important to note that the degree of participation in
projects is not necessarily constant throughout. For example, in Colombia's Biopacífico project (presented below), local participation was
initially very slight but over time increased significantly. In the following discussion, I refer to the different levels presented in Table 11.2
to discuss local participation in five conservation and development
projects. All but one of the projects incorporated some local participation, and three included communities in project decision making.
Although these three projects do not reflect general trends in participation, they suggest interesting directions in decision making and illustrate important conditions that facilitate local involvement.

Trends in Participation from Africa, Asia, and Latin America

In Thailand, the two lead organizations initially approached local
participation as a consultative process (Level 3). The PDA interviewed
Sup Tai villagers and used the information to design and launch the
project. Once the Sup Tai project was under way, the PDA and WFT encouraged community members to take on responsibility for the Environmental Protection Society. An elected group of community members partially administered the society. Economic incentives were the
primary means of encouraging behavior change. For the entire project, 60% of all households in six villages participated in the different
EPSs by taking out loans (PDA 1991). Thus, although the PDA consulted
villagers before designing the project and included them in decisions
regarding the application of project activities, they allowed only a minimal role for community members in determining the content of the

Table 11.2. Levels of Local Participation

Type of Participation	Description
Level 1: Passive participation	Local people are informed about a planned or completed intervention by project staff; people's responses are not taken into account.
Level 2: Participation in information giving	Local people answer questions posed by extractive researchers and project managers using surveys or similar approaches. People do not influence proceedings.
Level 3: Participation by consultation	Local people are consulted by outside agents and these agents use the information to define problems and design interventions. Local people do not take part in decision making.
Level 4: Participation for material incentives	Local people provide resources such as labor in return for cash, food, or other material incentives. They do not participate in decision making.
Level 5: Functional participation	Local people form groups to fulfill predetermined project objectives. This type of involvement typically occurs during application but not during problem definition and project design. The groups may take part in decision making. Groups initially depend on outside facilitators but may become independent.
Level 6: Interactive participation	Local people participate in joint analysis of problems and produce action plans, form new groups, or strengthen existing groups with other partner groups. The joint effort involves diverse methods that seek multiple perspectives and make use of systematic and structured learning processes. Local people participate as full partners in decision making.
Level 7: Self-mobilization	Local people organize themselves and initiate and undertake action on their own behalf with no external assistance.

Source: Adapted from M. P. Pimbert and J. N. Pretty, "Parks, People and Professionals: Putting Participation into Protected Area Management," United Nations Research Institute for Social Development (UNRISD) Discussion Paper #57. Geneva, Switzerland: UNRISD, 1995.

project. Perhaps the villagers continued to depend on the PDA to maintain the EPSs after the project's first seven years of operation because they had not participated in the project's construction and therefore felt little responsibility for project outcomes.

Similarly, lead organizations initially incorporated low levels of local participation at the Amber Mountain project in Madagascar and the Biopacífico project in Colombia (both Level 1). However, whereas the Amber Mountain project did not advance beyond the lowest level of local participation, conditions moved the Biopacífico project staff to eventually share significant amounts of decision-making power with local groups.

Madagascar: Amber Mountain

In addition to the Sup Tai case, Hough's (1994) account of the Amber Mountain Integrated Conservation and Development Project (ICDP) in Madagascar illustrates how participants operate to advance their own organizational goals and power. The WWF and the Malagasy Nature Conservation Service (SPN) led the Amber Mountain project. These two organizations emphasized conservation, causing local development organizations to withdraw from the project over disagreements about project objectives. In initiating the project, the WWF responded to the international donor community's strong interest in protecting Madagascar's unusually diverse natural areas.

Amber Mountain received generous funding, and in return, the WWF and SPN felt it necessary to demonstrate positive results rapidly because it was the first ICDP attempted in Madagascar. As a result, both organizations undertook activities that reflected their fields of expertise: protected areas management, tree planting, and the construction of small dams and irrigation systems. Local participation in decision making did not occur. Moreover, the WWF and SPN designed the project to provide significant infrastructure development for the SPN in the form of new offices and houses. Given a general scarcity of resources among government agencies in Madagascar, the SPN worked to increase its own power by hiring new staff rather than forming partnerships with other departments. In other words, the SPN was at least as interested in amassing more infrastructure and increasing its power vis-à-vis other state agencies as it was in undertaking conservation and development activities (see Hough 1994).

Initially, the SPN and WWF worked in conjunction with the development committee of the local Catholic church to plan activities that emphasized a participatory approach. However, before planning was finalized, representatives of the local Catholic church withdrew from the project because the WWF and SPN emphasized conservation research and enforcement over grassroots development. Without the involvement of the development committee members, local people distrusted the motives of the WWF and SPN. Although the two organizations worked with a regional religious leader, local communities did not participate directly in project decision making (Gezon 1997). As a result, the social and cultural "distance" between local people and project staff increased. According to Hough (1994:121), "Villagers generally perceived uniformed SPN personnel as hostile because of their traditional role as law enforcers, their authoritarian attitude, and, in the coastal and northern parts of Madagascar [the location of Amber Mountain], because government officials were perceived as agents of a hostile ethnic group, the Merina. That development workers were SPN staff reporting to a park warden severely handicapped development work in the field." Of the four case studies, Amber Mountain featured the least amount of local participation. Indeed local communities did not participate in any phase of project decision making.

Colombia: Biopacífico

United Nations Development Programme (UNDP) staff members designed Colombia's Biopacífico project without any input from the rural communities it was intended to benefit (Navajas 1995). Biopacífico is a six-year project financed in part by the Global Environment Facility (GEF) to conserve biodiversity and promote rural development in Colombia's Pacific Coast region. Initially, the predominantly black and indigenous communities of the region refused to participate in the project because community representatives were not included in project decision making (the project document approved by the GEF called for ample community participation) and because the project's activities did not address their goals of attaining land tenure and preserving their cultural identities. In the midst of increased local-level resistance to the project and followng a midterm external evaluation that heavily criticized the Biopacífico project for not involving communities in decision making, staff members opened discussions with community representatives. After several months of arduous negotiation, the black and indigenous representatives successfully negotiated with the project staff to reorient its goals to more heavily emphasize community interests (Wilshusen 1996). Consequently, local participation has increased substantially. Black and indigenous representatives now take part in a project planning team with staff members (Level 5).

The outside appraisal sponsored by the UNDP in 1995 concluded that Biopacífico had not met many of its goals because local communities had not participated (Hernández et al. 1995). The appraisal mission served as an important conduit for local people to further express their dissatisfaction about the project. The joint appraisal group changed the project's decision-making structure by creating a permanent technical committee including a broad range of representatives that have since rewritten the project's operational guidelines. So, whereas staff based in Bogotá controlled all aspects of decision making during the project's first three years, the technical committee took on decision-making responsibility following the joint appraisal.

Nepal: Annapurna

In Nepal, the King Mahendra Trust designed the Annapurna Conservation Area Project (ACAP) in collaboration with government officials to respond to the needs of local people and to conserve biodiversity. An initial survey of local inhabitants conducted by the trust found that people feared the designation of a new protected area because they believed it would cause them to be evicted or excluded from their traditional lands. The creation of Chitwan, Sagarmatha, and other Nepali national parks had involved relocation of local peoples. Because of the high degree of distrust, participation by local people in the Annapurna project came slowly. Project staff undertook consultations with communities and worked to convince them that the project would offer development benefits (Levels 1–3). ACAP took a second step to encourage greater participation by reviving centuries-old community forest management committees and allowing them to control decisions related to pasture and forest management (Level 5). The committees are responsible for enforcing hunting regulations, fining poachers, and authorizing timber harvests (Wells 1994).

The King Mahendra Trust approached the Annapurna project much like the PDA and WFT initiated the Sup Tai project where the lead organization surveyed local people and used the information to design the project. Unlike the PDA and WFT, however, the trust structured the Annapurna project to allow the community forest management committees to control all aspects of decision making for pasture and forest management. Thus, unlike the Sup Tai villagers, who administered the environmental protection society, communities in the Annapurna project performed all decision functions with respect to pasture and forest management. In other words, while community members executed activities for parts of both projects, only in the Annapurna project was full decision-making power granted to villagers to steer the course of some management practices.

Costa Rica: Osa Peninsula Forest Conservation and Management Project

Costa Rica's Osa Peninsula Forest Conservation and Management Project (BOSCOSA) emerged in 1987 in response to persistent, often violent, conflicts between gold miners, loggers, farmers, conservationists, biologists, foresters, and government administrators over land use in an area that includes the only remaining tracts of lowland rainforest on the Pacific coast of Central America (Donovan 1994). Because of its ecological and cultural importance, the Osa Peninsula is dominated by six protected areas. The government created most of these reserves in the mid- to late 1970s without consulting local populations. Local people resented government officials who claimed their lands, in some cases relocated families, and invariably failed to provide them any form of compensation.

The WWF and Neotropica Foundation (a Costa Rican NGO) designed BOSCOSA to ameliorate tensions with rural communities and improve their welfare as a means of reducing hunting, wood cutting, mining, and other pressures on remaining forested areas. The project's goal was to help organize local communities to undertake small-scale agriculture and forestry projects. Similar to the Annapurna case, local-level organizations in the communities of Rancho Quemado and La Palma made decisions regarding agricultural and forestry activities while BOSCOSA offered technical support (Level 5). In contrast to the other four case studies, BOSCOSA focused on local, regional, and national policy arenas in an attempt to reverse deforestation. It worked to increase interagency cooperation by creating the Osa Inter-Institutional Committee. Among other things, the BOSCOSA staff advocated reforms in land tenure policy at the national level for the 90% of Osa residents without land titles. Despite important organizational accomplishments, the project had yet to achieve any modifications in land tenure policy after three years of work (Donovan 1994).

Similar to the Annapurna project, BOSCOSA staff offered community-level organizations full decision-making power over certain aspects of the project's application. Presumably, the local organizations made all decisions with respect to agricultural and forestry activities and relied on the project's personnel to assist with intelligence, promotion, prescription, and appraisal activities. Unlike the redesigned Biopacífico project, however, neither Annapurna nor BOSCOSA shared decision-making power with community representatives in decisions regarding the overall course of project development. Therefore, even though Annapurna and BOSCOSA achieved high degrees of participation compared to Amber Mountain and Sup Tai, one could hypothesize that by limiting power sharing to the application of project prescriptions, lead organizations reduce the probability that local people will adopt and continue to apply the prescriptions beyond the project's duration. Conversely, one would expect community representatives to assimilate many of Biopacífico's objectives once the project is completed. Biopacífico is the only project reviewed in this paper in which power was shared explicitly with community representatives regarding the intelligence, promotion, prescription, and appraisal functions.

The Challenges of Participation

Under the circumstances described above, villagers take part in project activities when direct incentives are offered to them (e.g., a low-interest loan) but do not invest their energies in longer-term goals such as conservation (Table 11.3). My hypothesis is that villagers are more likely to support long-term goals when they have some control over decision making (Levels 5–7) and external factors permit. In cases such as Annapurna and BOSCOSA, where greater power sharing occurred during project application, lead organizations structured the degree and timing of participation by controlling the formative phases of project development.

Local participation occurs only superficially in many conservation and development projects because the different participants involved possess different levels and forms of power. Because participation among groups implies power sharing, which brings uncertainty, those groups with more power are not readily inclined to give much of it away. Thus, increased participation in decision processes decreases the amount of control that lead agencies have over outcomes. Many agencies that wish to gain credibility and power relative to their peer organizations, in addition to successfully completing a project, perceive greater involvement to be less efficient and riskier than strategies that direct interventions at communities.

In summary, lead organizations tend to structure participation so that local communities have higher amounts of decision-making power once a project is established and operating and less decision-making power during the formative phases of project development. Thus, whether advertently or inadvertently, lead organizations have excluded local people from the early phases of project decision making when

Table 11.3. Summary of Participant Analysis Categories

Category	Description	Primary Questions
Participants	Collective and individual actors	Who participates?
Perspectives	Demands, identifications, expectations	With what perspectives?
Situations	Contextual setting	In what arenas or situations?
Base values	Value assets, liabilities	Relying on what assets or base values?
Strategies	Coercive, persuasive, communicative, collaborative	Following which strategies?
Outcomes	Empirical events	Generating which outcomes?
Effects	Consequences of action; change–status quo	With what consequences?

Source: Adapted from H. D. Laswell, *A Pre-View of Policy Sciences.* New York: American Elsevier, 1971.

participants create the approach and structure the decision process of the project.[3] Regarding the appraisal function, these organizations have involved local communities in monitoring and evaluation but tend to rely most heavily on their own conclusions as well as those reached by project financiers to guide project development.

Conditioning Factors that Shape Local Participation in Conservation and Development Projects

At least five factors condition or shape the character of local participation in conservation and development projects: differences in goals and meanings; knowledge as power; histories of domination; class, ethnicity, and status differences; and external forces.

Differences in Goals and Meanings

Local people often value different outcomes than do conservation and development organizations. Moreover, because of professional training and institutional incentives, consultants or personnel tend to impose the goal of biodiversity conservation rather than attend to local peoples' interests. An environmental awareness campaign that focuses on the importance of protecting a park for future generations will seem irrational to groups whose most immediate concern is securing land tenure rights. Indeed, elements of a project can take on different meanings for participants depending on their values and perceptions. In the case of Colombia's Biopacífico project, rural communities' perceptions of biodiversity conservation had as much to do with the survival of cultural groups and their attachment to place as it did with the preservation of plants, animals, and habitats (Wilshusen 1996). Thus, negotiation of the project's goals represented not only power sharing and a discussion of resource allocation but also an exploration over meanings.

Knowledge as Power

Interventions often involve technical and other forms of knowledge that local people do not possess. Therefore, outside experts tend to have a disproportionate amount of power (in the form of specialized technical knowledge) compared to local people. As a result, the experts dominate decision making—particularly the formative phases of project development—to the point that local people participate marginally. Moreover, knowledge differences create communication barriers that alienate local people and cause them to become suspicious of activities carried out by experts. For example, the Biopacífico project initially emphasized biological research carried out by trained university scientists. Community members felt that the researchers extracted local knowledge about plants and animals and offered nothing in return. Scientists used methodologies and instruments that were unfamiliar to local people, and the outsiders made no attempt to include community members. Researchers excluded local people from field studies, and they, in turn, became suspicious that scientists were "stealing" valuable plants. As a result, communities began to prohibit outside researchers from performing studies on their lands (Wilshusen 1996).

Histories of Domination

Rural communities often deeply distrust the intentions of outside groups because of long histories of negative experiences with both conservation and development activities. For example, the Mexican government created the Calakmul Biosphere Reserve (located on the Yucatán Peninsula) by presidential decree without consulting local communities living within the designated area. As a result, local residents actively resisted the government's program for relocating communities (López and Boege 1995). Similarly in Nepal, local communities distrusted conservation groups' motives in establishing the Annapurna Conservation Area since other protected areas such as Chitwan and Sagarmatha had been created and local people had been relocated without prior consultation. As described above in the Nepali case, the King Mahendra Trust used careful planning to gain the trust of local communities in setting up the Annapurna Conservation Area (Wells 1994).

Class, Ethnicity, and Status Differences

Differences in class, ethnicity, and status represent social barriers among groups that restrict or even prevent interaction. Most analyses of conservation and development projects discuss social structural differences only peripherally. In the Amber Mountain case (Madagascar) local community distrust of government representatives was due, in part, to the fact that they belonged to a rival ethnic group (Hough 1994). Different cultural groups speak different languages, observe different customs, and possess different degrees of political power. As the ethnic civil war between the Hutus and the Tutsis of Rwanda illustrates, one group may historically oppress another. In Colombia, government representatives from Bogotá were reluctant to incorporate members of black and indigenous organizations into the Biopacífico project, in part because of class and status differences. Only after local groups had demonstrated their abilities as interlocutors in public meetings did project staff accept them as participants in decision making.

The social differences discussed above suggest that conservation and development discourse is distinct from local discourses. The language of conservation and development is replete with terms such as *protected area, buffer zone,* and *biodiversity* that have little or no meaning for local people. These people may value the natural world differently from representatives of conservation and development organizations. Moreover, local people often perceive the natural world differently compared to people trained in "Western" modes of thinking. While communities may share the basic goals or rationale of a project, outside groups' expression of these goals in project form or their treatment of nature may be conceptually foreign to local people who perceive different spatial and temporal realities.

External Forces

Political, economic, and social forces external to projects often impede, if not prevent, local participation. Economic factors include financial incentives generated by national economic policy and global market forces that encourage rapid extraction of timber, minerals, and other natural resources. In Brazil, the national congress maintained subsidies for large-scale cattle ranchers who "developed" land in the Amazon region by converting forests to pasture. Brazil later altered this policy in the early 1990s because of political pressure from domestic and international groups.

Migration within or between countries represents another process occurring widely in places like Thailand, Rwanda, and Brazil. National policies, wars, and resource disputes cause large numbers of people to relocate from one area to another. At Thailand's Khao Yai National Park, most of the 53,000 people living in the region in 1992 were migrants who illegally occupied land classified as forest reserves. One-third of Sup Tai villagers, most recent immigrants, possessed no legal land titles (Wells and Brandon 1992). Under circumstances of political repression, instability, or national emergency, local people may be unable to participate.

External political, economic, and social forces have varied impacts on projects and can both preclude local participation and interrupt ongoing projects. Policies that create environmentally exploitative subsidies or cause widespread migration and landlessness can be modified under certain circumstances, but other occurrences such as wars are usually uncontrollable. In Rwanda, the ethnic civil war forced hundreds of thousands of refugees to seek safety around Kahuzi Biega and Virunga national parks in neighboring Zaire (now the Democratic Republic of Congo). The rapid arrival of large numbers of people produced widespread land degradation as refugees cut trees, hunted wildlife, and established settlements to satisfy their basic needs (Hart and Hall 1996; Hart and Hart 1997). The ethnic civil war in Rwanda brought to a halt the mountain gorilla project at Volcanoes National Park, one of Africa's most celebrated conservation and development initiatives. The project was operated by the African Wildlife Foundation in conjunction with the Rwandan Office of Tourism and Nature Protection. It focused on tourism development as a means of generating revenue in a very poor region. The project brought in 6,000 tourists to view the mountain gorillas, increasing park income from a few thousand U.S. dollars in 1979 to a half million U.S. dollars annually by 1989 (Wells and Brandon 1992).

Elements of Successful Participation

Whereas the cases from Costa Rica (BOSCOSA), Nepal, and Colombia achieved high levels of community involvement (Level 5 and above) in some if not all decision-making functions, the examples from Thailand (Sup Tai) and Madagascar (Amber Mountain) represent typical levels of participation where outside organizations consult local people during the intelligence, promotion, and prescription phases. They attempt to involve communities in the application of prescriptions but grant them little or no decision-making power over the project itself. Fundamentally, those projects that have successfully shared decision-making power with local people have explicitly operationalized participation as a project goal and employed local organizations and support networks to facilitate all decision functions.

In Colombia (Biopacífico), project staff initially excluded local communities from all aspects of decision making, but black and indigenous representatives later negotiated a reformulation of the project that gave community representatives some power over project decisions and reflected local peoples' objectives more strongly. In Nepal (Annapurna) and Costa Rica (BOSCOSA), lead organizations dominated intelligence, promotion, and prescription but strengthened local organizations and encouraged local decision-making control over project components dealing with forest management. Thus, the Colombia project started with the lowest level of participation but, upon renegotiation, featured the highest level of involvement possible given outside intervention (Level 6). The cases from Nepal and Costa Rica incorporated high levels of participation in portions of the projects from the outset (Level 5). Local participation in these two projects succeeded in large part because the intervening organizations (the Neotropica Foundation in Costa Rica and the King Mahendra Trust in Nepal) made shared decision making a priority, channeled community involvement through existing local organizations, and constructed support networks. I consider each characteristic in turn.

Operationalizing Participation

The Annapurna and BOSCOSA projects both made local participation a primary objective, whereas other projects reviewed in this paper approached participation as a means for protecting biodiversity (see Little 1994). The difference between the two approaches is that once participation is stated as an objective, project organizers must assess how it can occur. When participation is pursued as a means to an end, lead organizations tend not to consider specific involvement strategies. In Nepal, the King Mahendra Trust initiated contact with local communities and consulted them about the project. The same process occurred in the Sup Tai project (Thailand). However, the King Mahendra

Trust sought mechanisms to encourage community decision making regarding forest management, whereas the Population and Community Development Association worked to increase local peoples' incomes. The PDA placed more emphasis on community empowerment through income generation but focused less attention on increasing local peoples' decision-making power. Like Annapurna, the BOSCOSA project evaluated potential mechanisms for local participation with the intent of promoting community decision making in forest management. In its initial formulation, Colombia's Biopacífico indicated the importance of local participation (it was not a formal objective) but did not define channels for involvement until communities' rejection of the project jeopardized its existence.

Guaranteeing Support Structures

Local organizations play a key role in community participation because they represent established conduits for decision making. In Nepal, the King Mahendra Trust relied on existing local forest management committees that traditionally made decisions regarding forest resources held in common by the communities. The committees worked under customary rules regarding harvesting and management such that the Annapurna project did little more than encourage communities to continue the pertinent practices. In Costa Rica, the Neotropica Foundation linked forest management activities to established agricultural cooperatives. Because the cooperatives already had a decision-making system in place, BOSCOSA staff channeled technical support through local organizations.

When the Biopacífico project began in 1992, a broad-based social movement composed of black and indigenous organizations from across the Pacific region was emerging. Initially, project staff made little attempt to join local organizations in decision making. By 1994, the collection of local organizations had coalesced to address land tenure and local autonomy issues in the national congress. Once established, this coalition of black and indigenous groups became the channel for community representation in the Biopacífico project. Whereas community organizations had been small and dispersed, the coalition represented a regional organization that matched the scale of project activities. Where the scale of organizational representation matches the project scope, it facilitates local participation by channeling decision making at compatible levels.

Support networks help lead organizations address the political, economic, and social processes occurring beyond the local context that influence how local participation occurs. The best example is the BOSCOSA project in Costa Rica, where the Neotropica Foundation has formed the Osa Inter-Institutional Committee including representatives from local organizations and regional and national government agencies to consider coordinated action on the Osa Peninsula. Through this interagency committee, project participants work to resolve local land tenure problems that prevent farmers from making long-term investments in forest stewardship. The Biopacífico project has incorporated an informal support network where a core planning team composed of project staff and representatives of the black and indigenous peoples' coalition interacts regularly with government agencies regarding actions in the country's Pacific Coast region. The Biopacífico project has not been as successful as BOSCOSA at fostering interagency cooperation but has influenced decision making on a regionwide economic development project that local communities had opposed (Wilshusen 1996).

The Future of Participation

In addition to the cases offered in this paper, appraisals such as the one completed by Wells and Brandon (1992) suggest that where political and economic factors and social differences create high potential for conflict, participation in conservation and development projects will be limited at all levels of decision making. Moreover, local involvement will be constrained in situations where there exist histories of external domination over rural communities, high levels of distrust, and wide differences in goals and knowledge between lead organizations and local people. The trends in participation for the Sup Tai (Thailand) and Amber Mountain (Madagascar) projects are representative of most conservation and development projects (see Wells and Brandon 1992; Western et al. 1994).

Conservation and development organizations will incorporate low levels of participation through consultation and avoid sharing power in decision making when they perceive the need to produce immediate results (Madagascar) or seek to advance their own power (Thailand and Madagascar). Lead organizations will also deemphasize participation when no strong local organizations are in place. Conversely, in instances where conservation and development organizations pursue local participation as an explicit goal (Nepal, Costa Rica) and external political, economic, and social forces permit, community involvement in decision making will occur successfully but mainly in the application of project prescriptions. In other words, even those organizations dedicated to local participation will tend to minimize power sharing during the formative phases of project development in order to control project content and procedures. The level and durability of local participation will depend on many factors including the intervening organization's ability to overcome differences in goals and knowledge as well as legacies of past injustices.

Strengthening Participation in Conservation and Development Projects

The conservation and development literature typically recommends that lead organizations increase the amount of community involvement in projects. Indeed, the case studies presented in this paper indicate that low levels of public participation can hinder and even derail projects. In Colombia, local people openly opposed the Biopacífico project initially because it was not being applied as originally designed. The project objectives emphasized community development and participation but, in practice, served specialized scientific research interests. As a result, the Biopacífico project gained a negative reputation among residents of Colombia's Pacific Coast region, and local communities refused to cooperate with project staff and researchers (Wilshusen 1996).

At the same time, examples suggest that high levels of local participation can occur despite social differences, external political and economic factors, distrust, and differences in goals and technical knowledge. Once black and indigenous representatives began to participate in all levels of decision making and gained the respect of members of the lead organizations, local communities started to receive direct support from Biopacífico and their acceptance of the project increased. Furthermore, the cases in this paper suggest that higher levels of local participation in projects may increase the impact and durability of conservation and development interventions. For example, community forest management committees existed in villages around Annapurna, and these customary practices had become institutionalized well before the creation of the modern Nepali state. By encouraging local participation through a well-established community institution, it appears that the Annapurna project's staff has increased the probability that its objectives will be pursued and amplified indefinitely.

As the cases from Colombia (Biopacífico) and Nepal (Annapurna) illustrate, higher levels of local participation in projects reduce conflict and may increase the durability of project results because communities become empowered to influence outcomes linked to their own interests. Strengthening local participation requires that lead organizations account for the political dynamics of projects that stem from differences in power. These differences manifest themselves in project goals, technical knowledge, historical events, social structure, and external political, economic, and social forces.

Although all the projects have involved local people in some way, none of them have included local communities in the initial problem identification and project design. In order for lead organizations to foster power sharing in decision making with local communities, they must understand participation dynamics. This is not simply a matter of increasing the amount of participation but knowing who will participate, when and how much they will participate, and in which phases of decision making they will participate. To help create an equitable and participatory decision-making environment that accounts for these political dynamics, I recommend three measures including clarifying standpoint and goals, completing participation assessments, and developing a participation strategy tailored to the local context. A key component of all three of these measures involves developing a monitoring process that allows all participants to receive feedback regarding both social process and progress toward project goals. I expect that by strengthening participation in all phases of decision making—especially during the project's formative stage when power relationships are structured—that conservation and development initiatives will be better able both to increase human dignity and to enhance biodiversity conservation.

Pimbert and Pretty's (1995:44) findings on participation in protected areas management summarize the change in approach that outside groups should adopt to strengthen local participation, improve project decision making, and encourage long-term project impacts. "Popular participation in defining what constitutes a 'protected area,' how it should be managed, and in whose interests, implies a shift from the more common passive, consultative and material-driven participation to more interactive and genuinely empowering forms of participation. Genuine people's participation in the conception, design, management and evaluation of protected areas implies new roles for conservation professionals and other outsiders. These new roles all require a new professionalism with new concepts, values, methods, and behaviour."

The main reason for strengthening participation relates directly to the twin concepts of authority and legitimacy in power relationships. Authority is often defined as *legitimate power* where *legitimacy* is "the property of a situation or behavior that is defined by a set of social norms as correct or appropriate" (Scott 1992:305). In all the cases presented in this paper, outside organizations such as state agencies or NGOs possess significant amounts of power. Further, these organizations often have the authority, which has been conferred upon them by the state, to manage natural resources or protected areas. Given historical trends where the invocation of state authority has eclipsed traditional land-use rights, local communities often have not recognized outside groups' authority to restrict access and use of natural resources. In Nepal, Thailand, Colombia, and Costa Rica, the creation of protected areas has resulted in the relocation of communities and generated conflict between local people and the state. As a result, although outside organizations have power and state-granted authority, local people often view their actions as illegitimate. In the absence of legitimacy, outside organizations will always have to invoke state-granted authority and use raw power (enforcement) to protect national parks and other areas. The other alternative is to understand and work with local people in a way that will increase outside organizations' legitimacy with the local community and encourage voluntary compliance under a negotiated set of norms regarding natural resources use that respects customary practices. The remainder of this section discusses three ways that lead organizations can increase their legitimacy with local communities during the formative phases of project development.

Clarifying Goals and Standpoint

Conservation and development projects pursue the goal of sustainable development, echoing a number of broad policy statements made by the World Commission on Environment and Development, the World Conservation Union, and others. The WCED defines sustainable development as "development that meets the needs of the present without compromising the ability of future generations to meet their own needs" (WCED 1987:43). Similarly, the IUCN, WWF, and United Nations Environment Programme (UNEP) made a joint policy statement called *Caring for the Earth*, which defines sustainable development as "improving the quality of human life while living within the carrying capacity of supporting ecosystems" (IUCN, UNEP, WWF 1991:10). However, by adopting these or similar declarations, conservation and development organizations tend not to clarify their standpoints in relation to specific projects. They make clear the intention to increase poor communities' economic and social welfare and improve nature protection but neither specify their roles vis-à-vis local people nor define how participation might occur *before* determining project goals. In the Sup Tai (Thailand), Amber Mountain (Madagascar), and Biopacífico (Colombia) projects, lead organizations involved local communities only after interventions were designed and activities were under way. In cases

where lead organizations fostered power sharing with local populations during the application of project prescriptions (Costa Rica and
Nepal), they established relationships early on with community representatives and made clear their goal of maximizing local participation
(albeit during specific phases of each project).

In order to strengthen local participation, lead organizations need
to clarify their intentions to local people and establish a working relationship with communities during the process of identifying problems
and producing project responses (intelligence, promotion, and prescription). Local people should have power in the process of project creation if they are to feel that it responds to their own development objectives. As Little (1994) observes, participation must be considered
not just as a means to an end but as an objective in and of itself. Therefore, conservation and development organizations should clearly establish and communicate their own stance on participation for a particular context before intervening.

Once defined, participation begins as a first meeting between local
and outside groups where both identify, discuss, and respond to problems collectively. This does not necessarily mean that outside organizations must always undertake co-management of projects with local
communities. These communities may decide to delegate administrative responsibility for projects to outside organizations. The main difference between this scenario and most projects is that communities
have the power to influence the course of projects from the outset; projects are not presented to communities as a *fait accompli*. This recommendation implies that conservation and development organizations
will need to dedicate more time and financial resources to the intelligence and promotion functions of decision making to create a dialogue
and an approach that satisfies all parties. By investing in participation
early on, these organizations increase their legitimacy in the eyes of
local populations, improving their ability to establish durable partnerships for conservation.

Outside organizations such as international lending agencies and
NGOs may resist openly structuring participation strategies with communities because the approach may seem to remove some of their
control over decision making. Those actors farthest removed geographically from the project context may be most susceptible to competing
interests at both individual and organizational levels.[4] For example,
upon examining the behavior of World Bank professionals in reject-
ing two high-potential development strategies—social rate-of-return
analysis and integrated rural development—Brunner and Ascher
(1992:313) found that "[d]espite the theoretical potential and com-

pelling normative bases of these approaches, they were rejected as World Bank strategies largely because of their professional, institutional, and careerist inconvenience, i.e., on the basis of considerations that are far removed from defensible appraisal criteria." Nonetheless, outside or lead organizations stand to benefit from expending more time and energy during project design because presumably it decreases costs related to reviewing and restructuring problematic projects during later phases.

Goal clarification is a fundamental step in establishing a relationship with local communities. As the Biopacífico case (Colombia) illustrates, local people often do not share the same goals as conservation and development organizations. In Colombia, Thailand, and Madagascar, local populations were not cognizant of project goals. Goal clarification requires that outside organizations communicate their goals to local communities and that these communities have some influence over their final expression. Regarding the goal of sustainable development, Lélé (1991:615) finds that the term should be defined in each context in order to be of practical use. Furthermore, Lélé suggests that "Any discussion of sustainability must first answer the questions 'What is to be sustained? For whom? How long?' The value of the concept, however, lies in its ability to generate an operational consensus between groups with fundamentally different answers to these questions, i.e., those concerned either about the survival of future human generations, or about the survival of wildlife, or human health, or the satisfaction of immediate subsistence needs (food, fuel, fodder) with a low degree of risks. It is therefore vital to identify those aspects of sustainability that do actually cater to such diverse interests, and those that involve tradeoffs."

Lélé's recommendation implies a process of negotiation that responds to the common interest as opposed to a set of specialized interests. In Colombia, land tenure was a priority for black communities that possessed no clear land titles. Although the designers of the Biopacífico project initially recognized land tenure as a problem, it was deemphasized in favor of ecological research activities. When community representatives gained a stronger voice in project decision making, project staff members rediscovered the importance of tenure and reallocated resources accordingly. Without this dialogue, the project would have continued to serve specialized research interests.

Assessing Participation Potential

Once conservation and development organizations clarify their goals and standpoints with regard to local participation, they should complete a comprehensive assessment of participation potential for sites targeted for intervention. A participation assessment would allow outside organizations to better understand the micropolitics of a given setting and decide how best to approach local participation, including who should participate. Blaikie and Jeanrenaud (1996:38) advocate an approach called "social mapping," stating that biodiversity conservation policy "demands that the whole 'cast of actors' concerned is identified along with actors' interests in the elements that comprise collectively the notion of biodiversity, how they go about pursuing their objectives and their source of power to reach them."

It is especially important to disaggregate actors such as "the state" or "local communities" because they are often analyzed as monolithic entities (Little 1994). Contrary to a preliminary assessment, Blaikie and Jeanrenaud (1996) note that conservation and development activities at Zambia's Lwangwa Valley National Park did not benefit the majority of local people. The initial report by Abel and Blaikie (1986) predicted that local communities could "get most of what they currently required from the park" while fulfilling outside organizations' conservation objectives. Instead, powerful actors such as local chiefs and project personnel pursued their own interests and thus benefited the most from development at the expense of the larger community. By their own estimation, Abel and Blaikie failed to understand the "prevailing unequal distribution of power" (Blaikie and Jeanrenaud 1996:38) at the local level because they analyzed the "community" as a whole and did not consider its internal politics. Similarly, for the Amber Mountain project in Madagascar, the World Wide Fund for Nature worked hard to gain the support of the regional religious leader but later discovered that he advocated project activities only as long as they generated tourism and other economic opportunities for him. Moreover, local villagers continued to remove wood from protected forests even when the religious leader made pronouncements in favor of the WWF and strict protection of nature reserves. As a result, the WWF found that the moral authority it had hoped to invoke through the religious leader was ineffective at altering local peoples' behavior in exploiting forest resources (Gezon 1997).

Assessing participation potential can best be carried out through a joint exercise in which lead organizations and local communities analyze social processes in a given setting. According to Lasswell (1971:24–26), analysis of social process accounts for participants, perspectives, situations, base values, strategies, outcomes, and effects. Participants are all actors who interact in a social context, and their perspectives are the subjective events experienced. Situations are the "zones in which interactions occur." Base values include all values available to an actor at a given time, and strategies are the means used to affect value outcomes. Outcomes and effects are the discrete, empirically identifiable events produced by interactions and their net repercussions. Lasswell (1971:16) states that human actors "participate selectively in what they do" and as a result are predisposed to complete acts in ways that they perceive will leave them better off than if they had completed them differently. Values are the means of differentiation for analyzing social process. In order to maintain a basis for reliable, empirical observation, Lasswell (1971:17) defines values as "culminating outcomes," meaning "events that are generally understood in a given situation to be very desirable (or undesirable)." He proposes eight values: power, enlightenment, wealth, well-being, skill, affection, respect, and rectitude. In the course of social interaction, some participants are indulged while others are deprived of certain desired outcomes.

Beyond the local context, a participation assessment should look at macropolitical processes such as the activities of regional and national governments that might influence patterns of local participation. In Costa Rica (BOSCOSA), Colombia (Biopacífico), and Thailand (Sup Tai), land tenure was an important issue. Local people apparently were less likely to invest their energies in conservation activities when they had no control over the land or other resources affected by conservation measures. While some external forces such as civil war cannot be controlled by project participants, other constraints such as land tenure can be changed through advocacy. In many cases external factors constrain but do not eclipse the possibility of local participation.

Conservation and development organizations should assess to what degree government policies, large-scale economic developments, or other broad social conditions inhibit or promote local participation. In Thailand, the PDA's assessment of Sup Tai village indicated that at least a third of the residents had no land titles and therefore no control over the means of production. In retrospect, this circumstance was an important constraint on project participation because many local people had little attachment to their surroundings, faced large debts, and did not have strong local support organizations.

Shaping Participation to Local Conditions

Finally, conservation and development organizations should develop a participation strategy as part of project design that reflects local conditions and power relationships. Although the projects from Nepal, Costa Rica, and Colombia successfully joined communities in decision making, none of them included local people in problem definition and project design and only Colombia's Biopacífico project featured local participation in project administration. The other two projects delegated decision-making power to community organizations on forest management activities, not the larger project. In other words, lead organizations successfully fostered power sharing with local people in the application of project prescriptions but did not include villagers in the initial carrying out of the intelligence, promotion, and prescriptive activities that structured each respective project and its inherent power dynamics. Therefore, I recommend that lead organizations strengthen local participation in problem definition and project design (intelligence, promotion, and prescription), work through and strengthen local organizations, and develop support networks.

Joint Problem Definition

In reference to local involvement in problem definition, Little (1994:359) argues that "[t]he critical questions are, whose definition of the problem is being invoked, and who shares in its meaning(s)? . . . The extent to which the local population shares in problem definition and participates in its identification is a prime factor affecting program success." The Annapurna case from Nepal suggests how lead organizations first define problems for project intervention—often with community input—and then present the results of the assessment to local people. Wells (1994:273) writes, "The perception of the crisis (the problem) and the need to address it originated from outside the region. But the early initiatives were worked out in face-to-face contact with people who would have to live with the results of any changes." Thus, despite negotiating the initiative directly with villagers, the Annapurna staff presented a version of the problem that did not necessarily coincide with local peoples' perception of the problem. One promising means for overcoming perceptual discrepancies regarding problem definition is through the use of participatory rural appraisal (PRA).

In the past ten years, both large and small conservation and development organizations have turned increasingly to PRA techniques to promote community-led problem identification. PRA is an approach to research and action that enables local people to share, enhance and analyze their own knowledge to act on their own behalf and in collaboration with other groups. This type of appraisal employs numerous methods including semistructured interviews, key informants, group work of various kinds, mapping and modeling, transect walks, timelines/trend analysis, oral histories, linkage diagramming, and seasonal calendars. Outside organizations often participate as facilitators while community members carry out the activities (see Chambers 1994). In Quintana Roo, Mexico, for example, a small NGO called *Yum Balam* worked with communities in the northern part of the state to produce murals that diagrammed the region's conservation and development problems. Based on this joint problem identification exercise, *Yum Balam* worked with communities to design interventions that responded to local interests and promoted natural resource conservation.

Strong Local Organizations

The degree to which communities feature strong organizations that are representative of local opinions and that can channel local decision making strongly influences the level of participation that takes place in conservation and development projects. Moreover, those communities with well-established organizations will be more capable of immediate project involvement compared to those that have weak or incipient organizations.

In the BOSCOSA case, project staff selected two villages with which to work—Rancho Quemado and La Palma—where the first had no viable community organizations while the second featured a well-established woodworking cooperative. The project intervened at Rancho Quemado by helping residents to form a producer association; it took nine months for community members to decide to form the organization. In contrast, local people formed the woodworkers' cooperative at La Palma a year before BOSCOSA began, and members had decided on specific goals and decision-making processes. In less than three months the cooperative and BOSCOSA established a tree nursery. In Rancho Quemado, it took a year to reach the same milestone (Donovan 1994). This example suggests that, to the extent possible, conservation and development organizations should choose intervention sites with strong local organizations with which to form partnerships. In communities with weak local organizations, outside groups should explicitly focus on organizational development. This last recommendation implies that outside groups should take the time to understand and support local forms of social organization as opposed to automatically promoting the creation of NGO replicas based on North American and European models.

Support Networks

Support networks allow participants in conservation and development projects to establish links to the broader external institutional environment, which has an impact on decision making at the local level. Lead organizations should work with local communities to construct either formal or informal lines of communication with outside agencies such as national or regional policy-making bodies. Such networks might allow project participants to exercise some influence over decisions regarding important issues such as land tenure or economic development in their geographic regions. The Osa Inter-Institutional Committee created under the BOSCOSA project in Costa Rica could serve as an organizational exemplar for other projects. The committee began as a series of informal meetings between representatives of different agencies and was formally created later when participants found it to be an effective arena for cooperation.

Support networks can also join local grassroots organizations, thus providing regular interchange of ideas. In Mexico's Yucatán Peninsula, funds from the Global Environment Facility's (GEF) Small Grants Programme allowed six small grassroots NGOs to establish communication links via electronic mail. In addition, each organization provides technical advice in its area of expertise to the other groups. The network of NGOs meets regularly to coordinate conservation and development activities for the entire Yucatán region. Where infrastructure to support more advanced technology such as computers does not exist, lead organizations could allocate resources for regular meetings between network participants.

Conclusions

Participation in decision making lies at the heart of democratic social processes that respond to the interests of all those involved in a given activity. In this sense sustainable development is a powerful concept when proponents explicitly define what is being sustained, for whom, and for how long. This paper has examined why local participation in conservation and development projects occurs only superficially in most cases despite the efforts of diverse projects to increase community involvement in decision making. Several barriers impede successful local participation including differences in goals and knowledge between communities and lead organizations; histories of domination of communities by the state; ethnicity, class, and other social differences; and external political, economic, and social forces.

Fundamentally, however, the problem of participation in conservation and development projects reverts to the politics of power sharing, especially during the formative phases of project development when participants decide on the structure of project responsibilities and activities. Of the five projects reviewed in this paper, even the ones that incorporated local people in decision making limited their participation to specific applications of project prescriptions as opposed to involvement in initial problem identification, project design, and project administration.

To establish a more equitable decision-making environment, I recommend, lead organizations should clarify their goals and standpoint regarding local participation to communities, assess participation potential, and develop participation strategies that are tailored to each project context. All these measures require that lead organizations adopt a new approach focused on power sharing where local communities have significant influence over project outcomes. Power sharing is important early on in project development in order to build trust. Communities could participate in joint problem identification exercises and take partial responsibility for project design. This new approach takes longer and is more resource intensive than project design based solely on consultations with local people, but it is the only way that lead organizations can increase their own legitimacy and that of their projects in the eyes of local people. Increased legitimacy, in turn, should raise the probability that communities will adapt conservation and development interventions to their own interests over the long term.

ACKNOWLEDGMENTS

I thank Tim Clark and Andy Willard for organizing the seminar "Foundations of Natural Resource Policy and Management" at the Yale School of Forestry and Environmental Studies, for which this paper was originally written. I am particularly grateful to Andy Willard for carefully reading multiple versions of this paper. His insightful comments helped me to rework and strengthen significant portions of the paper. I also thank Raúl Murguia, Karl Steyaert, and Vance Russell, who also offered constructive comments on the present version of this paper. The input allowed me to improve its contents, yet I bear full responsibility for any remaining factual errors or other shortcomings.

NOTES

1. Von Loebenstein et al. (1993) provide a slightly different version, stating that USAID discontinued funding the WFT's TEAM project in February 1991 when it terminated all activities in Thailand following a military coup. The report notes that the project was then funded by World Wildlife Fund—US.

2. Von Loebenstein et al. (1993) state that this conclusion was made based on the project staff's observations and conversations with villagers. They note that at the time of writing the report (August 1992) no system was in place for monitoring the conditions of the park and its resources relative to both the PDA and the WFT's projects (p. 46).

3. Harold Lasswell and Myres McDougal made the important distinction between "ordinary" and "constitutive" policy making. Ordinary policy making encompasses the day-to-day decisions regarding substantive policy content within a given context. Constitutive policy making embodies decisions regarding processual aspects of policy, that is, "*how* policy should by made, and by implication, *who* ought to be involved in choosing policies" (Healy and Ascher 1995:11). For further explanation on this distinction, refer to McDougal, Lasswell, and Reisman (1967); Lasswell (1971); and Lasswell and McDougal (1992). The distinction between ordinary and constitutive policy making is fundamental to this paper because local people tend not to participate during the formative or constitutive stage of project development when power relationships are structured or reinforced.

4. Referring particularly to the common misperception of state neutrality, Ascher and Healy (1990:177–178) write that "the interests of [agents] of the 'state' are varying and complex. In various combinations, their motivations are to: (a) enhance the standing of the agencies in which they work; (b) promote their own careers within these agencies (or elsewhere); (c) adhere to the highest professional standards, either for the sake of professionalism per se or to attain respect from professional peers; (d) pursue partisan political objectives; and (e) pursue a particular *a priori* policy objective (such as environmental protection at any cost). None of these objectives necessarily has anything to do with conservation and development yet they can significantly influence project outcomes."

LITERATURE CITED

Abel, N., and P. Blaikie. 1986. "Elephants, People, Parks and Development: The Case of the Luangwa Valley, Zambia," *Environmental Management* 10(6):735–751.

Ascher, W., and R. G. Healy. 1990. *Natural Resource Policy Making in Developing Countries: Environment, Economic Growth, and Income Distribution*. Durham, NC: Duke University Press.

Blaikie, P., and S. Jeanrenaud. 1996. "Biodiversity and Human Welfare," United Nations Research Institute for Social Development (UNRISD) Discussion Paper #72. Geneva, Switzerland: UNRISD.

Brunner, R. D., and W. Ascher. 1992. "Science and Social Responsibility," *Policy Sciences* 25:295–331.

Chambers, Robert. 1994. "The Origins and Practice of Participatory Rural Appraisal," *World Development* 22(7):953–969.

Dixon, J. A., and P. B. Sherman. 1990. *The Economies of Protected Areas: A New Look at Benefit and Costs*. Washington, DC: Island Press.

Dobias, R. 1988. *Integrating Park Conservation and Rural Development in Thailand*. Bangkok, Thailand: Population and Community Development Association.

Donovan, R. 1994. "BOSCOSA: Forest Conservation and Management Through Local Institutions (Costa Rica)," in D. Western and P. Wright, eds., *Natural Connections: Perspectives in Community-Based Conservation*. Washington, DC: Island Press.

Gezon, L. L. 1997. "Political Ecology and Conflict in Ankarana, Madagascar," *Ethnology* 36(2):85–100.

Griffin, J. G. 1994. "An Evaluation of Protected Area Management: A Case Study of Khao Yai National Park, Thailand." FAO Tiger Paper XXI (1):15–23.

Hart, J., and J. Hall. 1996. "Status of Eastern Zaire's Forest Parks and Reserves," *Conservation Biology* 10(2):316–324.

Hart, T., and J. Hart. 1997. "Zaire: New Models for an Emerging State," *Conservation Biology* 11(2):308–309.

Healy, R. G., and W. Ascher. 1995. "Knowledge in the Policy Process: Incorporating New Environmental Information in Natural Resource Policy Making," *Policy Sciences* 28(1):1–19.

Hernández, J., A. Bidoux, E. Cortés, and J. Tresierra. 1995. Proyecto Biopacífico: Primera evaluación externa. Informe Final. Proyecto Biopacífico, Bogotá, Colombia.

Hough, J. L. 1994. "Institutional Constraints to the Integration of Conservation and Development: A Case Study from Madagascar," *Society and Natural Resources* 7:119–124.

Lasswell, H. D. 1971. *A Pre-View of Policy Sciences*. New York: American Elsevier.

Lasswell, H. D., and M. S. McDougal. 1992. *Jurisprudence for a Free Society: Studies in Law, Science, and Policy*. 2 Vols. The Netherlands: Martinus Nijhoff Publishers.

Little, P. D. 1994. "The Link Between Local Participation and Improved Conservation: A Review of Issues and Experiences," in *Natural Connections: Perspectives in Community-Based Conservation*. Washington, DC: Island Press.

Lélé, S. M. 1991. "Sustainable Development: A Critical Review," *World Development* 19(6):607–621.

López, A., and E. Boege. 1995. "Case Study: Community Development Around the Biosphere Reserve at Calakmul, Mexico," Discussion Paper of the Social Science and Economics Program prepared for Workshop I of the ICDP Review. Washington, DC: World Wildlife Fund.

McDougal, M. S., H. D. Lasswell, and W. M. Reisman. 1967. "The World Constitutive Process of Authoritative Decision," in M. S. McDougal and W. M. Reisman, eds., *International Law Essays: A Supplement to International Law in Contemporary Perspective*. New York: Foundation Press.

Navajas, H. 1995. Informe de misión. Proyecto Biopacífico: UNDP COL/93/G31. Proyecto Biopacífico, Bogotá , Colombia.

Pimbert, M. P., and J. N. Pretty. 1995. "Parks, People and Professionals: Putting Participation into Protected Area Management," United Nations Research Institute for Social Development (UNRISD) Discussion Paper #57. Geneva, Switzerland: UNRISD.

Population and Community Development Association (PDA). 1990. "Sup Tai Project Evaluation, Phase II." Bangkok, Thailand: Population and Community Development Association.

———. 1991. "Fifth Project Progress Report, Rural Development for Conservation Project—Phase III, January–June, 1991." Bangkok, Thailand: Population and Community Development Association.

Scott, W. R. 1992. *Organizations: Rational, Natural, and Open Systems*. 3d edition. Englewood Cliffs, NJ: Prentice Hall.

von Loebenstein, K., A. Trux, and T. Welte. 1993. *Compensation and Reconciliation of Interests in the Field of Buffer Zone Management.* Volume 2: *Case Studies in Asia and Africa.* Bonn, Germany: GTZ.

Wells, M. 1994. "A Profile and Interim Assessment of the Annapurna Conservation Area Project, Nepal," in D. Western and P. Wright, eds., *Natural Connections: Perspectives in Community-Based Conservation.* Washington, DC: Island Press.

Wells, M., and K. Brandon. 1992. *People and Parks: Linking Protected Area Management with Local Communities.* Washington, DC: The World Bank.

West, P. C., and S. R. Brechin, eds. 1991. *Resident Peoples and National Parks: Social Dilemmas and Strategies in International Conservation.* Tucson: University of Arizona Press.

Western, D., P. Wright, and S. Strum, eds. 1994. *Natural Connections: Perspectives in Community-Based Conservation.* Washington, DC: Island Press.

Wilshusen, P. R. 1996. Case Study: Colombia Biopacífico Project—Biodiversity Conservation in the Chocó Biogeographic Region. Unpublished report. New York: UNDP.

World Commission on Environment and Development. 1987. *Our Common Future.* New York: Oxford University Press.

World Conservation Union (IUCN), United Nations Environment Programme (UNEP), and World Wide Fund for Nature (WWF). 1991. *Caring for the Earth: A Strategy for Sustainable Living.* Gland, Switzerland: IUCN, UNEP, and WWF.

World Wildlife Fund—US. 1995. "What's in a Name? Integrated Conservation and Development Projects (ICDPs)." Discussion Paper of the Social Science and Economics Program prepared for Workshop I of the ICDP Review: Linking Conservation to Human Needs: Creating Economic Incentives. Washington, DC: World Wildlife Fund—US.

12 Greening the United Nations High Commissioner for Refugees

Improving Environmental Management Practices in Refugee Situations

Gus Le Breton

The United Nations High Commissioner for Refugees (UNHCR) is the international United Nations agency responsible for providing care and assistance to refugees. Within recent years, and especially since the Rio Earth Summit in 1992, there has been a growing international awareness of the potentially adverse environmental impacts of hosting large numbers of refugees in geographically concentrated areas. As a result, considerable pressure has been placed on the UNHCR by the international community and nation donors to take measures to address issues of refugee-related environmental degradation within its programming. The UNHCR responded to this pressure by establishing a new unit within the organization: the Office of the Senior Coordinator for Environmental Affairs (OSCEA). The policy problem, therefore, is how the OSCEA can best promote and incorporate sustainable environmental management activities within the UNHCR's existing programs of refugee assistance.

The purpose of this paper is threefold: (1) to explore the social context within which the OSCEA operates; (2) to examine the decision processes through which the OSCEA will try to improve environmental management; and (3) to formulate recommendations as to how OSCEA can best improve UNHCR practices to minimize the adverse en-

Research for this project was conducted between 1993 and 1997.

vironmental effects of concentrating refugees in confined geographic areas.

My role in this analysis is as a consultant to the OSCEA. I am part of a team that has been recruited to advise the unit on how it can promote better environmental management practices within the organization. This is a role through which I am hoping to indulge my own enlightenment, skill, respect, power, and wealth.

My standpoint is that of both a critical outsider and a participant. My position as the director of a Zimbabwean nongovernmental organization (NGO) (Southern Alliance for Indigenous Resources, or SAFIRE) that addresses refugee-related environmental issues puts me at both an advantage and a disadvantage. The advantage is that, having worked in partnership with the UNHCR through SAFIRE, I have an understanding of how the UNHCR operates. I am also sufficiently independent of the commission to look critically at its policies. The disadvantage is that I come into the position of consultant with a series of preformed judgments and criticisms that I may be reluctant to shed.

I have conducted this analysis based on the following sources of information: my recently conducted series of interviews with senior UNHCR headquarters staff and several field staff; the extensive body of literature relating to the UNHCR, its operations, and its experiences in the field of environmental management; analysis of the OSCEA's recently formulated Guidelines for Environmental Management; my own experiences and understandings of the UNHCR's modus operandi.

The Policy Problem

Social Process and Context of Problem

A refugee is officially defined as: a person who, owing to a well-founded fear of being persecuted for reasons of race, religion, nationality, membership in a particular social group or holding a political opinion, is outside the country of his nationality and is unable, or, owing to such fear, is unwilling to avail himself of the protection of that country (1951 Convention Relating to the Status of Refugees).

The United Nations High Commissioner for Refugees was created in December 1950 by Resolution 428(V) of the United Nations General Assembly to provide international protection to refugees and to facilitate their voluntary repatriation or assimilation within new national communities. The UNHCR began its operations on 1 January 1951.

Within the past ten years, an average of 20 million refugees worldwide has existed at any one time, with an additional 30 million internally displaced people, many of whom also fall within the UNHCR's jurisdiction. Afghanistan alone spawned more than 6 million refugees throughout the 1980s and early 1990s. The UNHCR, with an annual operating budget of close to $1.3 billion, is currently assisting 27 million people in 140 different countries worldwide.

When refugees arrive in a new country, they often do so in large numbers. Typically they have traveled long distances over water or on foot and are tired, poorly nourished, and traumatized. Their immediate needs include food, water, medical care, and shelter. The UNHCR is often quick to respond to emerging refugee situations and has a long history of success saving large numbers of lives as a result of prompt intervention, proving that it is highly effective at fulfilling its mandate.

The UNHCR's procedures in dealing with refugees tend to be consistent. The refugees are sited in one or more convenient concentrations during what is referred to as the emergency phase, during which efforts are focused on providing for their basic needs. Once the population stabilizes, assistance moves to the care and maintenance phase. The UNHCR provides for less immediate but still pressing needs, including clothing, education, sanitation, and more permanent shelter. In many cases, this phase involves implementing steps toward the achievement of refugee self-sufficiency, including the introduction of agricultural activities, income-generating projects, and skills-training activities. Finally, once the political situation in the refugees' country of origin has stabilized, the program enters the "durable solutions" phase, in which refugees are either voluntarily repatriated to their own countries or resettled in the host or a third country.

While the programs have often been successful at helping to save lives, the arrival of a huge influx of refugees places enormous pressures on the environment in the host country. Localized deforestation, soil erosion, water contamination, and rapid depletion of groundwater supplies are common problems associated with refugee populations. Environmental concerns take a backseat, at least until the care and maintenance phase. As a result, there have been many cases where refugee populations have caused, and suffered from, intense localized degradation, the costs of which can be measured in millions of dollars.

Concerns, particularly those voiced by host governments, about the adverse environmental consequences of refugee hosting, have recently made it imperative that the UNHCR incorporate environmental considerations into its responses to refugee situations. Several steps have been taken, and the organization has expressed its commitment to protecting the environment of host countries. However, the realization of such a commitment has proven difficult, and the UNHCR is evaluating alternative strategies available for minimizing the environmental degradation of host countries.

Several different entities are participating in the process of "greening" the UNHCR, and it is important to understand the niche of each in the overall social context. A description of the social process is presented in Table 12.1.

Office of the Senior Coordinator for Environmental Affairs

The OSCEA is based at UNHCR headquarters in Geneva, Switzerland, and is staffed by environmental professionals. Created specifically to address environmental issues within the UNHCR, the OSCEA is devoted to improving the environmental management practices of the commission. The OSCEA has many motivations for improving environmental management: it must justify its continued existence and expansion within the UNHCR (power, respect), and it also has a desire to secure and expand its currently fragile funding base (wealth), as well as a genuine commitment to improve the environmental practices of the organization (rectitude). A personal commitment exists on the part of the (Japanese) senior environmental coordinator to fulfill his functions with honor and distinction (rectitude, respect). Finally, the OSCEA must address the need to secure refugee asylum in host countries by ensuring that those countries are not reluctant to accept refugees because of the potentially adverse ecological impacts associated with their hosting (rectitude).

United Nations High Commissioner for Refugees Field Staff

Field staff are all emergency relief professionals, many of whom have a broad depth of experience in emergency situations. Often working in extreme and traumatizing conditions, they are tough, pragmatic, and highly solution-oriented. In terms of environmental management activities, field staff are characteristically motivated by many factors. They desire to minimize the additional burden on what is already an intensive work load (well-being). In situations where staff are called upon to make rapid and important decisions, they resent what they see as unnecessary interference from headquarters (power, respect). Field staff also take a narrow, linear approach to emergency management that facilitates quick and effective action in emergencies but mitigates against broader strategic conceptualization and planning, which is required for successful environmental management (skill). Finally, their bounded rationality excludes action in any fields for

Table 12.1. Social Process for Participants in the "Greening of UNHCR"

Participants	Perspectives	Situations	Strategies	Preferred Outcomes	Preferred Effects
OSCEA	Mandated to address environmental issues within UNHCR	Mandate-based role in improving environmental management	Diplomatic pressure to field and PTSS staff, economic incentives to international, local partner agencies, host governments	OSCEA expands to have environmental coordinators within each country. Respect, power, wealth, and rectitude indulged.	OSCEA's role expanded
Field staff	Reluctant to take on environmental management because of increased work load and perceived lack of technical skills	Frontline staff in high-stress situations	Diplomatic resistance to additional work loads; diplomatic pressure for further training in environmental management skills, provision of technical staff	Minimal additions to work load for field staff, extra staff provided. Well-being, power, respect, skill, and enlightenment indulged.	OSCEA provides field staff with additional training and staff
Programme Technical Support Services	Desire to improve UNHCR's environmental management activities, but preferably through PTSS	Locked in territorial tension over environmental management	Diplomatic pressure on and resistance to OSCEA in favor of expanded role of PTSS in environmental management	OSCEA dissolved, PTSS given full responsibility for environmental management. Respect, power, wealth, rectitude and skill indulged.	PTSS given increased funding and resources

(continued)

Table 12.1. continued

Participants	Perspectives	Situations	Strategies	Preferred Outcomes	Preferred Effects
International partner agencies (multilateral lending)	Belief that NGOs may be best placed to undertake environmental management, but facilitated by UNHCR	Flexible, experienced, and capable of environmental management	Diplomatic and ideologic pressure on OSCEA for funding to undertake environmental management activities	International NGOs made fully responsible for environmental management. Wealth, power, skill, and respect indulged.	UNHCR provides funding and resources
International partner agencies (NGOs)	Belief that NGOs may be best placed to undertake environmental management, but facilitated by UNHCR	Flexible, experienced, and capable of environmental management	Diplomatic and ideologic pressure on OSCEA for funding to undertake environmental management activities	International NGOs made fully responsible for environmental management. Wealth, power, skill, and respect indulged.	UNHCR provides funding and resources
Local partner agencies (NGOs)	Belief that NGOs may be best placed to undertake environmental management activities, but facilitated by UNHCR	Locally knowledgeable and experienced	Diplomatic and ideologic pressure on OSCEA for funding to undertake environmental management activities	UNHCR hands full responsibility to local NGOs for environmental management activities. Wealth, skill, rectitude, and respect indulged.	UNHCR provides capacity building, funding, and resources to local NGOs

Note: NGOs = nongovernmental organizations; OSCEA = Office of the Senior Coordinator for Environmental Affairs; PTSS = Programme Technical Support Services; UNHCR = United Nations High Commissioner for Refugees.

which they do not feel themselves to be specifically trained (enlighten-ment, skill).

Programme Technical Support Services

A technical unit within the UNHCR headquarters, PTSS provides specific technical services to field operations at the request of field staff. Areas of expertise include health, sanitation, education, water supply, and agriculture. Prior to the formation of the OSCEA, PTSS had exclusive responsibility for environmental activities, even though it had no environmental specialists of its own. PTSS has four major motivating factors related to environmental management. First, the personnel have a strong belief that environmental activities should have remained within its jurisdiction, and that the formation of the OSCEA is, in some way, a threat to its authority and territory (respect, power). Second, the organization resents the OSCEA for the additional funding the latter has received that might have otherwise gone toward PTSS (wealth). Third, PTSS has a prior commitment to improve the UNHCR's environmental management activities (rectitude). Fourth, staff members take a positivist, technocentric approach to problem solving (skill).

International Partner Agencies (Multilateral Lending)

Environmental management is essentially a long-term developmental activity and falls within the mandates and experiences of several international multilateral agencies, including UN Development Program, Food and Agricultural Organization, World Bank, and UN Environment Programme. Without exception, these organizations work on huge annual operating budgets and their operations span the globe. All have field-based technical staff and are strongly territorial about the sectors in which they perceive they are mandated to work (e.g., forestry is seen as the FAO's jurisdiction). However, because most are also lending agencies, few have become involved in refugee-related environmental management activities; host governments would rarely take out loans for environmental management activities related to refugees in preference to their own citizens. Consequently, multilateral agencies have not had loan requests from host countries. Many factors motivate international agencies in relation to refugees and the environment. In an era of contracting funding, these agencies have a competitive desire to retain a strong environmental jurisdiction because of the perceived funding opportunities this provides (wealth, power). They also recognize their agencies' inherently superior abilities to the UNHCR to address environmental issues (skill). Finally, they genuinely accept the need to rationalize the activities of UN agencies to cut costs and enhance efficiency (wealth, rectitude, skill).

International Partner Agencies (Nongovernmental Organizations)

Regular implementing partners of the UNHCR, some NGOs including Cooperative for Assistance and Relief Everywhere (CARE) International and Lutheran World Federation (LWF) have been attempting to improve environmental management activities in refugee situations for some time. CARE has even produced its own set of environmental guidelines for emergency relief field staff. Five contributing factors motivate the UNHCR and environmental management. First, NGOs have experienced frustrations with the UNHCR, and they share a set of beliefs that the UNHCR's capacities as an emergency relief organization make it ill-equipped to address environmental issues because of its inappropriate bureaucratic procedures, its lack of technical knowledge in this field, and its tendency toward autocratic, exclusionary decision making (skill). Second, they believe they must compete with the UNHCR for funding perceived as being available for environmental activities (wealth). Third, NGOs want to expand into, and in some cases monopolize, what is seen as an emerging field within the aid industry (wealth, power). Fourth, they desire to alleviate the adverse environmental impacts of refugees (rectitude). Finally, a belief exists that NGOs, particularly those involved in relief and development activities, are better able to address environmental issues than is the UNHCR (respect, skill).

Local Partner Agencies (Nongovernmental Organizations)

Within many of the countries that it works, the UNHCR has formed partnerships with local NGOs. In the field of environmental management, these organizations are uniquely well suited to implementing appropriate programs because of their local knowledge and experience and the fact that they tend to be developmental in orientation. However, local NGOs typically have shortages of administrative, and sometimes technical, staff, and they are regularly constrained by limited funding. Motivations with respect to refugee-related environmental management activities include the opportunity for enhancing their funding and resource base (wealth); the same set of frustrations experienced by international NGOs concerning the UNHCR's limitations (skill); resentment toward the UNHCR for its perceived lack of commitment toward capacity-building within local NGOs (rectitude, skill); the desire to alleviate the adverse environmental impacts of refugees (rectitude); and the belief that NGOs, particularly those involved in relief and development activities, are better able to address environmental issues than is the UNHCR (respect, skill).

Host Governments

The governments of refugee-hosting countries have an especially strong vested interest in the improvement of the UNHCR's refugee-related environmental management capabilities. It is the local refugee-hosting communities, and through them their governments, that most keenly feel the adverse ecological impacts of refugees and whose livelihoods are affected by those impacts. Many have expressed strong reservations about hosting refugees as a result of these potential impacts, and many more have made strong demands on the UNHCR to address the issue. Host governments' motivations include the opportunity to expand their own funding and resource base and, often, their technical environmental management governmental departments (wealth, power, skill, respect) and the desire to avoid politically damaging environmental impacts among their own rural communities (respect, power).

Refugees

Although themselves ostensibly the cause of much of the environmental degradation associated with refugee-hosting, refugees do not intentionally or willingly cause degradation. Most are from rural communities that have long backgrounds of environmental management, forced into an extraordinary situation with the result that, in many cases, they have no alternatives for survival other than to degrade land. This is especially true in the case of fuelwood harvesting for cooking and heating. However, they also experience the consequences of environmental degradation. In Zimbabwe, for example, Mozambican refugees in the early 1990s had to walk in excess of fifteen kilometers to collect fuelwood because of the extent of local deforestation. Refugees are motivated by the desire to minimize the adverse effects of environmental degradation on their own lives (well-being), the need for assured asylum without the risk of denial on the grounds of potential environmental impacts (well-being), and the desire to avoid conflict with local communities over resource exploitation/degradation (well-being, respect).

Decision Process

The social process of the problem detailed above describes the context within which the OSCEA operates. It is also instructive to examine the problem's decision process. At its inception, the OSCEA developed a five-year plan intended to achieve its goals. An abridged version is presented in Table 12.2.

Most efforts to date involve the gathering and processing of intelligence. Central to intelligence is the study currently being undertaken to examine best practices of environmental management. Examination of the study reveals that it is flawed in three major ways. First, it looks at environmental problems not in terms of ecological conditions, but in terms of UNHCR planning cycles. Second, it assumes that case studies can themselves significantly improve field staff practices. If anything, case studies can only supplement broad problem-solving skills field staff may have and run the risk of linearly channeling their responses to replicate successful practices in other, and inevitably very different, contexts. Finally, the study examines not the processes that shape the environmental management practices of the UNHCR, but only the technical content of UNHCR's responses. (For a more detailed critique of the study, see the appendix to the chapter.)

Analysis of the Problem

Goals

The OSCEA's overall objective is to improve the UNHCR's environmental management activities within refugee situations such that the adverse ecological impacts associated with refugee hosting are prevented or, when they do occur, redressed. Specifically, the OSCEA seeks to ensure that the UNHCR adopts an ecologically sensitive approach to refugee assistance by explicitly including environmental concerns in planning and implementation; planning and implementing refugee assistance in a way that addresses the physical and social needs of refugees, while avoiding or minimizing a negative environmental impact on refugee-hosting areas; and playing a catalytic role in encouraging other institutions to address large-scale environmental problems in refugee-hosting areas.

Trends

The first steps toward the formation of the OSCEA were taken in 1993. PTSS was asked by the high commissioner (herself Japanese) to submit a

Table 12.2. Decision Process within the Office of Senior Coordinator for Environmental Affairs

Phase 1 (1993–1995)
• Strengthen environmental programming within UNHCR
• Identify cooperative partners and technical expertise
• Field-test environmental assessment methods
• Raise funds
• Consult with refugees, local communities, and national counterparts to design
• Use participatory approaches
• Study previous resettlement projects
• Establish a competitive UNHCR annual award system
• Identify a database system to store physical information

Phase 2 (1995–1997)
• Issue training documents and rapid environmental assessment packages
• Set up a global database for areas of refugee assistance
• Initiate long-term monitoring in camps and settlements
• Select promising approaches to siting and sizing of refugee camps
• Implement environmental rehabilitation projects in key countries

Phase 3 (1998 onward)
• Refine assessment and monitoring methods
• Select and implement a GIS database for all refugee camps and settlements
• Develop national capacities to deemphasize the role of UNHCR
• Continue the commitment to a unified approach to refugee assistance

Note: GIS = geographic information systems.

funding proposal to the Japanese government for the formation of an environmental unit within the UNHCR. This was done, and the Japanese government provided the initial funding (US$5 million) for the OSCEA's first two years of operations. It was agreed that, as part of the funding package, the senior coordinator would be seconded from the Japanese government. The first coordinator was duly appointed in June 1993.

It took some time for the unit to become fully established. The senior coordinator was new to the UNHCR, as was his position, and the first year was spent in raising the unit's profile within the UNHCR and familiarizing the senior coordinator with the UNHCR's operations, especially those involving environmental management activities.

During 1994, the OSCEA began its first small projects, allocating funds to various activities in response to requests from the field. It also began the process of preparing environmental guidelines, identifying a number of potential technical consultants who might be hired to assist in their writing.

At the same time, the unit recruited two full-time technical special-ists. One was an environmental educator, hired to boost the provision of environmental education to refugees, and the other was a geo-graphic information systems (GIS) specialist. The latter was given the task of developing a GIS database, using remote sensing and other data, of some of the refugee settlements in which the UNHCR was operational.

1994 was also the first year of the Rwandan crisis, in which several million refugees poured across the border from Rwanda into Zaire and Tanzania. The refugee-hosting area in eastern Zaire was of espe-cial environmental concern, incorporating the Virunga National Park, which contains one of the few remaining mountain gorilla populations in the world. Thus, this situation was an opportunity for the OSCEA to demonstrate its worth. This it did by initially appointing a physical planner from the technical staff in Zaire as the "environmental focal point," responsible for environmental issues relating to the refugee program. Subsequently, in 1995, the OSCEA funded a full-time environ-mental coordinator, based in Goma, for a one-year period.

In June 1995, the first senior environment coordinator's term ex-pired, and his replacement, also from the Japanese government, ar-rived. A more dynamic individual, he was quick to assert himself, and he rapidly began the process of consolidating the new environmental guidelines.

These guidelines were finally produced in 1996 and disseminated to all field staff. Although short on detail, they were intended to be sup-plemented by sectoral guidelines, covering household energy, forestry, environmental education, and site planning. The guidelines were not especially well received by field staff; many felt that adherence required a level of skill, time, and resources they did not possess. Although some field staff understood the importance of including environmental con-siderations within refugee assistance programs, the majority remained convinced that these considerations were of secondary importance to the immediate humanitarian concerns of any refugee situation. This was particularly manifested in their attitude toward the environmental focal point in Goma, whom many viewed as obstructive and unhelpful; they irreverently characterized the GIS images regularly being pro-duced to document deforestation around the Zairean refugee settle-ments as "grossly irrelevant satellite" images.

Nevertheless, the use of GIS at this time demonstrated its enormous potential as a tool for facilitating preventive planning. The OSCEA re-vised the job description of its GIS specialist, whose overall goal would be to prepare basic environmental planning data for a series of poten-tial flashpoint regions around the world. This revision was intended to facilitate the process of siting refugee settlements in an environmentally friendly manner in the event of a refugee situation.

In April 1996, the OSCEA moved the discussion on the UNHCR's role in environmental management to a new level by hosting a major international conference on the subject in Geneva. Over sixty experts from academia, host governments, local and international NGOs, international agencies, and the UNHCR itself were present. Refugees and environmental management became a recognized field of operations within the UNHCR, and the role of the OSCEA was greatly legitimized as a result. In July 1996, the conference was repeated in Dar es Salaam, Tanzania, aimed specifically at refugee-hosting countries in Africa, and this was the occasion on which the environmental guidelines were officially unveiled.

Meanwhile, the OSCEA had begun to receive requests, much in the way that PTSS does, from field staff for technical support and assistance. Many of these originated from the Great Lakes region of eastern Africa, where the new environment coordinator, and the Dar es Salaam workshop, had raised the unit's profile, and with it, the expectations of the field staff as to its capabilities. With its limited funding, the OSCEA was able to respond to only a few of the requests, primarily by hiring regional experts as consultants to field operations. Some of these consultancies resulted in substantive and cost-effective improvements in the UNHCR's operations.

The OSCEA launched its most ambitious project yet in 1997. The unit has recruited a team of consultants to examine environmental management activities in refugee situations in twenty-five countries around the globe. The team is expected to produce a series of case studies of "best practices" (i.e., prototyping) and from these to develop materials for use in training UNHCR field staff. Perhaps more significantly, the team has asked the OSCEA to identify broader ways it can improve environmental management activities in the field.

Conditions

Through the 1980s and early 1990s, an increasing number of refugee-related environmental management problems and programs emerged. This fact was brought to global attention when, in 1991, the Honduran government refused to grant asylum to Guatemalan refugees because of the perceived environmental damage they might cause.

In 1992, the year of the Rio Earth Summit, environmental concerns were placed firmly in the forefront of the international development discourse. It was agreed by all signatories to the Agenda 21 declaration that environmental impact assessments (EIAs) of development projects should be conducted prior to the giving of development assistance by any donor nation. Some donors began extending this principle to emergency relief assistance as well, and momentum rapidly built up among the UNHCR's donors for its refugee assistance programs to be seen as environmentally sensitive. It was largely as a result of this pressure that the OSCEA was established in 1993.

In 1994, the Rwandan refugee crisis, and the influx of over half a million refugees into the Virunga National Park in eastern Zaire, home to the threatened mountain gorilla, provoked a flurry of international media attention on refugee impacts. CNN showed satellite images that documented the rapid deforestation occurring around the refugee settlements. It was this coverage, more than anything else, that provoked the OSCEA to appoint an environmental focal point in Goma.

A further conditioning factor at this time was the perceived funding opportunities that hosting refugees and engaging in conservation efforts could offer. This aggravated the conflict between the OSCEA and PTSS, provoked a sudden interest from a range of other organizations in undertaking environmental management activities with refugees, and led to, at the local level, the emergence of a number of "briefcase" NGOs. These organizations, typically run by a single individual, suddenly appeared and submitted funding proposals to the UNHCR, which seriously undermined the credibility of legitimate local NGOs.

Despite the perceived financial opportunities, the OSCEA remained persistently short of funding, possibly because of the skepticism of the international donor community over the OSCEA's ability to reduce environmental degradation. This was not helped by the fact that the UNHCR's fundraising department, worrying that funds given for environmental management would detract from its overall program funding, refused to distinguish "environmental" proposals, preferring instead to "mainstream" environment into its other budget categories.

Finally, in terms of conditioning factors, one of the biggest constraints to the OSCEA's operations has been UNHCR funding cycles. Budgets are prepared a year at a time and are approved only shortly before they are released, with resultant difficulties in long-term planning. While this short cycle (which contrasts with the five-year cycles used by most development agencies) supports the flexibility required for emergency relief activities, it is extremely difficult in terms of environmental activities, which require long-term planning and commitment.

Projections

Based on the current five-year plan of the OSCEA toward the attainment of its objective, the following (speculative) projections can be made.

The current study will produce a plethora of additional training materials and documentation relating to environmental management in refugee situations. When disseminated among field staff (who already work with over 180 different training manuals), this will be viewed inevitably as an additional burden on their time and resources and will provoke strong antipathy toward the OSCEA.

PTSS, still determined to integrate environmental management into its normal sphere of operations, could capitalize on the disaffection of field staff, forming a coalition with them against the OSCEA. Field staff, recognizing that integration of environmental concerns into PTSS programming would alleviate any additional burden on them (because environmental activities would be the responsibility of PTSS specialists), would be willing partners in such a coalition.

The OSCEA, alienated from both parties, would come to be seen as increasingly irrelevant and toothless within the UNHCR. In line with current rationalization and budget cuts effective at the UNHCR, it could reasonably be projected that the OSCEA will then disband in 1999, at the conclusion of the next term of the senior environment coordinator.

PTSS is unlikely to be able to address environmental impacts successfully. Budgetary mainstreaming will inevitably result in a shortage of funds and, in turn, staff. PTSS staff are inherently linear and technocentric in their approach, characteristics that have been widely found to be inappropriate for environmental management. Furthermore, because PTSS acts only in response to specific requests from field staff, there would be no centralized strategic planning relating to environment, resulting in a series of reactive programs, many of which would not be applied until environmental damage already had reached an advanced stage.

PTSS's incapacity would therefore mitigate in favor of an approach to one or more international development agencies to assume a greater responsibility for environmental management. However, the lack of coordination between these agencies and the UNHCR, in addition to the former's difficulty raising funds for refugee-related environmental activities, would be likely to result in the UNHCR field staff repeatedly making decisions that compromise environmental sustainability and render the efforts of these agencies equally ineffective.

With little to show in terms of effective environmental management, host governments would become increasingly reluctant to host refugees, and donors would become increasingly reluctant to fund environmentally unfriendly relief programs. The cycle, probably starting with the formation of a new environmental unit similar to the OSCEA, would then be repeated.

Recommendations

Alternatives

The following alternatives can be considered in terms of improving UNHCR's environmental management practices:

Alternative 1: United Nations High Commissioner for
Refugees Recognizes Environment as a Sector

The UNHCR formally recognizes the importance of environmental considerations and accords the environment full sectoral recognition, with its own budget line. The OSCEA, because of the additional costs involved, and the fact that it is effectively outside the main UNHCR system, would probably be dissolved, and PTSS would hire its own environmental specialists to provide technical assistance to field staff on request.

Alternative 2: United Nations High Commissioner for Refugees
Mainstreams Environmental Management Activities

This is, in many ways, the direction toward which the UNHCR would like to see the OSCEA move. This would necessitate an ongoing role for the OSCEA, providing training and limited technical support to field staff who undertake environmental management activities within the course of their regular programming. The continued existence of the OSCEA would also enable some form of centralized management and preventive planning of environmental activities.

Alternative 3: United Nations High Commissioner for Refugees
Devolves Responsibility to International Agencies

The UNHCR recognizes its limitations and hands over responsibility for environmental management to one or more international agencies, with whom it collaborates to ensure that its own activities are not contrary to theirs. This would entail the dissolution of the OSCEA, and responsibility for ensuring that the UNHCR's activities are compatible with those of the international agencies would fall on the individual field offices.

Alternative 4: United Nations High Commissioner for Refugees Strengthens Its Catalytic Role

The UNHCR retains overall responsibility for environmental management but accepts its own limited capacity for carrying out such activities. Instead, it would aim to develop partnerships with host governments and local NGOs, focusing specifically on developing their capacity for environmental management activities, and assisting with securement of financing. The OSCEA would assume the responsibility centrally for identifying partners, capacity building, assisting in the preparation of funding proposals, ensuring that the UNHCR's overall activities do not conflict with environmental considerations, and undertaking preventative (GIS-based) planning.

Alternative 5: United Nations High Commissioner for Refugees Develops Partnerships with International Agencies

The UNHCR would develop a partnership with one or more international agencies, charged specifically with building the capacity of local NGOs and host governments to carry out environmental management activities. The OSCEA's role would be reduced to interacting with international partner agencies, guiding field staff in environmentally friendly operations, and preventative planning.

Evaluation of Alternatives

While continued success in saving the lives of refugees is vital to the operations of the UNHCR and OSCEA, the following evaluations focus on each alternative's ability to minimize environmental degradation and to secure funding.

Alternative 1: United Nations High Commissioner for Refugees Recognizes Environment as a Sector

Environmental issues would likely receive higher priority than currently, with consequent opportunities for improved practices. However, these opportunities would be contingent upon PTSS's ability to influence field operations, the caliber of PTSS staff, and field staff requesting PTSS input. PTSS staff, while technically competent, are not always imbued with the broad-based, problem-solving skills required of effective environmental managers. This option would be more likely to receive additional funding.

This alternative would indulge field staff in respect, enlightenment, and skill, and PTSS in respect, power, enlightenment, skill, and wealth. It would lead to value deprivation in all value categories for the OSCEA and in power, wealth, and respect for international agencies and local NGOs. The indulgence in and deprivation of values for host governments and refugees remain indeterminate.

Alternative 2: United Nations High Commissioner for Refugees Mainstreams Environmental Management Activities

This alternative is unlikely to improve significantly UNHCR practices to ensure environmental integrity, as environmental considerations will always remain a low priority to field staff. Similarly, development and application of environmental management skills among field staff will be sporadic. Funding could be enhanced overall, but again, when funding constraints exist, environmental considerations would be the first to be cut.

Alternative 2 would indulge the OSCEA in respect, power, enlightenment, skill, and wealth, and field staff in respect, enlightenment, and skill. It would lead to value deprivation in all value categories for PTSS, in well-being for field staff, in power, wealth, and respect for international agencies, local NGOs, and host governments, and in well-being for refugees.

Alternative 3: United Nations High Commissioner for Refugees Devolves Responsibility to International Agencies

International agencies are undoubtedly better equipped to undertake environmental management activities but likely to find themselves in constant conflict with local UNHCR environmentally unfriendly practices. Funding sources may also be uncertain, and the lack of any form of central coordination and planning within the UNHCR would certainly be detrimental to the overall quality of its environmental activities.

This alternative would indulge international agencies and local NGOs in respect, power, and wealth, and field staff in well-being. The OSCEA and PTSS would be deprived in all value categories, and international agencies and local NGOs would be deprived in the power, wealth, and respect categories. Value deprivation and indulgence remain indeterminate for host governments and refugees.

Alternative 4: United Nations High Commissioner for
Refugees Strengthens Its Catalytic Role

If the OSCEA is able to limit the UNHCR's environmentally unfriendly
practices by ensuring they are not in conflict with environmental management activities, this alternative is likely to be an effective option because of the inherently greater vested interests, and usually skills, of
local NGOs and host governments in carrying out effective environmental management activities. The OSCEA's capacity is likely to be enhanced by its limited role, making it more effective at the jobs it can do
(e.g., monitor UNHCR activities, conduct preventive planning, and identify and train local partners). The emphasis on local organizations,
along with the retention of some centralized role for the OSCEA, also increases the chances of funding. Respect, power, enlightenment, skill,
and wealth would be indulged for local NGOs, host governments, and
the OSCEA, and well-being would be indulged for field staff and refugees. PTSS would experience value deprivation in all value categories,
and international agencies would experience deprivation in the power,
wealth, and respect categories.

Alternative 5: United Nations High Commissioner for Refugees
Develops Partnership with International Agencies

International agencies are likely to be better than the OSCEA at developing local capacities. However, the addition of a further link in
the chain between the OSCEA and the local implementers is not necessarily desirable because it slows response time and increases the
likelihood of inefficiencies. Furthermore, the international agencies
would have a vested interest in not developing local capacities, because this could put them out of a job. The OSCEA's reduced role might
also have the effect of limiting its ability to control the UNHCR's environmentally unfriendly activities. This option, reflecting as it does the
attractive notion of interagency collaboration, along with that of local
capacity building, would probably be successful in terms of fundraising.

Alternative 5 would indulge international agencies, host governments, and local NGOs in respect, power, enlightenment, skill, and
wealth, field staff in well-being, and the OSCEA in respect. It would de-
prive the OSCEA of wealth, power, enlightenment, and skill, and PTSS of
wealth, power, enlightenment, skill, and respect. The value deprivation
and indulgence of refugees is indeterminate.

Selection of Strategy

Alternatives 4 and 5 are both attractive. Ultimately, however, there are several reasons for preferring alternative 4 relating to organizational characteristics, expectations, demands, and roles of the agencies involved. The OSCEA's motivation for developing local capacity is greater than the motivation of international agencies, because the OSCEA would have a genuine need and desire to relinquish responsibility to local partners. International agencies will have an inherent tendency toward goal substitution, because their involvement would be limited if they were effective in developing local capacity. With one less link in the chain between the UNHCR and partner agencies, alternative 4 would be more cost effective and, as a result, more likely to be funded.

Alternative 4 gives the OSCEA a larger role, which might serve to strengthen its overall capacity, particularly when it comes to regulating the UNHCR's activities. A small, two-person unit, as would probably be envisaged under alternative 5, would find it harder to be taken seriously within the UNHCR system than the larger, more powerful unit the OSCEA would constitute under alternative 4.

Conclusion

The OSCEA's overall objective is to improve the UNHCR's environmental management activities within refugee situations to prevent or redress adverse ecological impacts associated with refugee hosting. The policy problem, essentially, is how to achieve this goal. Analysis of the problem, using the policy sciences framework, shows that its definition is sound. The social process, though complex, is fairly clear, at least as to the OSCEA. Nevertheless, mapping it helps us understand areas that the OSCEA has perhaps deemphasized, such as the potential role local NGOs can play in sustainable environmental management. One point which the OSCEA should note within the social process, and an important factor in the selection of alternative 4 over alternative 5, is that there is a natural tendency toward goal substitution by all participants. Finally, as described, the decision process is deeply flawed, particularly in the design of the study currently being undertaken on behalf of the OSCEA. Further, the OSCEA does not seem to understand clearly how the decision process can and should be used to attain its goals.

The concluding recommendation is that the UNHCR should retain overall responsibility for environmental management, but should accept its limited capacity for carrying out such activities. It should therefore aim to develop partnerships with host governments and local NGOs, focusing specifically on developing their capacity for environmental management activities and assisting with securement of financing. This alternative would accord the OSCEA the central responsibility for doing this, and for monitoring the UNHCR's own refugee assistance activities to ensure they are not contradictory to the objectives of environmental management.

Appendix

Comments on Project Framework

Gus Le Breton
Team Leader, Study Group 5

I have gone through the project framework from a broad perspective and at a more focused level. There are several issues of concern to me, in particular relating to the overall structure of the study itself. In making these comments I have, I hope, accorded full recognition to the organizational complexities of the study, and to the wide levels of input and consultation that have gone into its design so far. However, the points I make are ones I feel to be important and are intended to improve the implementation of the study and the caliber of its final products.

Study Structure

The original plan, though loosely formulated, was to examine several cases of successful environmental management in refugee situations throughout Africa. One or more consultants would have visited each of the chosen countries and evaluated the environmental program from start to finish. In doing so, they would have attempted to draw replicable lessons for use in devising training materials for the UNHCR field staff. However, the subsequent division by the steering committee into five thematic study groups, aligned more closely to the sequence adopted in the UNHCR environmental guidelines, has made the task considerably more difficult for the following reasons.

To be effective, environmental management activities must take place throughout the five stages, from emergency preparedness and response through rehabilitation. Examination of each of these stages in isolation is virtually meaningless without relating it to activities that preceded or followed it. In many cases, the most important lessons are likely to relate to activities undertaken in the earliest phases of a refugee program, but because none of the study groups is able, or mandated, to examine the interrelationships between each phase, we lack the crucial link in the study. Thus, we will be examining bits of programs in a large number of countries. However, in no country will the refugee program be studied throughout all five phases.

Environmental management strategies are obviously highly contextual. If we are to develop broadly applicable guidelines, we need to identify commonalities among different situations. The fact that, for each phase of a refugee program, we will have examined only a very limited, and not necessarily diverse, range of countries reduces the likelihood that we will be able to identify commonalities. For example, guidelines for rehabilitation activities will be based on our study of rehabilitation in the dry miombo woodlands of southern Africa; their applicability to the moist forests of, say, Guinea, will obviously be inherently limited.

The integration of findings from five different study groups, looking at different phases in different ecozones, into one coherent and useful set of guidelines is going to be a mammoth task. Although there are many sound reasons UNHCR planning and operations are divided into these phases, it is equally possible that the very nature of this planning is itself a constraining factor to successful environmental management in refugee situations. By bracketing our study within these existing planning categories, the chances of accurately identifying these constraints are reduced.

While we may think in terms of sequential planning phases, ecosystems do not. To examine the human impact on an ecosystem, the ecosystem's response to such a disturbance, and the ways we can strengthen that ecosystem's resilience in the face of those disturbances, we need to look at the system in terms of its own time scale, not ours. By breaking the study into sequential phases allied to UNHCR programming, we have cut across the system's patterns and processes of change and development, with the concomitant risk that we might fail to see the "bigger picture."

Given the option, I would therefore strongly recommend that the study be restructured. Instead of breaking it into planning phases, I would divide it by ecozones. So, one group could look at refugee situations, from start to finish, in arid regions in Africa, another in the semi-arid Sahelian regions, another in the savanna woodlands of east and southern Africa, a fourth in the moist, tropical forests of west and central Africa, and the fifth, say, in moist highlands. Each group would examine one or two countries in as much detail as possible, throughout all the phases of the refugee program.

In addition to resolving all the issues highlighted above, this option would have the further advantage of fairly easy integration of information, because the same team would produce a set of guidelines that would, as far as possible, be applicable to a wide range of refugee situations in similar ecological conditions. I am convinced that the study could be restructured in this way, without necessarily altering the compositions of the teams or the countries to be visited.

The alternative, sticking to the existing structure, will require very careful attention to the integration of study results and their formulation into training guidelines. This will be by far the most challenging task of the study and will require meticulous attention to the details of each team's report, comparison of what may be different and contradictory conclusions, and some kind of process of evaluation, such that the final guidelines are rigorously evaluated for their usefulness and relevance to a wide variety of differing field conditions and situations.

In this respect, I am especially concerned that field staff deem the final guidelines useful and relevant, as I have a sense that they will be apprehensive about receiving another set of "rules and regulations" governing their field procedures and that this apprehension will translate into naked resentment.

Organizational Change

This study is intended to improve the UNHCR's practices. This will not be effected without some form of organizational change. However, an important area I think we have overlooked is the examination of how such organizational change can and should be brought about. This is likely to be especially difficult, in view of the fact that the UNHCR's mandate is to conduct short-term relief operations which, by their very nature, are inconsistent with the objectives of long-term environmental management. There is little in the organization's current institutional experience, operational procedures, and bureaucratic structure that suggests it is in any way equipped to cope with the vastly different demands of promoting and facilitating sustainable environmental management.

There are two important ways in which we can ensure that the lessons derived from this study are most likely to be effectively translated into improved practice. The first is to direct each study team to look not only at the specific content of each environmental management program within its case studies, but also to examine the decision-making processes by which that content was determined. Examining these processes will help us to derive broader lessons about the way the UNHCR's operating procedures inhibit or facilitate the development of effective environmental management strategies. The second way to ensure improved practice is to reexamine carefully the objective of providing guidelines to field staff. Theories of organizational learning suggest that imposed structures of action provide a framework for "single loop" learning but actively discourage "double loop" learning (which we can also call "adaptive management"). Yet it is precisely this double loop learning that an organization requires if it is to respond adaptively to the huge number of interactive variables inherent in any form of environmental management. Guidelines can be extremely conducive to double loop learning, if they are treated as standards to be constantly and critically reevaluated rather than as limits to action. This, of course, requires that field staff interpret them in that way, which may not necessarily be easy to achieve. I would recommend that whoever has the responsibility for preparing guidelines should be familiar with the extensive theories of organizational learning. This should then be incorporated into their outline and presentation, so that the final result is a set of guidelines conducive to double loop learning by field staff.

SUGGESTED READING

Arygris, C. 1992. *On Organization Learning*. Cambridge, MA: Blackwell Publishers.

Argyris, C., and D. A. Schon. 1978. *Organizational Learning: A Theory of Action Perspective*. Reading, MA: Addison-Wesley.

Clark, T. W. 1993. "Creating and Using Knowledge for Species and Ecosystem Conservation: Science, Organizations, and Policy," *Perspectives in Biology and Medicine* 36(3):497–525, appendices.

Clark, T. W., R. P. Reading, and A. L. Clarke. 1994. *Endangered Species Recovery: Finding the Lessons, Improving the Process*. Washington, DC: Island Press.

Etheredge, L. S. 1985. *Can Governments Learn?* New York: Pergamon Press.

Gunderson, L. H., C. S. Holling, and S. S. Light, eds. 1995. *Barriers and Bridges to the Renewal of Ecosystems and Institutions*. New York: Columbia University Press.

Morgan, G. 1986. *Images of Organization*. Beverly Hills, CA: Sage Publications.

Senge, P. M. 1990. *The Fifth Discipline: The Art and Practice of Learning Organization*. New York: Doubleday Books.

13 Improving Natural Resources Policy and Management

Epilogue and Prologue

Tim W. Clark, Andrew R. Willard,
and Christina M. Cromley

Natural resources policy and management challenges are diverse and pervasive. Part 1 of this book, "Improving Natural Resources Policy and Management," begins with the premise that meeting these challenges requires an integrative, genuinely interdisciplinary approach and a comprehensive set of concepts and propositions to investigate policy problems. Such a premise is not unique to this volume or to its authors (see, for example, Brewer 1995; Gunderson et al. 1995; Heberlein 1988; Ludwig et al. 1993). This book, however, introduces the problem-oriented, contextual, interdisciplinary framework found in the policy sciences as a rigorous, empirical, comprehensive, and practical method for understanding and solving problems within natural resources policy and management. It illustrates efforts to teach, learn, and apply this method. The policy sciences approach does not necessarily reject more conventional approaches to understanding and solving natural resources problems in biophysical, economic, sociological, or other disciplinary terms. Rather, the policy sciences approach and this volume attempt to answer the call for improved problem solving of natural resources issues and the need to examine these issues from an interdisciplinary and integrative perspective.

By focusing on the theory and application of the policy sciences, this book also offers a means to significantly improve the current understanding of the social and political dimensions of natural resources policy and management. Such an integrated and comprehensive understanding can also help improve natural resources use in a practical sense. In many ways, the underlying issue in natural resources policy—as in all public policy—is finding appropriate means to clarify and secure the common interest, finding ways of meeting the increasing demands of humans on natural resources while sustaining resources at a viable level. Readers of this book, whether interested in natural resources policy and management or other issues, can use the theory and cases as examples to learn the policy sciences and then apply to their own challenges the tools it outlines and illustrates.

The policy sciences were articulated and crystallized after World War II to help people understand and practically address diverse policy problems of growing complexity. In Chapters 1 and 2, we introduced the principal components of the policy sciences approach to problem solving and outlined the tools and comprehensive set of categories policy sciences recommends for mapping the social and decision processes that affect decision making in natural resources policy and management, or any public policy for that matter. The policy sciences emphasize that in problem solving, every detail is affected by interaction with the entire context of which the problem is a part. The social context includes participants with particular perspectives interacting in particular situations. The participants are motivated by different values, which affect their strategies to achieve desired outcomes. Problem solving is made up of five intellectual tasks that include clarifying the goals that people seek in relation to the problem(s) of concern; identifying historic trends to see if events are moving toward or away from clarified goals; understanding the factors or conditions that influence trends; projecting future trends; and creating, evaluating, and selecting potential solutions. The tasks are indispensable to integrative, interdisciplinary problem solving.

Teaching the policy sciences approach can be challenging, and part of the challenge is in shaping the presentation of the framework and theory to diverse audiences. One indispensable part of learning the policy sciences involves applying them to actual cases. The diversity of cases presented in Part 2 demonstrates the flexibility of the policy sciences approach in resolving almost any policy problem. The authors discuss human health and mortality in urban settings, urban parks, the development of energy projects, endangered species conservation, the relationship of national parks inside and outside of the United States to the human populations located near them, other issues of recreation and ecotourism, public participation in international development projects, and the relationship of human rights to environmental conservation in refugee situations. The standpoints taken by the authors are also diverse. Some authors take an activist view, others are less personally engaged, some sympathize with local communities, and many are participants in the policy processes examined.

In addition to showing how the policy sciences accommodate a wide range of problems in natural resources policy and management and permit the assumption of any standpoint in reference to particular problems, the ten case studies in this volume also demonstrate that by using a stable frame of reference, comparisons can be made across cases and patterns begin to emerge. All the cases reinforce the need for using a contextual, problem-oriented, integrative, multimethod approach. For example, Jessica Lawrence demonstrates that the relationships among parks, open space, and degraded natural areas in urban settings are intricately connected to urban social problems, including high crime rates, high rates of poverty, and drug dealing. Alejandro Flores emphasizes the need to consider both rates of ozone removal as well as human exposure—and behavior to reduce human exposure—in order to reduce the incidence of ozone-related health problems and mortalities. Eva Garen organizes her problem-oriented analysis by attempting to clarify the goals of the ecotourism industry and uses a literature review and field experience to identify the trends and conditions that have shaped how well participants in ecotourism have met the goals.

Many of the cases discuss specific problems with the decision process. In some cases, for example, the research used in intelligence emphasizes positivistic science with little understanding of the social and political context in which natural resources problems emerge. Christina Cromley, for example, discusses extensive intelligence activities conducted over a period of two and a half decades in grizzly bear conservation that resulted in limited information on social and decision process and which generated problems of access to intelligence-gathering activities. David Kaczka examines why an integrative and effective alternative promoted in sea turtle conservation—the use of turtle excluder devices—has not been prescribed or applied more extensively.

Many cases discuss the differential access and, accordingly, the differential influence of participants in decision-making processes overall. This is true in both urban and rural settings, in the United States and abroad. The common interest—meeting the demands of diverse stakeholders—is often sacrificed to and/or by special interest groups. This often involves the relationship among those with authority and

control in decision making and those with less power. Christopher Elwell discusses the exclusion of citizen groups from decisions to build
two pumped-storage hydroelectric power plants in the Sequatchie Valley, Tennessee. Peter Wilshusen's analysis focuses on the conditions
leading to the unsatisfactory degree of participation by local people in
the majority of conservation and development projects worldwide.

Other case studies discuss how the demands made by different participants in the policy process with unequal access in influencing policy processes affect natural resource integrity. Katherine Lieberknecht
discusses how demands for recreational activity, enjoyment of natural
settings, and commercial and residential growth have affected conservation efforts for the Barton Springs salamander. David Lyon seeks to
find a middle ground between conservation and development in a national park in Ghana that considers the needs and demands of local
residents while meeting conservation goals. Gus Le Breton conducts a
similar analysis, combining consideration of environmental issues of
sustainability with fundamental issues of human rights in refugee situations.

This brief summary provides an indication of how the policy sciences approach can facilitate the making of comparisons across diverse cases; however, even in reference to the ten cases in this volume,
the preceding discussion only begins to clarify commonalities and differences in natural resources cases. We need more applications of ever
more diverse natural resources policy and management challenges to
build and lay a foundation for a broader understanding of recurring
patterns in social and decision processes. More efforts in teaching,
learning, and applying the policy sciences to natural resources issues
can help lay such a foundation.

Although many students, scholars, and practitioners of natural resources policy and management are eager for the kind of approach the
policy sciences offer, resistance to the approach can be anticipated. In
an earlier and related study (Reisman and Willard 1988:264), it was
noted that "One of the difficulties in establishing a new focus that does
not fit neatly into existing disciplines and then fashioning special intellectual tools for dealing with the problems the focus casts in a new
light is that even those who understand and appreciate the purpose of
the exercise continue to feel that some of the traditional practices
should still be incorporated. What is resented is less the novel idea and
more the fact that it makes obsolescent beloved old ways of doing
things." Efforts to integrate the "beloved old ways of doing things"
should be attempted when it is feasible and appropriate. In many circumstances, however, clarifying and securing the common interest

will require the use of "new ways of doing things." The policy sciences approach provides a new and better way for those interested in improving how we use our natural resources. It is our conviction that the contents of this book more than amply support the validity of this observation.

LITERATURE CITED

Brewer, G. D. 1995. Environmental Challenges: Interdisciplinary Opportunities and New Ways of Doing Business. 1995 MISTRA Lecture, The Foundation for Strategic Environmental Research, Stockholm, Sweden.

Heberlein, T. A. 1988. "Improving Interdisciplinary Research: Integrating the Social and Natural Sciences," *Society and Natural Resources* 7:595–597.

Gunderson, L. H., C. S. Holling, and S. S. Light, eds. 1995. *Barriers and Bridges to the Renewal of Ecosystems and Institutions.* New York: Columbia University Press.

Ludwig, D., R. Hilbon, and C. J. Walters. 1993. "Uncertainty, Resource Exploitation and Conservation: Lessons from History," *Science* 260:17, 36.

Reisman, W. M., and A. R. Willard, eds. 1988. *International Incidents: The Law that Counts in World Politics.* Princeton, NJ: Princeton University Press.

Abbreviations

ACAP	Annapurna Conservation Area Project
ACE	Army Corps of Engineers
AER	Armstrong Engineering Resources
AQA	Air Quality Act
BCP	Balcones Canyonlands Conservation Plan
BMA	Baltimore metropolitan area
BOSCOSA	Osa Peninsula Forest Conservation and Management Project
BRSC	Black Rock/ Spread Creek allotment
BSSCAS	Barton Springs Salamander Conservation Agreement and Strategy
CAA	Clean Air Act
CAO	community-assisting organizations
CBO	community-based organization
CCWS	Center for Coastal and Watershed Studies
CI	Conservation International
CITES	Convention on International Trade in Endangered Species of Wild Fauna and Flora
CMC	Center for Marine Conservation
DPRT	Department of Parks, Recreation and Trees
EDF	Environmental Defense Fund
EII	Earth Island Institute
EIS	Environmental Impact Statement
EPA	United States Environmental Protection Agency
EPS	Environmental Protection Society

ERE	Elk Ranch East allotment
ESA	Endangered Species Act
FBPP	Friends of Beaver Ponds Park
FERC	Federal Energy Regulatory Commission
FMNP	Fazao-Malfacassa National Park
FWS	U.S. Fish and Wildlife Service
GATT	General Agreement on Tariffs and Trade
GEF	Global Environment Facility
GIS	Geographic information systems
GTNP	Grand Teton National Park
GYE	Greater Yellowstone Ecosystem
ICDP	Integrated Conservation and Development Project
IGBC	Interagency Grizzly Bear Committee
IGBST	Interagency Grizzly Bear Study Team
IUCN	World Conservation Union
KCP	Kyabobo Conservation Project
KRNP	New Kyabobo Range National Park
LWF	Lutheran World Federation
MDE	Maryland Department of the Environment
MGP	Mountain Gorilla Project
MS	management situation
MVAPC	Motor Vehicle Air Pollution Control Act
NAAQS	National Ambient Air Quality Standards
NAS	National Academy of Sciences
NEPA	National Environmental Policy Act
NFI	National Fisheries Institute
NGO	nongovernmental organization
NMFS	U.S. National Marine Fisheries Service
NOX	oxides of nitrogen
OAD	ozone action day
OSCEA	Office of the Senior Coordinator for Environmental Affairs
OSM	Office of Surface Mining
PDA	Community Development Association
PTSS	Programme Technical Support Services
SAFIRE	Southern Alliance for Indigenous Resources
SCSU	Southern Connecticut State University
SOS	Save Our Sequatchie
SOS	Save Our Springs Alliance
SPN	Malagasy Nature Conservation Service
STRP	Sea Turtle Restoration Project
TDEC	Tennessee Department of Environment and Conservation
TEAM	Environmental Awareness and Development Mobilization

TED	turtle excluder device
TNC	The Nature Conservancy
TNRCC	Texas Natural Resources Commission
TPWD	Texas Parks and Wildlife Department
TVA	Tennessee Valley Authority
TxDOT	Texas Department of Transportation
UN	United Nations
UNCLOS	United National Convention on the Law of the Sea
UNEP	United Nations Environment Programme
UNHCR	United Nations High Commissioner for Refugees
URI	Urban Resources Initiative
USFS	United States Forest Service
VEIP	Vehicle Emissions Inspection Program
VMT	vehicle miles traveled
VOC	volatile organic compound
VRNP	Vapor Recovery Nozzle Program
WCED	World Commission on Environment and Development
WD	Wildlife Department
WTO	World Tourism Organization
WTO	World Trade Organization
WYGF	Wyoming Game and Fish Department
WFT	Wildlife Fund Thailand
WWF	World Wide Fund for Nature
WWF	Word Wildlife Fund
YFES	Yale School of Forestry and Environmental Studies
YNP	Yellowstone National Park

Contributors

David Lyon completed a bachelor's degree in biological sciences at Oxford University in 1995 and a master's degree in environmental studies at Yale University in 1998, specializing in social ecology and environmental policy. He currently works as a social ecologist for the Baltimore Ecosystem Study, an urban long-term ecological research site.

Tim W. Clark is professor (adjunct) in the School of Forestry and Environmental Studies and fellow in the Institution for Social and Policy Studies, Yale University. He is also board president of the Northern Rockies Conservation Cooperative in Jackson, Wyoming. He received his Ph.D. in zoology from University of Wisconsin–Madison in 1973. He has written over 300 papers. His books and monographs include *Greater Yellowstone's Future: Prospects for Ecosystem Science, Management, and Policy* (1994) and *Averting Extinction: Reconstructing Endangered Species Recovery* (1997).

Christina M. Cromley received her bachelor's degree in English and environmental studies from Gettysburg College in 1995. She is a doctoral candidate at Yale School of Forestry and Environmental Studies. Her most recent work includes an assessment of a two-decades-long planning process and environmental assessment for bison on the National Elk Refuge in Wyoming. Cromley is currently writing her doctoral dissertation on the high-profile, highly contentious controversy over bison leaving Yellowstone National Park.

Christopher M. Elwell received his bachelor of arts degrees in natural resources and American studies from University of the South in 1995. He earned a Master of Forestry degree from Yale School of Forestry and Environmental Studies in 1998 and is the forester and manager of the Sustainable Forestry and Wood Products program of Appalachian Sustainable Development.

Alejandro Flores is a doctoral candidate in environmental policy at Yale School of Forestry and Environmental Studies. Before moving to the United States to pursue a master of science degree in ecology at University of California, Davis, and later to study at Yale, he graduated with a bachelor of science degree in industrial and systems engineering from Monterrey Institute of Technology in his native Monterrey, Mexico. His most recent publications have appeared in the 1998 *Journal of Urban and Landscape Planning* and in the occasional publication series of the Institute of Ecosystem Studies, in Millbrook, New York.

Eva J. Garen received a bachelor's degree in political science from Union College, New York, in 1994 and has worked for the legislative departments of the Sierra Club and Zero Population Growth. Her master's research at Yale School of Forestry and Environmental Studies focused on understanding ecotourism decision making on the island of Roátan off the northern coast of Honduras. Garen is a Ph.D. student at Yale School of Forestry and Environmental Studies, studying the use of ecotourism as a conservation and development strategy worldwide.

David Kaczka received his master's in environmental studies from Yale School of Forestry and Environmental Studies in 1997. He has published two articles on the shrimp–sea turtle case in the *Review of European Community and International Environmental Law: A Primer on the Shrimp–Sea Turtle Controversy*, and WTO's *Shrimp–Sea Turtle Decision*. Currently, Mr. Kaczka is a senior policy analyst for ISSI Consulting Group.

Jessica Lawrence received a bachelor of arts in biology from Vassar College in 1995 and a master of environmental studies degree in conservation biology from Yale University in 1998. She is interested in the social dynamics of ecosystem stewardship and initiatives that attempt to improve community well-being by sustainable and equitable use of natural resources. Her most recent work is in west Africa.

Gus Le Breton received a master's degree in social anthropology from Cambridge University, United Kingdom, and a master's in environmental studies at Yale School of Forestry and Environmental Studies.

Since 1992, he has directed the Southern Alliance for Indigenous Resources, a regional alliance of nongovernmental organizations based in Zimbabwe, dedicated to the economic development of communal areas based on sustainable and productive use of natural resources. He has led an international team recruited by the United Nations High Commissioner for Refugees in Geneva to develop environmental guidelines for refugee situations around the world.

Katherine Lieberknecht received her bachelor's degree in biology from the College of William and Mary in Virginia in 1996. She received a master's degree in environmental studies, studying conservation biology at Yale School of Forestry and Environmental Studies, in 1998. Her primary professional interest is in resolving land-use and biodiversity conservation conflicts.

Andrew R. Willard is associate research scholar and visiting lecturer at Yale Law School. He has also taught at Yale School of Forestry and Environmental Studies and in Yale College. He is the general editor of *New Haven Studies in International Law and World Public Order,* a monographic series. He also co-edited *International Incidents: The Law That Counts in World Politics* (1988). He has been a consultant to the United States Information Agency and serves as the president of both the Policy Sciences Center, Inc., and the Society for the Policy Sciences.

Peter R. Wilshusen completed his master's degree in forest science at Yale School of Forestry and Environmental Studies in 1996. He is currently a doctoral candidate at University of Michigan's School of Natural Resources and Environment. His research focuses on the impacts of changes in national forestry law on community-based forest management in Quintana Roo, Mexico. Wilshusen works as a consultant for the United Nations Development Programme and the Global Environmental Facility.

Index